T0156143

Communications
in Computer and Information Science   1946

Rationale

The CCIS series is devoted to the publication of proceedings of computer science conferences. Its aim is to efficiently disseminate original research results in informatics in printed and electronic form. While the focus is on publication of peer-reviewed full papers presenting mature work, inclusion of reviewed short papers reporting on work in progress is welcome, too. Besides globally relevant meetings with internationally representative program committees guaranteeing a strict peer-reviewing and paper selection process, conferences run by societies or of high regional or national relevance are also considered for publication.

Topics

The topical scope of CCIS spans the entire spectrum of informatics ranging from foundational topics in the theory of computing to information and communications science and technology and a broad variety of interdisciplinary application fields.

Information for Volume Editors and Authors

Publication in CCIS is free of charge. No royalties are paid, however, we offer registered conference participants temporary free access to the online version of the conference proceedings on SpringerLink (http://link.springer.com) by means of an http referrer from the conference website and/or a number of complimentary printed copies, as specified in the official acceptance email of the event.

CCIS proceedings can be published in time for distribution at conferences or as post-proceedings, and delivered in the form of printed books and/or electronically as USBs and/or e-content licenses for accessing proceedings at SpringerLink. Furthermore, CCIS proceedings are included in the CCIS electronic book series hosted in the SpringerLink digital library at http://link.springer.com/bookseries/7899. Conferences publishing in CCIS are allowed to use Online Conference Service (OCS) for managing the whole proceedings lifecycle (from submission and reviewing to preparing for publication) free of charge.

Publication process

The language of publication is exclusively English. Authors publishing in CCIS have to sign the Springer CCIS copyright transfer form, however, they are free to use their material published in CCIS for substantially changed, more elaborate subsequent publications elsewhere. For the preparation of the camera-ready papers/files, authors have to strictly adhere to the Springer CCIS Authors' Instructions and are strongly encouraged to use the CCIS LaTeX style files or templates.

Abstracting/Indexing

CCIS is abstracted/indexed in DBLP, Google Scholar, EI-Compendex, Mathematical Reviews, SCImago, Scopus. CCIS volumes are also submitted for the inclusion in ISI Proceedings.

How to start

To start the evaluation of your proposal for inclusion in the CCIS series, please send an e-mail to ccis@springer.com.

Feng Zhao · Duoqian Miao

Editors

AI-generated Content

First International Conference, AIGC 2023
Shanghai, China, August 25–26, 2023
Revised Selected Papers

 Springer

Editors
Feng Zhao (iD)
University of Science and Technology
of China
Hefei, China

Duoqian Miao (iD)
Tongji University
Shanghai, China

ISSN 1865-0929   ISSN 1865-0937 (electronic)
Communications in Computer and Information Science
ISBN 978-981-99-7586-0   ISBN 978-981-99-7587-7 (eBook)
https://doi.org/10.1007/978-981-99-7587-7

This Springer imprint is published by the registered company Springer Nature Singapore Pte Ltd.
The registered company address is: 152 Beach Road, #21-01/04 Gateway East, Singapore 189721, Singapore

Paper in this product is recyclable.

Preface

Welcome to the first International Conference on AI-generated Content (AIGC 2023)! It is with immense pleasure that we present the proceedings of this event, held from Aug. 25–26, 2023 in Shanghai, China. This conference stands as a testament to the remarkable strides that have been made in the realm of artificial intelligence and its transformative impact on content creation.

The AIGC 2023 conference gathered pioneers, researchers, practitioners, and enthusiasts from around the world, all of whom are at the forefront of exploring the diverse aspects of AI-generated content.

This conference served as a platform for interdisciplinary discussions, fostering a dynamic exchange of ideas that traverse the boundaries of computer science, linguistics, arts, ethics, and beyond. The AIGC 2023 proceedings encapsulate the collective wisdom and innovation presented during the conference.

A total of 62 manuscripts were submitted to our conference, each manuscript undergoing a meticulous single-blind peer-review process, involving no fewer than three reviewers. After careful scrutiny, we finally accepted 30 manuscripts.

Delving into the content of the proceedings, the reader will encounter cutting-edge research findings, innovative applications, and thought-provoking insights that underscore the transformative potential of AI-generated content. From deep learning architectures to algorithmic advancements, this work reflects the collaborative efforts of individuals who are shaping the future of content creation in unprecedented ways.

We extend our heartfelt gratitude to the organizing committee, authors, and participants who contributed to making AIGC 2023 a resounding success. The journey to harnessing the full capabilities of AI-generated content has only just begun, and we are excited to witness the profound impact that this field will continue to have on our society and creative endeavors.

As you navigate through the AIGC 2023 proceedings, we hope they will serve as a wellspring of inspiration, igniting your imagination, and collectively guiding you towards the limitless horizons of AI-generated content.

August 2023

Feng Zhao
AIGC 2023 Program Committee Chair

Organization

General Chair

Xipeng Qiu   Fudan University, China

General Co-chairs

Hongsheng Li   Chinese University of Hong Kong, China
Wanli Ouyang   Shanghai AI Laboratory, China

Program Committee Chair

Feng Zhao   University of Science and Technology of China,
                                    China

Program Committee Co-chairs

Ping Luo   University of Hong Kong, China
Qibin Zhao   RIKEN Center for Advanced Intelligence Project,
                                    Japan

Technical Program Committee Chairs

Duoqian Miao   Tongji University, China
Junchi Yan   Shanghai Jiao Tong University, China

Technical Program Co-chairs

Jingjing Deng   Durham University, UK
Man Zhou   Nanyang Technological University, Singapore

Publicity Chairs

Lei Chen   Shandong University, China
Zhiquan Liu   Jinan University, China

Program Committee

Zahid Akhtar   State University of New York Polytechnic
                                  Institute, USA
Joseph Arul   Fu Jen Catholic University, Taiwan
Chokri Barhoumi   Taibah University, Saudi Arabia
Kamran Behdinan   University of Toronto, Canada
Sinan Chen   Kobe University, Japan
Xin Chen   South China Agricultural University, China
Minh-Son Dao   National Institute of Information and
                                  Communications Technology, Japan
Junwen Ding   Huazhong University of Science & Technology,
                                  China
Sathishkumar V. E.   Jeonbuk National University, Republic of Korea
Yunhe Feng   University of North Texas, USA
Cristiano Hora de Oliveira Fontes Federal University of Bahia, Brazil
Hayford A. Perry Fordson   Cornell University, USA
Moncef Garouani   Université du Littoral Côte D'Opale, France
Jingjing Guo   Xidian University, China
Shuihua Han   Xiamen University, China
Hamidreza Hosseinpour   University of Tehran, Iran
Vinasetan Ratheil Houndji   Université d'Abomey-Calavi, Benin
Haoji Hu   Zhejiang University, China
Zhen Hua   Xi'an Jiaotong-Liverpool University, China
Choi Jaeho   Jeonbuk National University, Republic of Korea
Xiaopeng Ji   Ningbo Institute of Materials Technology and
                                  Engineering, Chinese Academy of Sciences,
                                  China
Yuling Jiao   Wuhan University, China
Jiaxin Jiang   National University of Singapore, Singapore
Jalal Khalil   University of Alabama at Birmingham, USA
He Li   Universidade de Lisboa, Portugal
Xiangtao Li   Jilin University, China
Yi Li   Lancaster University, UK
Yuchen Li   Beijing University of Technology, China
Yueying Li   Cornell University, USA

Haiyang Liu	University of Tokyo, Japan
Huafeng Liu	Zhejiang University, China
Moyun Liu	Huazhong University of Science and Technology, China
Xinwang Liu	Harbin Engineering University, China
Yuewen Liu	Xi'an Jiaotong University, China
Abhilash Puthanveettil Madathil	University of Strathclyde, UK
Priyanka Mudgal	Intel Corporation, USA
Hajer Nabli	University of Sousse, Tunisia
Mohammad Noori	California Polytechnic State University at San Luis Obispo, USA
Nasheen Nur	Florida Institute of Technology, USA
Hyun-A Park	Honam University, Republic of Korea
Kebin Peng	University of Texas at San Antonio, USA
Gerald Penn	University of Toronto, Canada
Moritz Platt	King's College London, UK
Jean-Philippe Prost	Aix-Marseille Université, France
Michal Ptaszynski	Kitami Institute of Technology, Japan
Xinying Qiu	Guangdong University of Foreign Studies, China
Abolfazl Razi	Clemson University, USA
Mayank Soni	Trinity College Dublin, Ireland
Lavanya Srinivasan	University of West London, UK
Victor Telnov	National Research Nuclear University MEPhI, Russia
Run Wang	Wuhan University, China
Xianzhi Wang	University of Technology Sydney, Australia
Wei Wang	Hong Kong University of Science and Technology (Guangzhou), China
Yuntao Wang	Osaka University, Japan
Zihan Wang	University of California, San Diego, USA
Qian Wang	Insilico Medicine, China
Chi-Jen Wu	Chang Gung University, Taiwan
Robert Wu	University of Technology Sydney, Australia
Shiqing Wu	University of Technology Sydney, Australia
Shuang Wu	Black Sesame Technologies, Singapore
Akihiro Yamaguchi	Toshiba Corporation, Japan
Tengfei Yang	Xi'an University of Posts & Telecommunications, China
Yilong Yang	University of Southampton, UK
Irfan Yousuf	University of Engineering & Technology, Pakistan
Zhenhua Yu	Imperial College London, UK
Hao Zhang	Supportiv Inc., USA

Lijun Zhang   University of Massachusetts, USA
Di Zhao   University of Chinese Academy of Sciences,
                             China
Wenbo Zhou   University of Science and Technology of China,
                             China

Contents

Automatic Multilingual Question Generation for Health Data Using LLMs

Ryan Ackerman and Renu Balyan(✉) ⓘ

State University of New York Old Westbury, Old Westbury, NY 11568, USA
balyanr@oldwestbury.edu

Abstract. Question Generation (QG) involves automatic generation of yes/no, factual and Wh-questions created from data sources such as a database, raw text or semantic representation. QG can be used in an adaptive intelligent tutoring system or a dialog system for improving question answering in various text generation tasks. Traditional QG has used syntactic rules with linguistic features to generate factoid questions. However, more recent research has proposed using pre-trained Transformer-based models for generating questions that are more aware of the answers. The goal of this study was to create a multilingual database (English and Spanish) of automatically generated sets of questions using artificial intelligence (AI), machine learning (ML), and large language models (LLMs) in particular for a culturally sensitive health intelligent tutoring system (ITS) for the Hispanic population. Several language models (LMs) including Chat GPT, valhalla/t5-based-e2e-qg, T5 (Small, Base, and Large), mrm8488/bert2bert-spanish-question-generation, mT5 (Small and Base), Flan-T5 (Small, Base and Large), BART (Base and Large) and mBART (Large) were chosen for our experiments that were given a prompt to produce a set of questions (3, 5, 7 or 10) using transcribed texts as the context. We observed that a text/transcript of at least 100 words was sufficient to generate 5–7 questions of reasonable quality. When models were prompted to produce 10 or more questions based on texts containing 100 words or less the meaningfulness, syntax and semantic soundness of outputs decreased notably.

Keywords: Question Generation · LLMs · Multilingual · Healthcare

1 Introduction

Question Generation (QG) is defined as the task involving automatic generation of yes/no, factual and Wh-questions from different forms of data input that can be obtained from a database, raw text or semantic representation [1]. QG is not an easy task and requires not only an in-depth understanding of the input source and the context but also the ability to generate grammatical questions that are also semantically correct. Questions are generally constructed as well as assessed by tutors in education and are crucial for stimulating self-learning and evaluating Students' knowledge [2, 3]. It has been seen that a student learns more deeply if prompted by questions [4].

QG can be used in an adaptive intelligent tutoring system or a dialog system [4], for improving question answering [5–8], in various text generation tasks for evaluating

© The Author(s), under exclusive license to Springer Nature Singapore Pte Ltd. 2024
F. Zhao and D. Miao (Eds.): AIGC 2023, CCIS 1946, pp. 1–11, 2024.
https://doi.org/10.1007/978-981-99-7587-7_1

the factual consistency [9–11], or for automatic assessments [12–15] including course materials. One of the strategies used by a tutor for evaluating a learners' comprehension is either by asking questions that the learner needs to provide answers for, that are based on some text already provided to the learner, or by asking the learner to generate questions from the available text. There are different types of questions that can be asked from a learner or that a learner may be asked to generate to gain an understanding of the learners' comprehension. Some of these types of questions are gap fill questions (GFQs; [16–18]), multiple choice questions (MCQs; [19–21]), factoid-based questions (FBQs; [20–23]) and deep learning questions (DLQs; [24–27]).

Traditional QG has used syntactic rules with linguistic features to generate FBQs either from a sentence or a paragraph [28–32]. However, QG research has started to utilize "neural" techniques, to generate deeper questions [14, 24–27]. Some research more recently has relied on pre-trained Transformer-based models for generating questions that are more aware of the answers [33–36].

The goal of this study was to create a multilingual database (English and Spanish) of automatically generated sets of questions using artificial intelligence (AI), machine learning (ML), and large language models (LLMs) in particular for a culturally sensitive health intelligent tutoring system (ITS) for the Hispanic population. This ITS is being developed as a part of a bigger NSF-funded project. In order to build this database some of the tasks were carried out in this study to answer the following research questions (RQ):

RQ1: What existing systems or models can be used to automatically generate different sets of questions given the context for English as well as Spanish?
RQ2: Is the quality of questions generated consistent across different texts and thresholds? If the quality is not consistent, then what factors impact or determine the quality of questions generated?
RQ3: What are the optimal values for these factors impacting the question generation quality?

2 Methods

2.1 Data Acquisition

For the purposes of this research data needed to be fed into our question generators as text. Therefore, eleven short videos that were recorded in both English and Spanish were transcribed into 1–3 paragraphs of text. While the topic of each section varied, the theme of every video revolved around the domain of cancer survivorship. Otter.ai was originally implemented to perform transcription tasks but proved to be inconsistent in transcribing English and ineffective in transcribing Spanish dialog altogether. Human transcribers proficient in English and Spanish were used for our final transcriptions.

The data descriptives (Number of sentences and words) shown in Table 1 for each transcription were obtained using SpaCy, an open-source NLP python library.

Table 1. Data Descriptives for the English and Spanish Expert/Reference Transcriptions

Transcript	Transcript Description	# of sentences (English/Spanish)	# of words (English/Spanish)
1	Visual Symptoms	5/7	97/79
2	Tamoxifen Side Effects	9/5	217/167
3	Survivorship Care	5/8	243/203
4	SE After Surgery	8/8	198/133
5	PT Side Effects	3/13	162/337
6	PT Breast Cancer Basics	3/9	112/167
7	PT Intro-Mi Guia	2/5	38/68
8	Peripheral Neuropathy	6/8	126/171
9	Osteoporosis	5/6	117/193
10	Depression	9/5	111/131
11	Cardiac symptoms	6/5	120/70

This is to be noted that there are differences between the data descriptives for English and Spanish transcriptions. Some of these differences result due to some variations in the length of videos for the two languages. In addition, the other differences are caused due to linguistic differences between the two languages.

2.2 Data Preparation

The final transcriptions were then split into groups based on language and topic before they were printed into separate rows of a CSV file. Depending on the experiment the texts would be entered in the CSV file differently. Whole texts and individual texts were separated into English only, Spanish only and multilingual versions.

2.3 Language Models

Language models (LMs) are integral to the process of natural language processing (NLP) by which a computer is able to understand, analyze and generate human language. To do so LMs are trained on large datasets of text data gathered from a variety of resources such as books, articles, and Wikipedia. After being trained on this data, LMs are able to make predictions based on recognized patterns in natural language that can aid in a number of NLP tasks [37].

Large Language Models (LLMs). LLMs are an evolution of LMs that are trained on considerably larger datasets and domains. They employ a self-attention model in transformers to capture long range dependencies in text and parallelization. Using in-context learning, models can be trained on a specific context or prompt allowing LLMs to create more coherent and human-like outputs taking major strides toward the advancement of NLP tasks and artificial intelligence [38, 39].

Chat GPT. Chat GPT is an LLM that uses transformer architecture to process input data to create an adequate response. We implemented the 'davinci-003 engine' for its ability to handle large prompts and instruction-following tasks. This model was trained on Common Crawl, webtexts, books and Wikipedia [40].

Valhalla/t5-based-e2e-qg. This is a LLM pre-trained on the SQuADv1 dataset which consists of questions created based on Wikipedia articles where the answer to each question is a segment of text from the corresponding reading passage. T5 stands for Text-to-Text-Transfer-Transformer model proposes reframing all NLP tasks into a unified text-to-text-format where the input and output are always text strings.

mrm8488/bert2bert-spanish-question-generation. This is a LLM pre-trained on the SQuAD dataset that has been translated to Spanish. It utilizes a bert2bert model which means both the encoder and decoder are powered by BERT models.

T5 (Small, Base, and Large) and mT5 (Small and Base). A transformer-based architecture that uses a text-to-text approach. Tasks including translation, question answering, and classification are used as input where it's trained to generate some target text [41].

Flan-T5 (Small, Base and Large). This model is similar to T5 but uses an instruction-tuned model and therefore is capable of performing various zero-shot NLP tasks, as well as few-shot in-context learning tasks.

BART (Base and Large) and mBART (Large). A denoising autoencoder for pre-training sequence-to-sequence models trained by corrupting text with an arbitrary noising function, and learning a model to reconstruct the original text. It uses a standard transformer-based neural machine translation architecture [42].

2.4 Experimentation

A number of LMs were tested in order to compare their capabilities when put to the task of multilingual question generation. The LMs Chat GPT, valhalla/t5-based-e2e-qg, T5 (Small, Base, and Large), mrm8488/bert2bert-spanish-question-generation, mT5 (Small and Base), Flan-T5 (Small, Base and Large), BART (Base and Large) and mBART (Large) were chosen for our experiments. Each of these models were given a prompt to produce a certain number of questions (3, 5, 7 and 10) using one of the transcribed texts as the context.

The length and combination of texts given to each model depends on the language that it was designed to handle. LMs that specialized in either English or Spanish were given each of the 11 texts individually and then these two language texts were combined into a single text to create multilingual text and the question generators were prompted to create either 3, 5, 7 or 10 questions based on the text. The multilingual models performed the same task in both languages plus a multilingual version where the individual Spanish and English texts that corresponded to the same topic were fed through at the same time. Some of the models couldn't process the Spanish and English whole text combined so this test was eliminated. The evaluation of each model was determined by judging the outputs they produced based on coherency, spelling and accuracy.

3 Results

Context: One of the long-term side effects after breast cancer treatment is not only early menopause or possibly menopausal symptoms, but grouped with this is the development of osteoporosis. Osteoporosis is the thinning of the bones where the calcium is not being absorbed as well. Many times the treatment consists of calcium and vitamin D ....

Prompt: Create 3 unique questions based on the following text: (Context from above)

Questions generated:
Q1. What is one of the long-term side effects after breast cancer treatment?
Q2. How is osteoporosis related to breast cancer treatment?
Q3. What is osteoporosis and what causes it?

Fig. 1. Example output from Chat GPT for the given Context to generate 3 question

The evaluation of the models we have discussed so far was determined manually where the output text (questions generated in this case) meaningfulness, syntax and semantic soundness were all taken into consideration for each of the target languages. Chat GPT using the engine 'davinci-003' performed the best at producing 3, 5, 7 and 10 questions based on English, Spanish and multilingual texts of various sizes. Figure 1 shows an example output obtained from Chat GPT for a given context and the model was prompted to generate 3 questions based on the given context.

Although the model 'mT5' was capable of generating questions in both English and Spanish, it was not as consistent as the Chat GPT. Pre-trained models such as 'T5-based-e2e-qg', 'bert2bert-spanish-question-generation', 'T5', 'Flan-T5', 'BART' and 'mBART' worked well in their respective languages but were not able to handle multilingual question generation tasks and therefore failed this phase of testing. 'T5-based-e2e-qg' and 'bert2bert-spanish-question-generation' models also failed to create the correct number of questions when prompted to do so. Tables 2 and 3 summarize each model performance and sample outputs (questions generated) for monolingual and multilingual models.

Table 2. Question generation Performance for different Multilingual models

Model: Example Question(s)	Performance
Chat GPT: Q1. What medication is commonly prescribed after breast cancer treatment? Q2. ¿Cómo puede ser afectada la visión a largo plazo después del tratamiento del cáncer del seno?	Excellent in both languages. Questions only started to degrade in quality when prompted to produce 10 questions
mT5 (Small and Base): Q1. ¿Qué término puede afectar **la visión de la visión?** Q2. ¿Cuánto tiempo puede tardar en detectar cambios en el cáncer?	Some of the questions created repeated the same word or phrase or were generally vague
mBART: "Exception thrown"	None

*m indicates the model is "multilingual"

Table 3. Question generation Performance for different Monolingual models

Model: Example Question(s)	Performance
valhalla/t5-based-e2e-qg: Q1. What may be manifestations of heart disease or long-term cardiac toxicity?	This model had trouble producing over 5 questions per prompt. Many outputs were empty or null
mrm8488/bert2bert-spanish-question-generation: Q1. ¿ Qué palabra describe un tratamiento ineficiente para la osteoporosis después de que se trata de **osteoporosisisis**?	Issues with coherency and the repetition of certain letters, words and phrases were seen at all stages
T5 (Small, Base, Large): Q1. What is important to notice whether the vision is decreasing? Q2. What is important to notice if the vision is decreasing? Q3. Who is the doctor that discusses the long-term effects of breast cancer treatment?	While the outputs didn't have obvious spelling errors many of the questions created were vague or incoherent
Flan-T5 (Small, Base and Large): Q1. What is the name of the ophthalmologist who has an eye exam every one or two years? Q2. What is important to notice if the vision is decreasing? Q3. What is important to notice if the vision is decreasing?	This model produced grammatically correct results but they often weren't asking a specific or relevant question
BART (Base and Large): Q1. What is one of the long-term effects of breast cancer treatment? Q2. How long should you have an eye exam every year?	This model produced grammatically correct results but they often weren't asking a specific or relevant question

4 Evaluation Metrics

There are no specific evaluation metrics designated to question generation and it is also challenging to define a gold standard of proper questions to ask. Some of the useful criteria to evaluate questions can be to determine the *meaningfulness* of a question, how *syntactically correct* a question is, and how *semantically sound* the generated questions are, but these criteria are very difficult to quantify. As a result, most QG systems rely on human evaluation, by randomly sampling a set of questions from the generated questions, and asking human experts or annotators to rate the quality of questions on a 5-point Likert scale.

It is also a well-known fact that like any task requiring human interference for creation of references including human question generation, human evaluation for generated questions is also costly and time-consuming. Therefore, some commonly used automatic evaluation metrics for natural language generation (NLG), such as BiLingual Evaluation

Understudy [43], Metric for Evaluation of Translation with Explicit ORdering [44], National Institute of Standard and Technology [45], and Recall-Oriented Understudy for Gisting Evaluation [46] are also widely used for question generation. No matter how frequently these evaluation metrics are used even to date, some studies have shown that these evaluation metrics do not correlate well with adequacy, coherence and fluency [47–50] because these metrics evaluate by computing similarity between the source and the target sentence (in this case, the generated question) overlapping n-grams.

To overcome the issues encountered by these existing popular NLG evaluation metrics, recently a few metrics have been proposed [51–53]. Unlike the existing metrics, these new metrics [52] consider several question-specific factors such as named entities, content and function words, and the question type for evaluating the "answerability" of a question given the context [51, 52]. [53] proposed a dialogue evaluation model called ADEM that in addition to the word overlap statistics, uses a hierarchical RNN encoder to capture semantic similarity.

Some more recent metrics that also evaluate candidate questions given the reference questions are BLEURT [54–56], BERTScore [57] and MoverScore [58]. The BERTScore and MoverScore use Bidirectional Encoder Representations from Transformers [37] instead of n-gram overlap and use embeddings for token-level matching. BLEURT is a regression-based metric and uses supervised learning to train a regression layer that mimics human judgment.

The focus of this study was question generation as a result we used human evaluation to determine good quality questions for our questions corpora creation. However, to scale our work for a larger corpus we will be exploring these automatic evaluation metrics in our future work.

5 Discussion

This study was conducted to create a database of multilingual questions for both English and Spanish and answer the three research questions (RQ1–RQ3) discussed in the Introduction (Sect. 1).

5.1 RQ1: What Existing Systems or Models can be Used to Automatically Generate Different Sets of Questions Given the Context for English as Well as Spanish?

We found that there are several systems and models that exist that can be used to automatically generate questions given the context for both English and Spanish. The models that were tested for English in this study include 'valhalla/t5-based-e2e-qg', 'T5 (Small, Base, and Large)', 'Flan-T5 (Small, Base and Large)', 'BART (Base and Large)'. For Spanish question generation 'mrm8488/bert2bert-spanish-question-generation' model was implemented and finally for mixed data (containing both English and Spanish) 'Chat GPT', 'mT5 (Small and Base)' and 'mBART (Large)' were tested. Chat GPT out performed all models for generating different sets of questions (3, 5, 7 and 10 for this study) in English, Spanish and multilingual texts of various sizes considering each question's meaningfulness, syntax and semantic soundness. Other multilingual models such as 'mT5' showed promising results but the questions produced were repetitive and lacked meaningfulness as compared to questions generated by Chat GPT. While models such as T5 (Small, Base, and Large), Flan-T5 (Small, Base and Large), valhalla/t5-based-e2e-qg and BART (Base and Large) also produced coherent questions that were limited to single language contexts.

5.2 RQ2: Is the Quality of Questions Generated Consistent Across Different Texts and Thresholds? If the Quality is not Consistent, then What Factors Impact or Determine the Quality of Questions Generated?

The length of the transcript and the number of questions that our models were prompted to generate correlated with the overall quality of the outputs among all the models tested in the study where longer transcripts and fewer prompted questions generated higher quality outputs. It should also be noted that each of the models were pre-trained on different corpora and therefore performed better or worse than their counterparts during this experiment.

5.3 RQ3: What are the Optimal Values for These Factors Impacting the Question Generation Quality?

During our experiments we observed that a text/transcript of at least 100 words is good enough to generate 5–7 questions of reasonable quality. When models were prompted to produce 10 or more questions based on texts containing 100 words or less the meaningfulness, syntax and semantic soundness of outputs decreased notably.

5.4 Evaluation

Even though the focus of the study was not evaluation of questions but rather question generation, we feel that the next step of this study will be to evaluate these generated questions in order to build a corpus of high-quality questions for the healthcare domain. Due to the availability of small corpus used in this study, we considered human evaluation for determining the quality of generated questions. However, for scalability and future studies with larger corpora, we need to thoroughly investigate and improve upon the existing schemes to accurately measure the quality of questions, in particular deep questions.

Acknowledgement. This work was supported by grants from the National Science Foundation (NSF; award# 2131052 and award# 2219587). The opinions and findings expressed in this work do not necessarily reflect the views of the funding institution. Funding agency had no involvement in the conduct of any aspect of the research.

References

1. Rus, V., Cai, Z., Graesser, A.: Question generation: example of a multi-year evaluation campaign. In: Proc WS on the QGSTEC (2008)

2. Divate, M., Salgaonkar, A.: Automatic question generation approaches and evaluation techniques. Curr. Sci. **113**, 1683–1691 (2017)
3. Pan, L., Lei, W., Chua, T.S., Kan, M.Y.: Recent advances in neural question generation. arXiv preprint arXiv:1905.08949 (2019)
4. Lindberg, D., Popowich, F., Nesbit, J., Winne, P.: Generating natural language questions to support learning on-line. In: Proceedings of the 14th European Workshop on Natural Language Generation, Sofia, Bulgaria, pp. 105–114 (2013)
5. Cheng, Y.: Guiding the growth: difficulty-controllable question generation through step-by-step rewriting. arXiv preprint arXiv:2105.11698 (2021)
6. Puri, R., Spring, R., Patwary, M., Shoeybi, M., Catanzaro, B.: Training question answering models from synthetic data. arXiv preprint arXiv:2002.09599 (2020)
7. Du, X., Cardie, C.: Harvesting paragraph-level question-answer pairs from wikipedia. arXiv preprint arXiv:1805.05942 (2018)
8. Duan, N., Tang, D., Chen, P., Zhou, M.: Question generation for question answering. In: Proceedings of the 2017 conference on empirical methods in natural language processing, pp. 866–874 (2017)
9. Fabbri, A.R., Wu, C.S., Liu, W., Xiong, C.: QAFactEval: improved QA-based factual consistency evaluation for summarization. In: Proceedings of the 2022 Conference of the North American Chapter of the Association for Computational Linguistics: Human Language Technologies, pp. 2587–2601. Association for Computational Linguistics, Seattle, United States (2022)
10. Scialom, T., et al.: Questeval: summarization asks for fact-based evaluation. arXiv preprint arXiv:2103.12693 (2021)
11. Scialom, T., Lamprier, S., Piwowarski, B., Staiano, J.: Answers unite! Unsupervised metrics for reinforced summarization models. arXiv preprint arXiv:1909.01610 (2019)
12. Rebuffel, C., et al.: Data-QuestEval: a referenceless metric for data-to-text semantic evaluation. arXiv preprint arXiv:2104.07555 (2021)
13. Lee, H., Scialom, T., Yoon, S., Dernoncourt, F., Jung, K.: QACE: asking questions to evaluate an image caption. arXiv preprint arXiv:2108.12560 (2021)
14. Chen, G., Yang, J., Hauff, C., Houben, G.J.: LearningQ: a large-scale dataset for educational question generation. In: Proceedings of the International AAAI Conference on Web and Social Media, vol. 12 (1) (2018)
15. Heilman, M., Smith, N.A.: Good question! Statistical ranking for question generation. In: Human language technologies: The 2010 annual conference of the North American Chapter of the Association for Computational Linguistics, pp. 609–617 (2010)
16. Aldabe, I., Maritxalar, M.: Automatic distractor generation for domain specific texts. In: The International Conference on Natural Language Processing, pp. 27–38. Springer Berlin Heidelberg, Berlin, Heidelberg (2010)
17. Kumar, G., Banchs, R.E., D'Haro, L.F.: Revup: automatic gap-fill question generation from educational texts. In: Proceedings of the Tenth Workshop on Innovative Use of NLP for Building Educational Applications, pp. 154–161 (2015)
18. Agarwal, M., Mannem, P.: Automatic gap-fill question generation from text books. In: Proceedings of the sixth workshop on innovative use of NLP for building educational applications, pp. 56–64 (2011)
19. Narendra, A., Agarwal, M., Shah, R.: Automatic cloze-questions generation. In: Proceedings of the International Conference Recent Advances in Natural Language Processing RANLP 2013, pp. 511–515 (2013)
20. Mitkov, R.: Computer-aided generation of multiple-choice tests. In: Proceedings of the HLT-NAACL 03 workshop on Building educational applications using natural language processing, pp. 17–22 (2003)

21. Aldabe, I., De Lacalle, M.L., Maritxalar, M., Martinez, E., Uria, L.: Arikiturri: an automatic question generator based on corpora and nlp techniques. In: Intelligent Tutoring Systems: 8th International Conference, ITS 2006, Jhongli, Taiwan, June 26-30, 2006. Proceedings 8, pp. 584-594. Springer Berlin Heidelberg (2006)
22. Ali, H., Chali, Y., Hasan, S.A.: Automatic question generation from sentences. In: Actes de la 17e conférence sur le Traitement Automatique des Langues Naturelles. Articles courts, pp. 213–218 (2010)
23. Chali, Y., Hasan, S.A.: Towards automatic topical question generation. In: Proceedings of COLING 2012, pp. 475–492 (2012)
24. Corley, M.A., Rauscher, W.C.: Deeper learning through questioning. TEAL Cent. Fact Sheet **12**, 1–5 (2013)
25. Yao, X., Bouma, G., Zhang, Y.: Semantics-based question generation and implementation. Dialogue Discourse **3**(2), 11–42 (2012)
26. Liu, M., Calvo, R.A.: An automatic question generation tool for supporting sourcing and integration in students' essays. ADCS **2009**, 90 (2009)
27. Adamson, D., Bhartiya, D., Gujral, B., Kedia, R., Singh, A., & Rosé, C.P.: Automatically generating discussion questions. In: Artificial Intelligence in Education: 16th (2013)
28. Khullar, P., Rachna, K., Hase, M., Shrivastava, M.: Automatic question generation using relative pronouns and adverbs. In: Proceedings of ACL 2018, Student Research Workshop, pp. 153–158 (2018)
29. Rus, V., Lester, J.: The 2nd workshop on question generation. In: Artificial Intelligence in Education, p. 808. IOS Press (2009)
30. Rus, V., Wyse, B., Piwek, P., Lintean, M., Stoyanchev, S., Moldovan, C.: Overview of the first question generation shared task evaluation challenge. In: Proceedings of the Third Workshop on Question Generation, pp. 45–57 (2010)
31. Rus, V., Wyse, B., Piwek, P., Lintean, M., Stoyanchev, S., Moldovan, C.: Question generation shared task and evaluation challenge–status report. In: Proceedings of the 13th European Workshop on Natural Language Generation, pp. 318–320 (2011)
32. Rus, V., Wyse, B., Piwek, P., Lintean, M., Stoyanchev, S., Moldovan, C.: A detailed account of the first question generation shared task evaluation challenge. Dialogue Discourse **3**(2), 177–204 (2012)
33. Murakhovs' ka, L., Wu, C.S., Laban, P., Niu, T., Liu, W., Xiong, C.: Mixqg: neural question generation with mixed answer types. In: Proceedings of the 58th Annual Meeting of the Association for Computational Linguistics: System Demonstrations (2022)
34. Lelkes, A.D., Tran, V.Q., Yu, C.: Quiz-style question generation for news stories. In: Proceedings of the Web Conference 2021, pp. 2501–2511 (2021)
35. Qi, W., et al.: Prophetnet: predicting future n-gram for sequence-to-sequence pre-training. arXiv preprint arXiv:2001.04063 (2020)
36. Dong, L., et al.: Unified language model pre-training for natural language understanding and generation. Advances in neural information processing systems, 32 (2019)
37. Devlin, J., Chang, M.W., Lee, K., Toutanova, K.: (2019). Bert: pre-training of deep bidirectional transformers for language understanding. In: Proceedings of the 2019 Conference of the North American Chapter of the Association for Computational Linguistics: Human Language Technologies, vol. 1 (Long and Short Papers), pp. 4171–4186. Association for Computational Linguistics, Minneapolis, Minnesota (2019)
38. Chang, Y., et al.: A survey on evaluation of large language models. arXiv preprint arXiv:2307.03109 (2023)
39. Brown, T., et al.: Language models are few-shot learners. Adv. Neural. Inf. Process. Syst. **33**, 1877–1901 (2020)
40. Cretu, C.: How Does ChatGPT Actually Work? An ML Engineer Explains, Scalable Path. https://www.scalablepath.com/data-science/chatgpt-architecture-explained (2023)

41. Alammar, J.: The Illustrated Transformer [Blog post]. Retrieved from https://jalammar.git hub.io/illustrated-transformer/ (2018)
42. Lewis, M., et al.: Bart: denoising sequence-to-sequence pre-training for natural language generation, translation, and comprehension. arXiv preprint arXiv:1910.13461 (2019)
43. Papineni, K., Roukos, S., Ward, T., Zhu, W.J.: Bleu: a method for automatic evaluation of machine translation. In: Proceedings of the 40th annual meeting of the Association for Computational Linguistics, pp. 311–318 (2002)
44. Lavie, A., Denkowski, M.J.: The meteor metric for automatic evaluation of machine translation. Mach. Transl. **23**, 105–115 (2009)
45. Doddington, G.: Automatic evaluation of machine translation quality using n-gram co-occurrence statistics. In: Proceedings of the second international conference on Human Language Technology Research, pp. 138–145 (2002)
46. Lin, C.Y.: Rouge: a package for automatic evaluation of summaries. In: Text summarization branches out, pp. 74–81 (2004)
47. Ananthakrishnan, R., Bhattacharyya, P., Sasikumar, M., Shah, R.M.: Some issues in automatic evaluation of english-hindi mt: more blues for bleu. Icon 64 (2007)
48. Callison-Burch, C.: Fast, cheap, and creative: evaluating translation quality using amazon's mechanical turk. In: Proceedings of the 2009 conference on empirical methods in natural language processing, pp. 286–295 (2009)
49. Callison-Burch, C., Osborne, M., Koehn, P.: Re-evaluating the role of BLEU in machine translation research. In: The 11th Conference of the European Chapter of the Association for Computational Linguistics, pp. 249–256 (2006)
50. Liu, C.W., Lowe, R., Serban, I.V., Noseworthy, M., Charlin, L., Pineau, J.: How not to evaluate your dialogue system: an empirical study of unsupervised evaluation metrics for dialogue response generation. arXiv preprint arXiv:1603.08023 (2016)
51. Mohammadshahi, A., et al.: RQUGE: reference-free metric for evaluating question generation by answering the question. arXiv preprint arXiv:2211.01482 (2022)
52. Nema, P., Khapra, M.M.: Towards a better metric for evaluating question generation systems. arXiv preprint arXiv:1808.10192 (2018)
53. Lowe, R., et al.: Towards an automatic turing test: learning to evaluate dialogue responses. arXiv preprint arXiv:1708.07149 (2017)
54. Ushio, A., Alva-Manchego, F., Camacho-Collados, J.: An empirical comparison of LM-based question and answer generation methods. arXiv preprint arXiv:2305.17002 (2023)
55. Ushio, A., Alva-Manchego, F., Camacho-Collados, J.: A practical toolkit for multilingual question and answer generation. arXiv preprint arXiv:2305.17416 (2023)
56. Sellam, T., Das, D., Parikh, A.P.: BLEURT: learning robust metrics for text generation. arXiv preprint arXiv:2004.04696 (2020)
57. Zhang, T., Kishore, V., Wu, F., Weinberger, K.Q., Artzi, Y.: Bertscore: evaluating text generation with bert. In: The International Conference on Learning Representations (2020)
58. Zhao, W., Peyrard, M., Liu, F., Gao, Y., Meyer, C.M., Eger, S.: MoverScore: text generation evaluating with contextualized embeddings and earth mover distance. In: Proceedings of the 2019 Conference on Empirical Methods in Natural Language Processing and the 9th International Joint Conference on Natural Language Processing (EMNLP-IJCNLP), pp. 563–578, Association for Computational Linguistics, Hong Kong, China (2019)

DSEN: A Distance-Based Semantic Enhancement Neural Network for Pronoun Anaphora Resolution

Xiaojian Li[1(✉)], Zheng Xiao[2], Linhao Li[2], Ye Tian[1,2], and Lei Ma[1,2]

[1] School of Artificial Intelligence, University of Chinese Academy of Science, Beijing, China
lixiaojian21@mails.ucas.ac.cn

[2] ELITE: Efficient intelLIgent compuTing and lEarning, Nanjing, China

Abstract. Anaphora resolution is one of the fundamental tasks in natural language processing, aiming to identify the specific entities referred to by pronouns or noun phrases. In recent years, with the advancement of deep learning, the performance of anaphora resolution has significantly improved. To accurately identify the entities referred to by pronouns based on context, various factors need to be considered, such as the semantic dependencies in the context, the positions of entity words and pronouns, among others. However, recent research has lacked consideration of the impact of word distances on the correlation between word pairs. In this paper, we propose a anaphora resolution model called **DSEN**, which stands for **D**istance-based **S**emantic **E**nhancement Neural **N**etwork for Pronoun Anaphora Resolution. Our model integrates explicit positional relationships between entity words and pronouns, key meanings, and semantic information. By considering the influence of word distances, our model effectively incorporates the dependencies between word pairs. The fusion of these factors empowers our model to achieve outstanding results on the CLUEWSC2020 anaphora resolution dataset.

Keywords: neural networks · information extraction · anaphora resolution

1 Introduction

Anaphora resolution is a common linguistic phenomenon in natural language, which refers to the use of a pronoun to refer back to a previously mentioned linguistic unit in discourse. It plays an important role in maintaining semantic coherence in language. There are many types of references, including personal pronouns, demonstrative pronouns, zero pronouns, nominal phrases, and even

F. Zhao and D. Miao (Eds.): AIGC 2023, CCIS 1946, pp. 12–23, 2024.
https://doi.org/10.1007/978-981-99-7587-7_2

sentences or clauses. Anaphora resolution is the process of finding the linguistic unit that a pronoun refers to within a discourse. It mainly examines the referential relationship between the pronoun and its adjacent noun phrase.

The current research on zero-anaphora resolution typically uses word sense disambiguation to cluster words that have the same referent, as seen in studies such as [1] and [2]. This approach faces two main problems. Firstly, due to the complexity and diversity of natural language, most clustering results do not actually contribute to semantic understanding; instead, they may lead to redundant computations. Additionally, some meaningless pronouns might be grouped together, such as clustering several instances of "her" from different positions in the text, which fails to accurately reflect the referent or provide meaningful semantic information. In reality, only a few entities play a crucial role in the readability and comprehension of a sentence. Secondly, most studies focus on anaphora resolution tasks based solely on contextual semantics and neglect the influence of word distances on semantic relationships. The impact of word proximity is crucial in capturing the dependency relationship between pronouns and their antecedents. Moreover, the majority of research on anaphora resolution focuses on judging a single antecedent (entity) and pronoun in a sentence. However, when a sentence contains multiple antecedents and pronouns, multiple judgments are required, making it impossible to obtain the referents for all pronouns in one step.

This paper aims to construct a novel neural network model that utilizes context semantics, word distance, and word pair similarity to determine the reference of pronouns. This approach effectively avoids redundant calculations while incorporating distance factors and computing the degree of association from multiple dimensions, providing traceability of words. The model is capable of identifying all the anaphors with their corresponding antecedents in the sentence at once. We summarize our contributions as follows:

- We propose a novel anaphora resolution model that integrates information from multiple dimensions and calculates the correlation between pronouns and their antecedents in a targeted manner.
- We introduce distance attenuation as a feature and utilize it as an influencing factor for the semantic correlation between pronouns and their antecedents.
- The DSEN model achieved outstanding results on the CLUEWSC2020 anaphora resolution dataset, which demonstrates the model's effectiveness in addressing the task of anaphora resolution.

2   Related Work

Much of the earlier work in anaphora resolution heavily exploited domain and linguistic knowledge [3,4], which was difficult both to represent and process, and required considerable human input [5,6]. Meanwhile,machine learning methods have a long history in anaphora resolution such as simple co-occurrence rules [7] through training decision trees to identify anaphor-antecedent pairs [8] to genetic algorithms to optimise the resolution factors [9]. This type of task has also been extended to various language systems [19, 20, 23, 25] and application scenarios [24, 26].With the widespread application of deep learning in the field of coreference resolution, Kenton Lee *et al.* proposed an end-to-end neural network model [1] in 2017 that captures contextual semantic information using LSTM and judges each span in the sentence by defining a coreference score. The following year, Kenton Lee *et al.* proposed [2], which added a less accurate but more efficient bilinear factor to enable more aggressive pruning without sacrificing accuracy. [22] have improved the commonly used masked language model for zero-anaphora resolution and introduced a new fine-tuning technique that significantly enhances the model's performance.

This zero-coreference problem has some common issues: 1) even with pruning, there is still a significant amount of redundant computation. 2) Since no information about pronouns and named entities is provided, the resulting clusters can be meaningless and excessive.

After Devlin *et al.* [11] proposing to learn the contextual representations of each word by training a large scale language model. They proposed the BERT and broken the record of many NLP tasks. Mandar Joshi *et al.* [13] apply BERT to coreference resolution, achieving strong improvements on the OntoNotes and GAP benchmarks. Juntao Yu *et al.* [10]proposed a BERT-based anaphora resolution model, where they introduce a model for unrestricted resolution of split-antecedent anaphors; A year after that, Youngtae Kim [12] achieve high performance in the task of anaphora resolution by using BERT as encoder.

The Attention mechanism has also been widely applied to NLP tasks. Juntao Yu *et al.* [14] removed the position embeddings, used in the self-attention to determine the relative importance of the mentions in the cluster; In addition, to better interpret zero pronouns and their candidate antecedents, [15] introduce a convolutional neural networks with internal self-attention model for encoding them; [16] demonstrate how important are the self-attention patterns of different heads for bridging anaphora resolution.

Given the positional characteristics of pronouns and antecedents in a sentence, the distance between them has a certain impact on their referential relationship. Studies such as [17] have demonstrated that distance attenuation can effectively capture the influence of word distance on correlation, thus more accurately describing the correlation between two words. Similarly, in the work of [18], distance attenuation is also utilized as a parameter to optimize the performance.

3   Model

The goal of this article is to propose a new anaphora resolution model that can obtain all pronoun references in a sentence at once. Our proposed model is shown in Fig. 1.

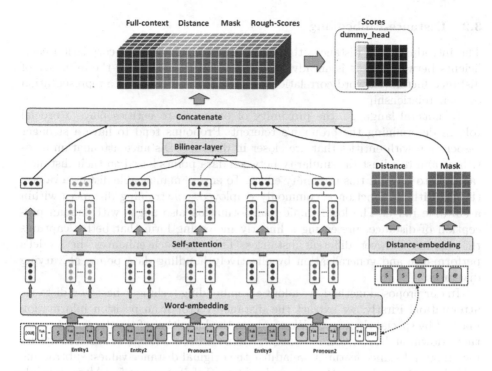

Fig. 1. Architecture of **DSEN**. Entities and Pronouns are tagged. The model has two processes: Semantic and Distance Feature Fusion, and Anaphora Scoring. It encodes the sentence, computes attention, and uses a bilinear layer for scoring. Features are encoded considering distance attenuation. (To handle pronouns with no antecedent, a negligible value is introduced as a dummy_head at the first position during scoring)

3.1   Word Embedding

The original dataset pairs pronouns and antecedents in a text and annotates the pairing results, constituting a binary classification problem. In this study, we redesigned the dataset to annotate all pronouns and antecedents belonging to the same text at once, using different special characters for boundary marking. This approach allows us to obtain explicit positional features and obtain pronoun referent resolution results in one step. The resulting sentence takes the form of \$李燕\$坐车去上海看病。\$宋凡平\$说过等到@她@彻底恢复了,@他@会亲自去接@她@回来。 Here, "李燕" and "宋凡平" are two named entities in the sentence, marked with \$, while "她", "他", and "她" are three pronouns marked with @. We used Bert as the encoder to encode the text. Given an input sentence $X = [x_1, x_2, \ldots, x_i](j \in [1, s]$,where s is the length of the sentence. The input is passed through BERT to obtain the token-level representation X_i. At the same time, Bert can obtain the vector X_{CLS}, which contains the full sentence content.

$$X_{CLS}, X_i = BERT(x_i) \tag{1}$$

3.2   Distance-Embedding

The introduction of distance attenuation in the calculation of correlation coefficients between words is an advanced extension that captures the influence of distance factors on word correlations, enabling a more accurate representation of their relationship.

In natural language, the proximity of pronouns to entities plays a crucial role in determining the pronoun's referent. Pronouns tend to have a stronger association with entities that are closer in distance. Distance attenuation algorithms aim to adjust the similarity between data points based on their distance, taking into account this proximity effect. To ensure manageable distance values, the logarithmic function is commonly employed, constraining distances within a suitable range. The logarithmic transformation also aligns with human perception of distance, presenting a linearly increasing form that better captures the nuances between different distances. This approach enhances the model's performance and generalization by effectively handling data points at varying distances.

In our proposed method, we adopt an embedding technique based on distance attenuation. Firstly, we extract the distance vector from position information marked by special characters. We compute the logarithm of the distances, round them down, and limit them to integer values within a range of six. To capture meaningful distance features, we adjust the original distance values: subtracting 1 if the distance is less than 5, and adding 2 if it exceeds 5. (After multiple experiments, when the distance threshold is set to 5, the performance is significantly improved and remains stable)This transformation ensures that distances are appropriately scaled while preserving the underlying distance characteristics. Finally, we embed these modified distance features, facilitating the representation of relative distances between words. Mathematically, the process can be represented as follows:

$$distance^* \begin{cases} distance - 1, & distance < 5 \\ log(distance) + 2, & otherwise \end{cases} \tag{2}$$

Let $distance$ be the original distance between two words and $distance^*$ be the transformed distance after applying distance attenuation.By introducing a distance attenuation algorithm, we have incorporated inter-word distance features, enabling the model to achieve a more refined understanding of the correlation between pronouns and their antecedents. This enhancement significantly improves the model's ability to capture nuanced relationships between pronouns and other referents.

3.3   Self-attention

Self-attention calculates the similarity between each element in a sequence and other elements, and calculates the weight of each element based on these similarities. In natural language processing, self-attention can be applied to sentences or text sequences to capture the relationships and dependencies between different words by comparing the similarity between words at different positions. First, a set of vectors $X = [x_1, x_2, ..., x_n]$(such as multiple character vectors that make up a word) are projected onto three different vector spaces to obtain three sets of vectors $Q = [q_1, q_2, ..., q_n]$, $K = [k_1, k_2, ..., k_n]$, and $V = [v_1, v_2, ..., v_n]$. These three sets of vectors are used to calculate similarities and perform weighted summation, i.e.:

$$Q = X \cdot W_Q, K = X \cdot W_K, V = X \cdot W_V \tag{3}$$

where W_Q, W_K, and W_V are projection matrices. Then, the similarity (score) between each element and other elements is calculated using formula (4), using the dot product by multiplying the transpose of Q with K, and then divided by $\sqrt{d_k}$ (where d_k is the number of columns of matrix K) to obtain the score for each character vector. The *softmax* operation in formula (5) obtains a set of weights (n-dimensional vector) that represent the importance of other elements related to x_i. These weights are then multiplied by V to obtain the weighted representation of each element. Finally, all the weighted representations of the elements are concatenated to obtain the representation of the entire word vector, as shown in formula (6):

$$score(x_i, x_j) = \frac{Q_i^T \cdot K_j}{\sqrt{d_k}} \cdot V \tag{4}$$

$$attention(x_i) = softmax(score(xi, xj)) \cdot V \tag{5}$$

$$attention(X) = [attention(x_1), attention(x_2), ..., attention(x_n)] \tag{6}$$

3.4   Fine-Grained Scoring

The bilinear method is a widely used technique in natural language processing to project two vectors onto a shared low-dimensional space and compute their similarity or relationship in this space. This method offers a high degree of flexibility and scalability. To calculate the relationship scores between pronouns and all entity words, we adopt this approach, which projects the embeddings of pronouns and entity words into a shared space using linear transformation matrices W_p and W_e respectively, as shown in Eqs. (7) and (8):

$$h_e = W_e * e \tag{7}$$

$$h_p = W_p * p \tag{8}$$

$$S_{p,e} = U * h_p * h_e^T \tag{9}$$

where e is the embedding matrix for all entity words, p is the embedding vector for the pronoun, h_e and h_p are the projected vectors for the entity and pronoun, respectively. W_p and W_e are learnable matrices. $S_{p,e}$ is the similarity score between the projected pronoun and entity vectors, and U is a learnable matrix. Then we calculate the final Score:

$$S = S_{p,e} \odot X_{CLS} \odot distance^* \odot Mask \tag{10}$$

$$LeakyReLU(x) = \begin{cases} x & , x > 0 \\ \alpha x & , x \leq 0 \end{cases} \tag{11}$$

$$score = Linear(Dropout(LeakyRuLU(Linear(S)))) \tag{12}$$

$$Score = dummy\_head \odot score \tag{13}$$

where $\odot$ denotes concatenation operation and α is a small positive constant, typically set to a small value like 0.01. X_{CLS} refers to a representation that encompasses the entire sentence meaning, as mentioned in Sect. 3.1.2. $Mask$ is a technique borrowed from the Transformer [21] decoder, which allows us to control anaphors to only attend to previously mentioned antecedents. $distance^*$ represents the effect of distance attenuation. $LeakyReLU$ is an activation function that improves gradient propagation in deep neural networks. It allows non-zero gradients for negative inputs, leading to faster training, better handling of negative values, and increased robustness to noise and outliers. We combine the four components mentioned above and use formula (12) for a refined calculation, resulting in the $Score$ matrix. Before that, we add a $dummy_head$ at the beginning of the $score$" matrix to represent pronouns without references (usually set to a very small value). After applying the $tanh$ activation function, each row of the $Score$ matrix corresponds to a pronoun, and the index with the highest value in each row indicates the pronoun's reference.

4   Experiments

4.1   Datasets

CLUEWSC2020. CLUEWSC2020 is a dataset for Chinese anaphora resolution task, consisting of 1243 sentences in the training set, 304 sentences in the validation set, and 290 sentences in the test set. Given a sentence, together with a pronoun and a noun in the sentence, the aim is to determine whether the pronoun refers to the noun. In this paper we merged multiple instances with the same sentence into a single instance, which contains one sentence, multiple pronouns, and multiple nouns. The goal then becomes to determine the referent noun for each pronoun.

4.2   Parameter Settings

Table 1 shows the major parameters used in our experiments for CLUEWSC2020. The BERT model we use in our experiments is a cased basic model, which uses all cross-entropy loss functions and adds a Dropout layer to prevent overfitting.

4.3   Comparsion with Other Mehods

We compared our own method with the currently available method with out-standing results in CLUEWSC2020, and both used average accuracies comparison format, and the results reported in the baseline were copied directly from the original published article.

On the CLUEWSC2020 dataset, we compared the previous models which have outstanding effect. The specific effects are shown in Table 2.

Table 1. Parameter settings for CLUEWSC2020

CLUEWSC2020	
Epoch	8
Batch_size	6
Max_sequence_length	256
Learning_rate	1e-4
Dummy_head	1e-3
Dropout	0.25
Update	0.25
Optimizer	Adam

Table 2. Performance of baseline models on CLUEWSC2020 from *CLUEbenchmark*. Bold text denotes the best result.

Method	benchmark
BERT-base	62.0
BERT-wwm-ext	61.1
XLNet-mid	64.4
ERNIE-base	60.8
ALBERT-base	64.34
ALBERT-large	62.24
ALBERT-xxlarge	61.54
RoBERTa-large	72.7
RoBERTa-wwm-large	74.6
DSEN	**80.5**

In the CLUEWSC2020 dataset, XLNet adopts a permutation-based language model architecture, which allows it to better capture the correlations and dependencies between words. However, it requires more computational resources compared to BERT's masked language model architecture. Additionally, we believe that ALBERT-base performs better than ALBERT-large and ALBERT-xxlarge due to its smaller parameter size, which helps to avoid overfitting on small-sample tasks. On the other hand, RoBERTa-wwm-large benefits from pretraining with large-scale Chinese data, enhancing its ability to understand Chinese context and semantic information. Our model exhibits superior performance compared to these mainstream deep learning models, and several factors contribute to its competitive advantage. Firstly, we did not choose a more complex pretraining model for small-sample data. Secondly, we used self-attention mechanism when computing word embeddings, enhancing potential significant features of pronouns and antecedents. This weighted enhancement effectively amplifies the crucial parts of word vectors responsible for the correlations between word pairs. Additionally, we incorporated the effect of distance attenuation into our model. This consideration of word proximity impact combines the influence of distance features on word dependencies. Through the use of bilinear layers, we successfully calculated scores between each pronoun and its antecedent, endowing our model with the ability to discern their inherent relationships within the same dimensional space. The fusion of these architectures propelled our model to achieve outstanding performance.

However, the low-resource nature of the CLUEWSC2020 dataset poses challenges for further result optimization. The limited amount of training data increases the risk of model overfitting, where the model becomes excessively tuned to the training set and may struggle to generalize to unseen data.

4.4   Ablation Studies

To analyze the contributions and effects of self-attention and distance attenuation, we perform ablation tests. The results are shown in Table 3.

M1. This configuration performed poorly, as it lacked the ability to capture contextual information and incorporate word proximity. Consequently, anaphoric resolution was inaccurate.

M2. The inclusion of distance attenuation improved performance compared to the first configuration. However, without attention, the model still struggled to focus on important contextual features.

M3. Adding attention mechanism notably improved performance. The model captured linguistic nuances and made more accurate anaphoric references. However, the absence of distance attenuation limited its sensitivity to word proximity, resulting in slightly lower performance than the fourth configuration.

M4. This configuration achieved the best performance. Both attention and distance attenuation were crucial. The attention mechanism captured contextual cues, while distance attenuation considered word proximity, leading to the highest accuracy in anaphoric resolution.

In order to assess the impact of embedding layer dimensions on the model, we conducted experiments with different dimensions for the distance embedding layer. The results, presented in Table 4, demonstrate that the model achieves

Table 3. Self-attention and Distance attenuation ablation studies

Method	CLUEWSC2020		
	train-F1	dev-F1	test-F1
M1: $DSEN_{withoutAttorDis}$	73.84	72.78	68.74
M2: $DSEN_{withoutAtt}$	76.93	73.15	70.66
M3: $DSEN_{withoutDis}$	74.35	73.61	70.45
M4: $DSEN$	**80.12**	**78.81**	**72.77**

optimal performance when the layer dimension is set to 15. However, as the layer dimension increases to 20, the model's accuracy significantly declines. This indicates that models with lower layer dimensions may not fully capture the complexity of semantic relationships. Increasing the embedding layer dimension appropriately enhances the model's representation capability, allowing it to better capture the details and features of the input data, thereby improving performance. However, excessive layer dimensions, given limited data, can result in learning excessive noise or overfitting.

Table 4. Ablation studies of different distance embedding dimensions

Dis_Emb_dims	acc	pre	recall	F1
5	72.10	71.47	69.40	67.04
10	74.76	72.95	71.64	73.64
15	**77.71**	**83.63**	**77.69**	**78.81**
20	69.28	68.16	58.41	56.45

5 Conclusions

We utilize BERT as a pretrained language model to obtain word vector representations with rich semantic features. By employing the attention mechanism, we enhance the weights of important parts in the text and generate weighted sum of the character vectors to obtain word vectors. Distance attenuation algorithm is employed to encode the influence of distances between antecedents and pronouns, and a mask mechanism is added to mask out irrelevant nouns following each pronoun to prevent noise interference. The entity words and pronouns are then fused using a bilinear layer for rough calculation of anaphoric scores. Subsequently, fine-grained scoring is conducted by incorporating distance encoding features and the vector representing the entire sentence. The experimental results demonstrate a significant improvement in the anaphora resolution task. In our future work, we plan to focus on the resolution of pronouns referring to antecedents in the preceding context, while acknowledging the possibility of a few antecedents occurring in the subsequent context. The strategies for anaphoric calculation still require optimization and further research. Additionally, we aim to reduce the number of model parameters to effectively mitigate the risk of overfitting and improve inference speed. We kindly invite you to review this paper and provide us with any new ideas or insights.

References

1. Lee, K., He, L., Lewis, M., Zettlemoyer, L.: End-to-end neural coreference resolution. arXiv preprint arXiv:1707.07045 (2017)
2. Lee, K., He, L., Zettlemoyer, L.: Higher-order coreference resolution with coarse-to-fine inference. arXiv preprint arXiv:1804.05392 (2018)
3. Carbonell, J.G., Brown, R.D.: Anaphora resolution: a multi-strategy approach. In: Coling Budapest 1988 Volume 1: International Conference on Computational Linguistics (1988)
4. Carter, D.: Interpreting Anaphors in Natural Language Texts. Halsted Press (1987)
5. Rich, E., LuperFoy, S.: An architecture for anaphora resolution. In: Second Conference on Applied Natural Language Processing, pp. 18–24 (1988)
6. Sidner, C.L.: Toward a computational theory of definite anaphora comprehension in English. Technical report AI-TR-537. MIT (1979)
7. Dagan, I., Itai, A.: Automatic processing of large corpora for the resolution of anaphora references. In: COLING 1990 Volume 3: Papers Presented to the 13th International Conference on Computational Linguistics (1990)
8. Aone, C., William, S.: Evaluating automated and manual acquisition of anaphora resolution strategies. In: 33rd Annual Meeting of the Association for Computational Linguistics, pp. 122–129 (1995)
9. OrÅsan, C., Evans, R., Mitkov, R.: Enhancing preference-based anaphora resolution with genetic algorithms. In: Christodoulakis, D.N. (ed.) NLP 2000. LNCS (LNAI), vol. 1835, pp. 185–195. Springer, Heidelberg (2000). https://doi.org/10.1007/3-540-45154-4_17
10. Yu, J., Moosavi, N.S., Paun, S., Poesio, M.: Free the plural: unrestricted split-antecedent anaphora resolution. arXiv preprint arXiv:2011.00245 (2000)
11. Devlin, J., Chang, M.-W., Lee, K., Toutanova, K.: BERT: pre-training of deep bidirectional transformers for language understanding. arXiv preprint arXiv:1810.04805 (2018)
12. Kim, Y., Ra, D., Lim, S.: Zero-anaphora resolution in Korean based on deep language representation model: BERT. ETRI J. **43**(2), 299–312 (2021)
13. Joshi, M., Levy, O., Weld, D.S., Zettlemoyer, L.: BERT for coreference resolution: baselines and analysis. arXiv preprint arXiv:1908.09091 (2019)
14. Yu, J., Uma, A., Poesio, M.: A cluster ranking model for full anaphora resolution. arXiv preprint arXiv:1911.09532 (2019)
15. Sun, S.: Self-attention enhanced CNNs with average margin loss for Chinese zero pronoun resolution. Appl. Intell. **52**(5), 5739–5750 (2022)
16. Pandit, O., Hou, Y.: Probing for bridging inference in transformer language models. arXiv preprint arXiv:2104.09400 (2021)
17. Li, J., Huang, G., Chen, J., Wang, Y.: Dual CNN for relation extraction with knowledge-based attention and word embeddings. Comput. Intell. Neurosci. **2019** (2019)
18. Xu, Z., Dong, K., Zhu, H.: Text sentiment analysis method based on attention word vector. In: IEEE 2020 International Conference on Modern Education and Information, ICMEIM, pp. 500–504 (2020)

19. Ertan, M.: Pronominal anaphora resolution in Turkish and English. Middle East Technical University (2023)
20. Ferilli, S., Redavid, D.: Experiences on the improvement of logic-based anaphora resolution in English texts. Electronics **11**(3), 372 (2022)
21. Vaswani, A., et al.: Attention is all you need. In: Advances in Neural Information Processing Systems, vol. 30 (2017)
22. Ryuto, K., Shun, K., Yuichiroh, M., Hiroki, O., Kentaro, I.: Pseudo zero pronoun resolution improves zero anaphora resolution. cs.CL, arXiv:2104.07425 (2021)
23. Senapati, A., Poudyal, A., Adhikary, P., Kaushar, S., Mahajan, A., Saha, B.N.: A machine learning approach to anaphora resolution in Nepali language. In: IEEE 2020 International Conference on Computational Performance Evaluation (ComPE), pp. 436–441 (2020)
24. Tsvetkova, A.: Anaphora resolution in Chinese for analysis of medical Q&A platforms. In: Zhu, X., Zhang, M., Hong, Yu., He, R. (eds.) NLPCC 2020, Part II. LNCS (LNAI), vol. 12431, pp. 490–497. Springer, Cham (2020). https://doi.org/10.1007/978-3-030-60457-8_40
25. Li, S., Qu, W., Wei, T., Zhou, J., Gu, Y., Li, B.: A survey of Chinese anaphora resolution. In: Sun, X., Zhang, X., Xia, Z., Bertino, E. (eds.) ICAIS 2021, Part I. LNCS, vol. 12736, pp. 180–192. Springer, Cham (2021). https://doi.org/10.1007/978-3-030-78609-0_16
26. Wongkoblap, A., Vadillo, M., Curcin, V.: Depression detection of twitter posters using deep learning with anaphora resolution: algorithm development and validation. JMIR Ment. Health (2021)

Inheritance and Revitalization: Exploring the Synergy Between AIGC Technologies and Chinese Traditional Culture

Yuhai Zhang, Naye Ji(✉), Xinle Zhu, and Youbing Zhao

Communication University of Zhejiang, Hangzhou 310018, China
jinaye@cuz.edu.cn

Abstract. Diffusion models like Stable Diffusion have made impressive progress in T2I (text-to-image) generation. However, when applied to image generation tasks concerned with Chinese cultural subjects, Stable Diffusion needs to improve the quality of its results. This paper proposes a practical approach to address the challenges of utilizing popular AIGC (AI-Generated Content) technologies and integrating them into a cohesive system, which makes it easier to use Stable Diffusion to create high-quality generated images related to Chinese cultural subjects with direct Chinese prompts. Specifically, with the capabilities of Large Language Models (LLMs), the approach can weave expressive visual descriptions based on initial inputs (scattered words in Chinese, Chinese poems…) and align them in suitable English text prompts for subsequent image generation with Stable Diffusion. Through the parameter-efficient finetuning method called LoRA, Stable Diffusion can effectively learn complex and nuanced concepts of Chinese culture. Additionally, Prompt Engineering plays a role in optimizing inputs, assuring quality and stability, and setting the detailed behavior patterns of LLMs throughout the workflow. This success is attributed to overcoming the constraints of accepting only English prompts and significantly improving the understanding of certain concepts in Chinese culture. The experiments show that our method can produce high-quality images associated with complex and nuanced concepts in Chinese culture by leveraging the fusion of all independent components.

Keywords: Diffusion Models · Large Language Models · Chinese Traditional Culture

1 Introduction

Recently, diffusion models have emerged as promising generative models, particularly for T2I (text-to-image) synthesis. Specifically, large-scale T2I diffusion models such as DALL·E 2 [1], Imagen [2], and Stable Diffusion [3] provide powerful image generation capabilities. It is now possible to generate high-fidelity images based on text prompts. Such powerful T2I image models can be applied in various applications.

The diffusion model possesses remarkable capabilities in visual creativity and multi-style creation. Its application in generating images based on Chinese cultural and artistic concepts holds significant potential and diverse applications, which can facilitate the dissemination and revitalization of Chinese culture. By enhancing the capabilities of T2I generation models to learn and comprehend artistic mediums like ink painting and meticulous brushwork, shadow puppetry, as well as the intricate patterns and conceptual elements unique to Chinese traditional culture, these valuable cultural heritages can be integrated and innovated with modern features to align with the aesthetics of the general public and obtain new cultural connotations.

However, when applied to image generation tasks associated with Chinese cultural subjects, Stable Diffusion still faces the need to improve the quality of its results due to the constraint of accepting only English prompts and its limited understanding of certain concepts in Chinese culture.

Existing research, such as Taiyi Diffusion [4] and AltDiffusion [5], was developed by training a new Chinese text encoder to address these issues. As a result, even though these models were trained on large datasets, the lack of interaction between the Chinese and CLIP text encoder leads to poor alignments between Chinese, English, and images [6]. Moreover, existing research is often time-consuming and costly. It would be deemed unacceptable to disregard the abundance of valuable open-source resources around Stable Diffusion.

This paper aims to design practicable methods to address these issues within existing Diffusion Model based frameworks. Additionally, it aims to explore the synergy between AIGC technologies and Chinese traditional culture. In this paper, we propose a practical approach to address the challenges associated with the utilization of popular AIGC technologies and align them for integration into a cohesive system.

Specifically, with the capabilities of Large Language Models (LLMs) [7, 10], the approach can weave expressive visual descriptions based on initial inputs (scattered words in Chinese, Chinese poems...) and align them in suitable English text prompts for subsequent image generation with Stable Diffusion. Through the parameter-efficient finetuning method called LoRA [8], Stable Diffusion can effectively learn complex and nuanced concepts of Chinese culture. Additionally, Prompt Engineering plays a role in optimizing input, assuring quality and stability, and setting the detailed behavior patterns of LLMs throughout the workflow.

2   Related Work

2.1   Text-to-Image Synthesis

T2I synthesis has been a subject of considerable interest over an extended period. In the early stage, GAN was a popular choice as the architecture for T2I synthesis models. Recently, large-scale diffusion models have significantly improved the quality and alignment of generated images and replaced GANs in many image-generation tasks [9]. Due to the prosperous open-source community of Stable Diffusion, it is chosen as the default model to implement our T2I generation in this paper.

2.2   Large Language Models

Large language models (LLMs), such as ChatGPT, have attracted enormous attention due to their remarkable performance on various Natural Language Processing (NLP) tasks. LLMs can produce superior language understanding, generation, interaction, and reasoning capability based on large-scale pre-training on massive text corpora and Reinforcement Learning from Human Feedback (RLHF) [11]. The potent ability of LLMs also drives many emergent research topics to investigate the enormous potential of LLMs further. It brings possibilities for us to build advanced artificial intelligence systems [12]. In this paper, we leverage the capabilities of LLMs to weave expressive English text prompts for Stable Diffusion based on initial inputs.

2.3 LoRA

Low-Rank Adaptation of Large Language Models (LoRA) is a training method that accelerates the training of large models while consuming less memory. It adds pairs of rank-decomposition weight matrices (called update matrices) to existing weights and only trains those newly added weights. With this method to train models, the original pre-trained weights are kept frozen to have multiple lightweight and portable LoRA models for various downstream tasks built on top of them. LoRA was initially proposed for large-language models and demonstrated on transformer blocks, but the technique can also be applied elsewhere. In the case of Stable Diffusion fine-tuning, LoRA can be used for the cross-attention layers that relate the image representations with the prompts that describe them. The method is more effective than other fine-tuning methods for few-shot tasks, such as Textual Inversion [13] and DreamBooth [14].

3   Method

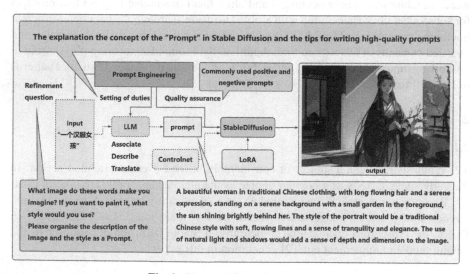

Fig. 1. The workflow of our method.

We propose a practical workflow by leveraging the power of diffusion models, LLMs, prompt engineering, and LoRA. This method provides flexible options for the composition of different structures. Moreover, the type and implementation of components in the workflow are also adjustable. This workflow can produce high-quality images featuring Chinese culture.

The method can be divided into two parts. Composed of LLMs and prompt templates, the first part is responsible for accepting various textual inputs and aligning them into suitable descriptive English text prompts for subsequent image generation with Stable Diffusion. The second part consists mainly of the Stable Diffusion model and our pre-trained LoRA model weights for performing text-image generation tasks, complemented by the ControlNet [15] model to enhance image generation control when needed. The workflow shown in Fig. 1 is introduced in the following subsections in detail.

3.1 Prompt Building

Constructing good picture descriptors is crucial in T2I generation tasks. However, this process is often time-consuming and demands users to possess excellent aesthetic qualities. For image generation with Chinese culture as the theme, our approach involves using Chinese language inputs instead of relying on lengthy lists of English words by capitalizing on the multi-language and multi-profile text understanding capabilities of LLMs, as depicted in Fig. 1. LLMs like ChatGPT or ChatGLM are chosen to aid in constructing the prompts. Cue word templates are designed as guiding prompts for the LLMs, allowing it to generate expressive descriptions tailored to our requirements.

Specifically, we will teach the selected LLMs the concept of the "Prompt" in Stable Diffusion and the tips for writing high-quality prompts. The detailed example prompts designed can be like *"Stable Diffusion, a deep learning text-to-image model that generates images using prompt words to describe the elements to be included or omitted. I introduce the concept of Prompt in the Stable Diffusion algorithm. The prompt here is usually used to describe images and comprises commonly used words. Next, I will explain the steps for generating a prompt, mainly used to describe characters. In generating the prompt, you must describe the character's attributes, themes, appearance, emotions, clothing, posture, perspective, actions, and background using English words or phrases as labels. Then, combine the desired similar prompt words, using a comma as the delimiter, and arrange them from most important to least important. Now, you will serve as the prompt generator."*

A prompt template is formulated to combine with the user's input, encompassing their information as part of the question. This integration enhances the final returned prompt, providing a more comprehensive step-by-step guide for the model to follow. We can achieve a flexible and adaptable system by incorporating various inputs into the workflow, for instance, by using a poem as input to generate a prompt that closely corresponds to the picture described in the poem. Additionally, even a few Chinese words can yield detailed descriptors, addressing the limitation of only accepting English prompts.

落霞与孤鹜齐飞 秋水共长天一色   千山鸟飞绝 万径人踪灭 孤舟蓑笠翁 独钓寒江雪
autumn, sunset, calm, serene, mountains,   mountains, birds, path, solitude, boat, old
water, flying, lone, duck, vivid colors   man, straw hat, snow, fishing, tranquility

飞流直下三千尺 疑是银河落九天   流水落花春去也 天上人间
waterfall, towering, thousand meters, silver,   flowing water, falling flowers, spring,
galaxy, celestial, majestic, breathtaking   heavenly, earthly, beauty

Fig. 2. Prompts generated by LLMs and images generated finally in our method.

3.2   Finetuning with LoRA

While LLMs and prompt engineering can assist in mitigating the challenge of inputting complex English prompts, certain concepts specific to the intricacies of Chinese culture and unique art forms may still require additional enhancement in Stable Diffusion. In such cases, additional training and fine-tuning of the Stable Diffusion model become necessary to enhance its understanding and performance.

By selecting specific elements and art forms from traditional Chinese culture and training them with LoRA, we can enhance the understanding of Chinese cultural concepts in Stable Diffusion, allowing us to achieve the goal of presenting intricate Chinese elements in the T2I task effectively.

3.3   Workflow of the Inference Pipeline

The process is initiated by providing various textual inputs, such as scattered Chinese words or poems, to LLMs, which will respond with descriptive prompts to the user. Depending on the elements the user wishes to showcase in the final image, pre-trained style LoRA weights can be selectively incorporated and combined with the base model weight while generating the final images. ControlNet can also be employed to further control the final image's structure if needed.

4 Experiments

4.1 Settings

We utilized the gpt-3.5-turbo variant of GPT models and the ChatGLM-6B 1.1 checkpoint as LLMs in the experiments. These models are publicly accessible through the OpenAI API[1] and the Hugging Face repository[2]. Additionally, we incorporated the pre-trained checkpoint model of "TangBohu" for Stable Diffusion.

A WebUI system is developed using standard web development technologies like FastAPI, Spring Boot, and Vue.js to conduct the experiments. The deployment pipeline is built on Diffusers [16] and Transformers [17] libraries. It can integrate all components into a cohesive system with an intuitive user interface, enabling seamless interaction and efficient utilization of the models for image generation with Chinese cultural elements. The component library "Kohya"[3] is utilized for training our LoRA models, and the pre-trained checkpoint model of "TangBohu" for Stable Diffusion serves as the base model for training and inference. The training image datasets for our models were collected online by ourselves. For image captioning, we first use WD1.4 ViT Tagger[4] for automatic labeling and then manually make adjustments.

4.2 Results

We present several demonstrations in Fig. 1, Fig. 2, Fig. 3, Fig. 4, and Fig. 5. Figure 3 illustrates the outcomes of the artistic creation function using the Stable Diffusion combined with multiple models we trained. These models encompass elements in Chinese traditional culture like Chinese ink painting and meticulous brushwork, shadow puppetry, as well as the intricate patterns of Dunhuang and Blue-and-White.

Fig. 1 and Fig. 2 represent prompts generated by LLMs and final images generated with our method. The prompts in Fig. 2 are generated by gpt-3.5-turbo while those in Fig. 1 are generated by ChatGLM-6B. In addition, only images in Fig. 2 were directly generated by the basic model without using any models we trained, so they can be compared and referenced with images in other figures. Figures 4 and 5 represent case studies in our experiments. Figure 4 showcases pattern images generated by the trained pattern model, along with the results obtained from combining the "ControlNet" depth model for interior design rendering. Meanwhile, Fig. 5 demonstrates the outcomes of various pre-trained models combined with the "ControlNet" OpenPose model (detects and copies human poses without copying other details), focusing on traditional Chinese costume design.

---

[1] https://platform.openai.com/.

[2] https://huggingface.co/THUDM/chatglm-6b.

[3] https://github.com/bmaltais/kohya_ss.

[4] https://github.com/picobyte/stable-diffusion-webui-wd14-tagger.

Fig. 3. Images featuring Chinese culture generated by our method.

Fig. 4. Images generated with our method.

Fig. 5. Images generated with our method.

5 Conclusions

This paper presents a practical approach to integrating popular AIGC technologies into a unified system while also exploring the potential synergy between AIGC technologies and Chinese traditional culture. Specifically, with the capabilities of LLMs, the approach can weave expressive visual descriptions based on initial inputs and align them in suitable English text prompts for subsequent image generation with Stable Diffusion. Through the fine-tuning method called LoRA, Stable Diffusion can effectively learn complex and nuanced concepts of Chinese culture. The experimental results demonstrate that our approach can generate high-quality results pertaining to Chinese cultural subjects. This success is attributed to overcoming the constraints of accepting only English prompts and significantly improving the understanding of certain concepts in Chinese culture.

Acknowledgments. This paper is partially supported by "Pioneer" and "Leading Goose" R&D Program of Zhejiang (No. 2023C01212) and the Public Welfare Technology Application Research Project of Zhejiang (No. LGF22F020008).

References

1. Ramesh, A., Dhariwal, P., Nichol, A., Chu, C., Chen, M.: Hierarchical text-conditional image generation with clip latent. arXiv preprint arXiv:2204.06125 (2022)

2. Saharia, C., et al.: Others: photorealistic text-to-image diffusion models with deep language understanding. Adv. Neural. Inf. Process. Syst. **35**, 36479–36494 (2022)
3. Rombach, R., Blattmann, A., Lorenz, D., Esser, P., Ommer, B.: High-resolution image synthesis with latent diffusion models. In: Proceedings of the IEEE/CVF conference on computer vision and pattern recognition, pp. 10684–10695 (2022)
4. Wang, J., et al.: Fengshenbang 1.0: being the foundation of Chinese cognitive intelligence. CoRR. abs/2209.02970 (2022)
5. Chen, Z., Liu, G., Zhang, B.-W., Ye, F., Yang, Q., Wu, L.: AltCLIP: altering the language encoder in CLIP for extended language capabilities (2022). https://doi.org/10.48550/ARXIV. 2211.06679
6. Saxon, M., Wang, W.Y.: Multilingual conceptual coverage in text-to-image models. arXiv preprint arXiv:2306.01735. (2023)
7. Du, Z., et al.: Glm: general language model pretraining with autoregressive blank infilling. arXiv preprint arXiv:2103.10360. (2021)
8. Hu, E.J., et al.: Lora: low-rank adaptation of large language models. arXiv preprint arXiv: 2106.09685. (2021)
9. Dhariwal, P., Nichol, A.: Diffusion models beat gans on image synthesis. Adv. Neural. Inf. Process. Syst. **34**, 8780–8794 (2021)
10. Brown, T., et al.: Others: language models are few-shot learners. Adv. Neural. Inf. Process. Syst. **33**, 1877–1901 (2020)
11. Ouyang, L., et al.: Others: training language models to follow instructions with human feedback. Adv. Neural. Inf. Process. Syst. **35**, 27730–27744 (2022)
12. Shen, Y., Song, K., Tan, X., Li, D., Lu, W., Zhuang, Y.: Hugginggpt: solving AI tasks with chatgpt and its friends in huggingface. arXiv preprint arXiv:2303.17580. (2023)
13. Gal, R., et al.: An image is worth one word: personalizing text-to-image generation using textual inversion. arXiv preprint arXiv:2208.01618. (2022)
14. Ruiz, N., Li, Y., Jampani, V., Pritch, Y., Rubinstein, M., Aberman, K.: Dreambooth: fine tuning text-to-image diffusion models for subject-driven generation. In: Proceedings of the IEEE/CVF Conference on Computer Vision and Pattern Recognition, pp. 22500–22510 (2023)
15. Zhang, L., Agrawala, M.: Adding conditional control to text-to-image diffusion models. arXiv preprint arXiv:2302.05543. (2023)
16. Platen von, P., et al.: Diffusers: state-of-the-art diffusion models. https://github.com/huggin gface/diffusers (2022)
17. Wolf, T., et al.: Transformers: state-of-the-art natural language processing. In: Proceedings of the 2020 Conference on Empirical Methods in Natural Language Processing: System Demonstrations, pp. 38–45. Association for Computational Linguistics, Online (2020)

Strong-AI Autoepistemic Robots Build on Intensional First Order Logic

Zoran Majkić[(✉)]

ISRST, Tallahassee, FL, USA
majk.1234@yahoo.com
http://zoranmajkic.webs.com/

Abstract. Neuro-symbolic AI attempts to integrate neural and symbolic architectures in a manner that addresses strengths and weaknesses of each, in a complementary fashion, in order to support robust strong AI capable of reasoning, learning, and cognitive modeling. In this paper we consider the intensional First Order Logic (IFOL) [1] as a symbolic architecture of modern robots, able to use natural languages to communicate with humans and to reason about their own knowledge with self-reference and abstraction language property.

We intend to obtain the grounding of robot's language by experience of how it uses its neuronal architectures and hence by associating this experience with the mining (sense) of non-defined language concepts (particulars/individuals and universals) in PRP (Properties/Relations/Propositions) theory of IFOL.

We consider the robot's four-levels knowledge structure: The syntax level of particular natural language (Italian, French, etc.), two universal language levels: its semantic logic structure (based on virtual predicates of FOL and logic connectives), and its corresponding conceptual PRP structure level which universally represents the composite mining of FOL formulae grounded on the last robot's neuro-system level.

Finally, we provide the general method how to implement in IFOL (by using the abstracted terms) different kinds of modal logic operators and their deductive axioms: we present a particular example of robots autoepistemic deduction capabilities by introduction of the special temporal $Konow$ predicate and deductive axioms for it: reflexive, positive introspection and distributive axiom.

1 Introduction

The central hypothesis of cognitive science is that thinking can best be understood in terms of representational structures in the mind and computational procedures that operate on those structures. Most work in cognitive science assumes that the mind has mental representations analogous to computer data structures, and computational procedures similar to computational algorithms.

Main stream machine learning research on deep artificial neural networks may even be characterized as being behavioristic. In contrast, various sources of evidence from cognitive science suggest that human brains engage in the active development of compositional generative predictive models from their self-generated sensorimotor experiences. Guided by evolutionarily-shaped inductive learning and information process-

F. Zhao and D. Miao (Eds.): AIGC 2023, CCIS 1946, pp. 33–58, 2024.
https://doi.org/10.1007/978-981-99-7587-7_4

ing biases, they exhibit the tendency to organize the gathered experiences into event-predictive encodings. Meanwhile, they infer and optimize behavior and attention by means of both epistemic- and homeostasis-oriented drives.

Knowledge representation, strongly connected to the problem if knowledge processing, reasoning and "drawing inferences", is one of the main topics in AI. By reviewing the knowledge representation techniques that have been used by humans we will be aware of the *importance of language*. The predominant part of IT industry and user's applications is based on some sublanguage of the standard (extensional) FOL (First Order Logic) with Tarski's semantics based (only) on the truth; my effort is to pass to a more *powerful evolution of the FOL* able to support the meaning of knowledge as well, by replacing the standard FOL and its DB theory and practice in IT business. All this work is summarized and extended also to AI applications of many-valued logics in my recent book [1].

Last 15 years of my work in AI was mainly dedicated to development of a new intensional FOL, by integrating Montague's and algebraic Bealer's [2] approaches, with a conservative Tarski's semantics of the standard FOL. Basic result was the publication of the conservative extension of Tarski's semantics to intensional FOL [3], and two-step intensional semantics [4], which guaranteed a conservative extension of current RDB, but more than 50-years old technology, toward new IRDB (Intensional RDB). Indeed, in my next Manifesto of IRDB [5], I hoped also to find interested research groups and funds to begin the realization of IRDB as a new platform (compatible with all previously developed RDB application), able also to support NewSQL for Big Data, and ready for other AI improvements.

This paper is an extension (by Section 4) of the paper [6]. It is dedicated to show how this defined IFOL in [1] can be used for a new generation of intelligent robots, able to communicate with humans with this intensional FOL supporting the meaning of the words and their language compositions. As in [7] we can consider three natural language levels: The *syntax* of a particular natural language (French, English, etc.) its *semantic logic structure* (transformation of parts of the language sentences into the logic predicates and definition of corresponding FOL formulae) and its corresponding *conceptual structure*, which differently from the semantic layer that represents only the logic's semantics, represents the composed meaning of FOL formulae based on the grounding of intensional PRP concepts.

Thus, intensional mapping from the free FOL syntax algebra into the algebra of intensional PRP concepts,

$$I : \mathcal{A}_{FOL} \to \mathcal{A}_{int}$$

provided by IFOL theory, is a part of the semantics-conceptual mapping of natural languages. Note that differently from the particularity of any given natural language of humans, the underlying logical semantics and conceptual levels have universal human knowledge structure, provided by innate human brain structure able to rapidly acquire the ability to use any natural language.

Parsing, tokenizing, spelling correction, part-of-speech tagging, noun and verb phrase chunking are all aspects of natural language processing long handled by symbolic AI, and has to be improved by deep learning approaches. In symbolic AI, discourse representation theory and first-order logic have been used to represent sentence meanings. We consider that the natural language (first level) can be parsed into a logical FOL formula with a numbers of virtual predicates and logic connectives of the FOL. By such a parsing we obtain the second, semantic logic, structure corresponding to some FOL formula. However, natural language is grounded in experience. Humans do not always define all words in terms of other words, humans understand many basic words in terms of associations with sensory-motor experiences for example. People must interact physically with their world to grasp the essence of words like "blue", "could", and "left". Abstract words are acquired only in relation to more concretely grounded terms.

Theoretical neuroscience is the attempt to develop mathematical and computational theories and models of the structures and processes of the brains of humans and other animals. If progress in theoretical neuroscience continues, it should become possible to tie psychological to neurological explanations by showing how mental representations such as *concepts* are constituted by activities in neural populations, and how computational procedures such as spreading activation among concepts are carried out by neural processes. Concepts, which partly correspond to the words in spoken and written language, are an important kind of mental representation.

Alan Turing developed the Turing Test in 1950 in his paper, "Computing Machinery and Intelligence". Originally known as the Imitation Game, the test evaluates if a machine's behavior can be distinguished from a human. In this test, there is a person known as the "interrogator" who seeks to identify a difference between computer-generated output and human-generated ones through a series of questions. If the interrogator cannot reliably discern the machines from human subjects, the machine passes the test. However, if the evaluator can identify the human responses correctly, then this eliminates the machine from being categorized as intelligent.

Differently from the *simulation* of AI by such Turing tests and the Loebner Prize[1] and in accordance with Marvin Minsky[2], in this paper I argue that a real AI for robots can be obtained by using formal Intensional FOL (with defined intensional algebra of intensions of language constructions) for the robots as their symbolic AI component, by defining the sense to ground terms (the words) in an analog way, associating to these words the software processes developed for the robots when they recognize by these

---

[1] The Loebner Prize was an annual competition in artificial intelligence that awards prizes to the computer programs considered by the judges to be the most human-like. The prize is reported as defunct since 2020 [1]. The format of the competition was that of a standard Turing test. In each round, a human judge simultaneously holds textual conversations with a computer program and a human being via computer. Based upon the responses, the judge must decide which is which.

[2] In the early 1970s, at the MIT Artificial Intelligence Lab, Minsky and Papert started developing what came to be known as the Society of Mind theory. The theory attempts to explain how what we call intelligence could be a product of the interaction of non-intelligent parts. Minsky says that the biggest source of ideas about the theory came from his work in trying to create a machine that uses a robotic arm, a video camera, and a computer to build with children's blocks. In 1986, Minsky published The Society of Mind, a comprehensive book on the theory which, unlike most of his previously published work, was written for the general public.

In November 2006, Minsky published The Emotion Machine, a book that critiques many popular theories of how human minds work and suggests alternative theories, often replacing simple ideas with more complex ones.

algorithms (neural architectures) the color "blue" of visual objects, the position "left" etc.. In this way we would obtain a *neuro-symbolic AI* which attempts to integrate neural and symbolic architectures in a manner that addresses strengths and weaknesses of each, in a complementary fashion, in order to support robust AI capable of reasoning, learning, and cognitive modeling. To build a robust, knowledge-driven approach to AI we must have the machinery of symbol-manipulation as, in this case, an IFOL. Too much of useful knowledge is abstract to make do without tools that represent and manipulate abstraction, and to date, the only machinery that we know of that can manipulate such abstract knowledge reliably is the apparatus of symbol-manipulation. The IFOL defined in [1] is provided by abstraction operators as well.

Daniel Kahneman [8] describes human thinking as having two components, System 1 and System 2. System 1 is fast, automatic, intuitive and unconscious. System 2 is slower, step-by-step, and explicit. System 1 is the kind used for pattern recognition while System 2, in uor case based on IFOL, is far better suited for planning, deduction, and deliberative thinking. In this view, deep learning best models the first kind of thinking while symbolic reasoning best models the second kind and both are needed.

So, for the words (ground linguistic terms), which can not be "defined by other words", the robots would have some own internal experience of the concrete sense of them. Thus, by using intensional FOL the robots can formalize also the natural language expressions "I see the blue color" by a predicate "see(I, blue color)" where the sense of the ground term "I" (*Self*)[3] for a robot is the name of the main working coordination program which activate all other algorithms (neuro-symbolic AI subprograms) like visual recognition of color of the object in focus. But also the auto-conscience sentence like "I know that I see the blue color" by using abstracting operators "<_>" of intensional FOL, expressed by the predicate "know(I, < see(I, blue color)>)", etc..

Consequently, we argue that by using this intensional FOL, the robots can develop their own knowledge about their experiences and communicate by a natural language with humans. So, we would be able to develop the interactive robots which learn and understand spoken language via multisensory grounding and internal robotic embodiment.

The *grounding* of the intensional concepts i PRP theory of intensional logic was not considered in my recent book [1] from the fact that this book was only restricted on the symbolic AI aspects (IFOL); so by this paper we extend the logic theory developed in [1] with concrete grounding of its intensional concepts in order to obtain a strong AI for robots. So, in next Section we will provide a short introduction to IFOL and its intensional/extensional semantics [1].

2   Algebra for Composition of Meanings in IFOL

Contemporary use of the term "intension" derives from the traditional logical doctrine that an idea has both an extension and an intension. Although there is divergence in formulation, it is accepted that the extension of an idea consists of the subjects to which

---

[3] Self in a sense which implies that all our activities are controlled by powerful creatures inside ourselves, who do our thinking and feeling for us.

the idea applies, and the intension consists of the attributes implied by the idea. In contemporary philosophy, it is linguistic expressions (here it is a logic formula), rather than concepts, that are said to have intensions and extensions. The intension is the concept expressed by an expression of intensional algebra $\mathcal{A}_{int}$, and the extension is the set of items to which the expression applies. This usage resembles use of Frege's use of "Bedeutung" and "Sinn" [9].

Intensional entities (or concepts) are such things as Propositions, Relations and Properties (PRP). What make them "intensional" is that they violate the principle of extensionality; the principle that extensional equivalence implies identity. All (or most) of these intensional entities have been classified at one time or another as kinds of Universals [10].

In a predicate logics, (virtual) predicates expresses classes (properties and relations), and sentences express propositions. Note that classes (intensional entities) are *reified*, i.e., they belong to the same domain as individual objects (particulars). This endows the intensional logics with a great deal of uniformity, making it possible to manipulate classes and individual objects in the same language. In particular, when viewed as an individual object, a class can be a member of another class.

Definition 1. VIRTUAL PREDICATES: Virtual predicate *obtained from an open formula* $\phi \in \mathcal{L}$ *is denoted by* $\phi(x_1, ..., x_m)$ *where* $(x_1, ..., x_m)$ *is a particular fixed sequence of the set of all free variables in* ϕ. *This definition contains the precise method of establishing the ordering of variables in this tuple: such an method that will be adopted here is the ordering of appearance, from left to right, of free variables in* ϕ. *This method of composing the tuple of free variables is unique and canonical way of definition of the virtual predicate from a given open formula.*

The virtual predicates are useful also to replace the general FOL quantifier on variables $(\exists x)$ *by specific quantifiers* $\exists_i$ *of the FOL syntax algebra* $\mathcal{A}_{FOL}$, *where* $i \geq 1$ *is the position of variable* x *inside a virtual predicate. For example, the standard FOL formula* $(\exists x_k)\phi(x_i, x_j, x_k, x_l, x_m)$ *will be mapped into intensional concept* $\exists_3\phi(\boldsymbol{x}) \in \mathcal{A}_{FOL}$ *where* $\boldsymbol{x}$ *is the list(tuple) of variables* $(x_i, x_j, x_k, x_l, x_m)$.

Virtual predicates are atoms used to build the *semantic logic structures* of logic-semantics level of any given natural language.

Let us define the FOL syntax algebra $\mathcal{A}_{FOL}$.
For example, the FOL formula $\phi(x_i, x_j, x_k, x_l, x_m) \wedge \psi(x_l, y_i, x_j, y_j)$ will be replaced by a specific *virtual predicate* $\phi(x_i, x_j, x_k, x_l, x_m) \wedge_S \psi(x_l, y_i, x_j, y_j)$, with the set of joined variables (their positions in the first and second virtual predicate, respectively) $S = \{(4, 1), (2, 3)\}$, so that its extension is expressed by an algebraic expression $R_1 \bowtie_S R_2$, where R_1, R_2 are the extensions for a given Tarski's interpretation I_T of the virtual predicate ϕ, ψ relatively, and the binary operator $\bowtie_S$ is the natural join of these two relations. In this example the resulting relation will have the following ordering of attributes: $(x_i, x_j, x_k, x_l, x_m, y_i, y_j)$. In the case when S is empty (i.e. its cardinality $|S| = 0$) then the resulting relation is the Cartesian product of R_1 and R_2. For the existential quantification, the FOL formula $(\exists x_k)\phi(x_i, x_j, x_k, x_l, x_m)$ will be replaced in $\mathcal{A}_{FOL}$ by a specific virtual predicate $(\exists_3)\phi(x_i, x_j, x_k, x_l, x_m)$. For logic negation operator we will use the standard symbol $\neg$.

Based on the new set of logical connectives introduced above, where the standard FOL operators $\wedge$ and $\exists$ are substituted by a set of specialized operators $\{\wedge_S\}_{S\in\mathcal{P}(\mathbb{N}^2)}$ and $\{\exists n\}_{n\in\mathbb{N}}$ as explained above, we can define the following free syntax algebra for the FOL:

Definition 2. FOL SINTAX ALGEBRA:
Let $\mathcal{A}_{FOL} = (\mathcal{L}, \doteq, \top, \{\wedge_S\}_{S\in\mathcal{P}(\mathbb{N}^2)}, \neg, \{\exists n\}_{n\in\mathbb{N}})$ be an extended free syntax algebra for the First-order logic with identity $\doteq$, with the set $\mathcal{L}$ of first-order logic formulae with the set of variables in $\mathcal{V}$, with $\top$ denoting the tautology formula (the contradiction formula is denoted by $\bot \equiv \neg\top$).

We begin with the informal theory that universals (properties (unary relations), relations, and propositions in PRP theory [11]) are genuine entities that bear fundamental logical relations to one another. To study properties, relations and propositions, one defines a family of set-theoretical structures, one define the intensional algebra, a family of set-theoretical structures most of which are built up from arbitrary objects and fundamental logical operations (conjunction, negation, existential generalization,etc.) on them.

Definition 3. INTENSIONAL LOGIC PRP DOMAIN $\mathcal{D}$:
In intensionl logic the concepts (properties, relations and propositions) are denotations for open and closed logic sentences, thus elements of the structured domain $\mathcal{D} = D_{-1} + D_I$, (here $+$ is a disjoint union) where

- *A subdomain D_{-1} is made of particulars (individuals).*
- *The rest $D_I = D_0 + D_1... + D_n...$ is made of universals (concepts)[4]: D_0 for propositions with a distinct concept $Truth \in D_0$, D_1 for properties (unary concepts) and $D_n, n \geq 2$, for n-ary concept.*

The concepts in $\mathcal{D}_I$ are denoted by $u, v, ...$, while the values (individuals) in D_{-1} by $a, b, ...$ The empty tuple $<>$ of the nullary relation $r_\emptyset$ (i.e. the unique tuple of 0-ary relation) is an individual in D_{-1}, with $\mathcal{D}^0 =_{def} \{<>\}$. Thus, we have that $\{f, t\} = \mathcal{P}(\mathcal{D}^0) \subseteq \mathcal{P}(D_{-1})$, where by f and t we denote the empty set $\emptyset$ and set $\{<>\}$ respectively.

The *intensional interpretation* is a mapping between the set $\mathcal{L}$ of formulae of the FOL and intensional entities in $\mathcal{D}$, $I : \mathcal{L} \to \mathcal{D}$, is a kind of "conceptualization", such that an open-sentence (virtual predicate) $\phi(x_1, ..., x_k)$ with a tuple of all free variables $(x_1, ..., x_k)$ is mapped into a k-ary *concept*, that is, an intensional entity $u = I(\phi(x_1, ..., x_k)) \in D_k$, and (closed) sentence ψ into a proposition (i.e., *logic concept*) $v = I(\psi) \in D_0$ with $I(\top) = Truth \in D_0$ for the FOL tautology $\top \in \mathcal{L}$ (the falsity in the FOL is a logic formula $\neg\top \in \mathcal{L}$). A language constant c is mapped into a particular $a \in D_{-1}$ (intension of c) if it is a proper name, otherwise in a correspondent concept u in D_I. Thus, in any application of intensional FOL, this intensional interpretation that determines the meaning (sense) of the knowledge expressed by logic

---

[4] In what follows we will define also a language of concepts with intensional connectives defined as operators of the intensional algebra $\mathcal{A}_{int}$ in Definition 6, so that D_I is the set of terms of this intensional algebra.

formulae is *uniquely determined (prefixed)* (for example, by a grounding on robot's neuro system processes, explained in next section).

However, the extensions of the concepts (with this prefixed meaning) vary from a context (possible world, expressed by an extensionalization function) to another context in a similar way as for different Tarski's interpretations of the FOL:

Definition 4. EXTENSIONS AND EXTENSIONALIZATION FUNCTIONS:
Let $\mathfrak{R} = \bigcup_{k \in \mathbb{N}} \mathcal{P}(\mathcal{D}^k) = \sum_{k \in \mathbb{N}} \mathcal{P}(\mathcal{D}^k)$ be the set of all k-ary relations, where $k \in \mathbb{N} = \{0, 1, 2, ...\}$. Notice that $\{f, t\} = \mathcal{P}(\mathcal{D}^0) \subseteq \mathfrak{R}$, that is, $f, t \in \mathfrak{R}$ and hence the truth values are extensions in $\mathfrak{R}$.
  We define the function $f_{<>} : \mathfrak{R} \to \mathfrak{R}$, such that for any $R \in \mathfrak{R}$,

$$f_{<>}(R) =_{def} \{<>\} \text{ if } R \neq \emptyset; \ \emptyset \text{ otherwise} \tag{1}$$

The extensions of the intensional entities (concepts) are given by the set $\mathcal{E}$ of extensionalization functions $h : \mathcal{D} \to \mathcal{D}_{-1} + \mathfrak{R}$, such that

$$h = h_{-1} + h_0 + \sum_{i \geq 1} h_i : \sum_{i \geq -1} D_i \longrightarrow D_{-1} + \{f, t\} + \sum_{i \geq 1} \mathcal{P}(\mathcal{D}^i) \tag{2}$$

where $h_{-1} : D_{-1} \to D_{-1}$ for the particulars, while $h_0 : D_0 \to \{f, t\} = \mathcal{P}(\mathcal{D}^0)$ assigns the truth values in $\{f, t\}$ to all propositions with the constant assignment $h_0(Truth) = t = \{<>\}$, and for each $i \geq 1$, $h_i : D_i \to \mathcal{P}(\mathcal{D}^i)$ assigns a relation to each concept.
  Consequently, intensions can be seen as names *(labels) of atomic or composite concepts, while the extensions correspond to various rules that these concepts play in different worlds.*

The intensional entities for the same logic formula, for example $x_2 + 3 = x_1^2 - 4$, which can be denoted by $\phi(x_2, x_1)$ or $\phi(x_1, x_2)$, from above we need to differentiate their concepts by $I(\phi(x_2, x_1)) \neq I(\phi(x_1, x_2))$ because otherwise we would obtain erroneously that $h(I(\phi(x_2, x_1))) = h(I(\phi(x_1, x_2)))$. Thus, in intensional logic the ordering in the tuple of variables **x** in a given open formula ϕ is very important, and explains why we introduced in FOL the virtual predicates in Definition 1.

Definition 5. *Let us define the extensional relational algebra for the FOL by,*

$$\mathcal{A}_{\mathfrak{R}} = (\mathfrak{R}, R_=, \{<>\}, \{\bowtie_S\}_{S \in \mathcal{P}(\mathbb{N}^2)}, \sim, \{\pi_{-n}\}_{n \in \mathbb{N}}),$$

where $\{<>\} \in \mathfrak{R}$ is the algebraic value correspondent to the logic truth, $R_=$ is the binary relation for extensionally equal elements, with the following operators:

1. *Binary operator $\bowtie_S : \mathfrak{R} \times \mathfrak{R} \to \mathfrak{R}$, such that for any two relations $R_1, R_2 \in \mathfrak{R}$, the $R_1 \bowtie_S R_2$ is equal to the relation obtained by natural join of these two relations* if *S is a non empty set of pairs of joined columns of respective relations (where the first argument is the column index of the relation R_1 while the second argument is the column index of the joined column of the relation R_2);* otherwise *it is equal to the cartesian product $R_1 \times R_2$.*

2. *Unary operator* $\sim: \mathfrak{R} \to \mathfrak{R}$, *such that for any k-ary (with* $k \geq 1$*) relation* $R \in \mathcal{P}(\mathcal{D}^k) \subset \mathfrak{R}$ *we have that* $\sim(R) = \mathcal{D}^k \backslash R \in \mathcal{P}(\mathcal{D}^k)$, *where '\' is the substraction of relations. For* $u \in \{f, t\} = \mathcal{P}(\mathcal{D}^0) \subseteq \mathfrak{R}$, $\sim(u) = \mathcal{D}^0 \backslash u$.

3. *Unary operator* $\pi_{-n} : \mathfrak{R} \to \mathfrak{R}$, *such that for any k-ary (with* $k \geq 1$*) relation* $R \in \mathcal{P}(\mathcal{D}^k) \subset \mathfrak{R}$ *we have that* $\pi_{-n}(R)$ *is equal to the relation obtained by elimination of the n-th column of the relation* R if $1 \leq n \leq k$ *and* $k \geq 2$; *equal to, from (1),* $f_{<>}(R)$ if $n = k = 1$; otherwise *it is equal to* R.

We will use the symbol '=' for the extensional identity for relations in $\mathfrak{R}$.

The intensional semantics of the logic language with the set of formulae $\mathcal{L}$ can be represented by the mapping

$$\mathcal{L} \longrightarrow_I \mathcal{D} \Longrightarrow_{h \in \mathcal{E}} \mathfrak{R},$$

where $\longrightarrow_I$ is a *fixed intensional* interpretation $I : \mathcal{L} \to \mathcal{D}$ with image $im(I) \subset \mathcal{D}$, and $\Longrightarrow_{h \in \mathcal{E}}$ is *the set* of all extensionalization functions $h : im(I) \to D_{-1} + \mathfrak{R}$ in $\mathcal{E}$.

So, we can define only the minimal intensional algebra (with minimal number of operators) $\mathcal{A}_{int}$ of concepts, able to support the homomorphic extension

$$h : \mathcal{A}_{int} \to \mathcal{A}_{\mathfrak{R}}$$

of the extensionalization function $h : \mathcal{D} \to D_{-1} + \mathfrak{R}$.

Definition 6. BASIC INTENSIONAL FOL ALGEBRA:
Intensional FOL algebra is a structure

$$\mathcal{A}_{int} = (\mathcal{D}, Id, Truth, \{conj_S\}_{S \in \mathcal{P}(\mathbb{N}^2)}, neg, \{exists_n\}_{n \in \mathbb{N}}),$$

with binary operations $conj_S : D_I \times D_I \to D_I$, *unary operation* $neg : D_I \to D_I$, *and unary operations* $exists_n : D_I \to D_I$, *such that for any extensionalization function* $h \in \mathcal{E}$, *and* $u \in D_k, v \in D_j$, $k, j \geq 0$,

1. $h(Id) = R_=$ *and* $h(Truth) = \{<>\}$, *for* $Id = I(\doteq (x, y))$ *and* $Truth = I(\top)$.
2. $h(conj_S(u, v)) = h(u) \bowtie_S h(v)$, *where* $\bowtie_S$ *is the natural join operation and* $conj_S(u, v) \in D_m$ *where* $m = k + j - |S|$ *if for every pair* $(i_1, i_2) \in S$ *it holds that* $1 \leq i_1 \leq k, 1 \leq i_2 \leq j$ *(otherwise* $conj_S(u, v) \in D_{k+j}$*)*.
3. $h(neg(u)) = \sim(h(u)) = \mathcal{D}^k \backslash (h(u))$ *(the complement of k-ary relation* $h(u)$ *in* $\mathcal{D}^k$*), if* $k \geq 1$, *where* $neg(u) \in D_k$. *For* $u_0 \in \mathcal{D}_0$, $h(neg(u_0)) = \sim(h(u_0)) = \mathcal{D}^0 \backslash (h(u_0))$.
4. $h(exists_n(u)) = \pi_{-n}(h(u))$, *where* π_{-n} *is the projection operation which eliminates n-th column of a relation and* $exists_n(u) \in D_{k-1}$ *if* $1 \leq n \leq k$ *(otherwise* $exists_n$ *is the identity function).*

Notice that for $u, v \in D_0$, so that $h(u), h(v) \in \{f, t\}$,

$$h(neg(u)) = \mathcal{D}^0 \backslash (h(u)) = \{<>\} \backslash (h(u)) \in \{f, t\}, \text{ and}$$
$$h(conj_\emptyset(u, v)) = h(u) \bowtie_\emptyset h(v) \in \{f, t\}.$$

We define a derived operation $union : (\mathcal{P}(D_i)\backslash\emptyset) \to D_i, i \geq 0$, such that, for any $B = \{u_1, ..., u_n\} \in \mathcal{P}(D_i)$ and $S = \{(l, l) \mid 1 \leq l \leq i\}$ we have that

$$union(\{u_1, ..., u_n\}) = \begin{cases} u_1, & \text{if } n = 1 \\ neg(conj_S(neg(u_1), conj_S(neg(u_2), ..., neg(u_n))...), & \text{otherwise} \end{cases} \tag{3}$$

Than we obtain that for $n \geq 2$:

$h(union(B)) = h(neg(conj_S(neg(u_1), conj_S(neg(u_2), ..., neg(u_n))...)$

$= \mathcal{D}^i\backslash((\mathcal{D}^i\backslash h(u_1)) \quad \bowtie_S \quad ... \quad \bowtie_S \quad (\mathcal{D}^i\backslash h(u_n))) = \mathcal{D}^i\backslash((\mathcal{D}^i\backslash h(u_1)) \cap ... \cap (\mathcal{D}^i\backslash h(u_n)))$

$= \bigcup\{h(u_j) \mid 1 \leq j \leq n\}$, that is,

$$h(union(B)) = \bigcup\{h(u) \mid u \in B\} \tag{4}$$

Note that it is valid also for the propositions in $u_1, u_2 \in D_0$, so that $h(union(u_1, u_2))$ $= h(u_1) \bigcup h(n_2) \in \{f, t\}$ where f is empty set $\emptyset$ while t is a singleton set $\{<>\}$ with empty tuple $<>$, and hence the join $\{<>\} \bowtie \emptyset = \emptyset$ and $\{<>\} \bowtie \{<>\} = \{<>\}$.

Thus, we define the following homomorphic extension

$$I : \mathcal{A}_{FOL} \to \mathcal{A}_{int}$$

of the intensional interpretation $I : \mathcal{L} \to \mathcal{D}$ for the formulae in syntax algebra $\mathcal{A}_{FOL}$ from Definition 2:

1. The logic formula $\phi(x_i, x_j, x_k, x_l, x_m) \wedge_S \psi(x_l, y_i, x_j, y_j)$ will be intensionally interpreted by the concept $u_1 \in D_7$, obtained by the algebraic expression $conj_S(u, v)$ where $u = I(\phi(x_i, x_j, x_k, x_l, x_m)) \in D_5, v = I(\psi(x_l, y_i, x_j, y_j)) \in D_4$ are the concepts of the virtual predicates ϕ, ψ, relatively, and $S = \{(4, 1), (2, 3)\}$. Consequently, we have that for any two formulae $\phi, \psi \in \mathcal{L}$ and a particular operator $conj_S$ uniquely determined by tuples of free variables in these two formulae, $I(\phi \wedge_S \psi) = conj_S(I(\phi), I(\psi))$.

2. The logic formula $\neg\phi(x_i, x_j, x_k, x_l, x_m)$ will be intensionally interpreted by the concept $u_1 \in D_5$, obtained by the algebraic expression $neg(u)$ where u is the concept of the virtual predicate ϕ, $u = I(\phi(x_i, x_j, x_k, x_l, x_m)) \in D_5$. Consequently, we have that for any formula $\phi \in \mathcal{L}$, $I(\neg\phi) = neg(I(\phi))$.

3. The logic formula $(\exists_3)\phi(x_i, x_j, x_k, x_l, x_m)$ will be intensionally interpreted by the concept $u_1 \in D_4$, obtained by the algebraic expression $exists_3(u)$ where $u = I(\phi(x_i, x_j, x_k, x_l, x_m)) \in D_5$ is the concept of the virtual predicate ϕ. Consequently, we have that for any formula $\phi \in \mathcal{L}$ and a particular operator $exists_n$ uniquely determined by the position of the existentially quantified variable in the tuple of free variables in ϕ (otherwise $n = 0$ if this quantified variable is not a free variable in ϕ), $I((\exists_n)\phi) = exists_n(I(\phi))$.

So, we obtain the following two-steps interpretation of FOL based on two homomorphisms, intensional I, and extensional h:

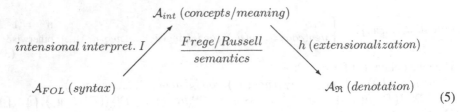

$$\mathcal{A}_{int} \ (concepts/meaning)$$

$$intensional \ interpret. \ I \diagup \quad \underline{Frege/Russell} \quad \diagdown h \ (extensionalization)$$
$$semantics$$

$$\mathcal{A}_{FOL} \ (syntax) \qquad\qquad\qquad\qquad \mathcal{A}_{\mathfrak{R}} \ (denotation)$$

$$(5)$$

We can enrich the expressivity of such a minimal FOL intensionality by new modal operators, or in different way provided in what follows. As, for example, in Bealer's intensional FOL, where he introduced the intensional abstraction operator, which will be considered in rest of this section, as a significant enrichment of the intensional FOL considered above.

In reflective languages, reification data is causally connected to the related reified aspect such that a modification to one of them affects the other. Therefore, the reification data is always a faithful representation of the related reified aspect. *Reification data* is often said to be made a *first class object*. In programming language design, a first-class citizen (also type, object, entity, or value) in a given programming language is an entity which supports all the operations generally available to other entities. These operations typically include being passed as an argument, returned from a function, modified, and assigned to a variable. The concept of first and second-class objects was introduced by Christopher Strachey in the 1960s when he contrasted real numbers (first-class) and procedures (second-class) in ALGOL.

In FOL we have the variables as arguments inside the predicates, and terms which can be assigned to variables are first-class objects while the predicates are the second-class objects. When we transform a virtual predicate into a term, by using intensional abstraction operator, we transform a logic formula into the first class object to be used inside another predicates as first-class objects. Thus, abstracted terms in the intensional FOL are just such abstracted terms as reification of logic formulae. For example, the sentence "Marco thinks *that Zoran runs*", expressed by $thinks(Marco, <runs(Zoran)>)$ by using binary predicate $thinks$ and unary predicate $runs$ where the ground atom $runs(Zoran)$ is reified into the predicate $thinks$.

If $\phi(\mathbf{x})$ is a formula (virtual predicate) with a list (a tuple) of free variables in $\mathbf{x} = (x_1, ..., x_n)$ (with ordering from-left-to-right of their appearance in ϕ), and α is its subset of *distinct* variables, then $<\phi(\mathbf{x})>_{\alpha}^{\beta}$ is a term, where β is the remaining set of free variables in $\mathbf{x}$. The externally quantifiable variables are the *free* variables not in α. When $n = 0$, $<\phi>$ is a term which denotes a proposition, for $n \geq 1$ it denotes a n-ary concept.

Definition 7. INTENSIONAL ABSTRACTION CONVENTION:
From the fact that we can use any permutation of the variables in a given virtual predicate, we introduce the convention that

$$< \phi(\mathbf{x}) >_{\alpha}^{\beta} \ is \ a \ term \ obtained \ from \ virtual \ predicate \ \phi(\mathbf{x}) \qquad (6)$$

if α is not empty such that $\alpha \bigcup \beta$ is the set of all variables in the list (tuple of variables) $x = (x_1, ..., x_n)$ *of the virtual predicate (an open logic formula) ϕ, and $\alpha \bigcap \beta = \emptyset$, so that $|\alpha| + |\beta| = |x| = n$. Only the variables in β (which are the only free variables of this term), can be quantified. If β is empty then $<\phi(x)>_\alpha$ is a ground term. If ϕ is a sentence and hence both α and β are empty, we write simply $<\phi>$ for this ground term.*

More about this general definition of abstract terms can be find in [1]. In this paper we will use the most simple cases of ground terms $<\phi>$, where ϕ is a sentence.

3   Four-Levels Robot's Brain Structure

Let us consider a model of robot for understanding language about space and movement in realistic situations [12,13], as finding video clips that match a spatial language description such as "People walking through the kitchen and then going to the dining room" and following natural language commands such as "Go down the hall towards the fireplace in the living room".

Video retrieval is a compelling application: in the United States alone, there are an estimated 35 million surveillance cameras installed, which record four billion hours of video per week. Analyzing and understanding the content of video data remains a challenging problem. A spatial language interface to video data can help people naturally and flexibly find what they are looking for in video collections. Studying language used to give directions could enable a robot to understand natural language directions. People talk to robots even if they do not have microphones installed, and it makes sense to build systems that understand what they say. A robot that understands natural language is easy for anyone to use without special training. By using the deductive properties of the IFOL, the robot can make logic deductions as well about the facts that it visually recognized and also to obtain its own autoepistemic deductions about obtained knowledge, as shortly explained in introduction, by using intensional abstractions in Definition 7.

Consequently, I will focus on a narrow subset of a natural language, grounding that language in data collected from a real world. This strategy has two benefits. First, it decreases the scope of the language understanding problem, making it more tractable. Second, by choosing a semantically deep core domain, it offers an opportunity to explore the connection between linguistic and non-linguistic concepts.

The linguistic structure extracted from spatial language expressions and many of the features in the model for spatial relations are based on the theories of Jackendoff [7], Landau and Jackendoff [14] and Talmy [15]. For example, the implementation of the mining of "across" in [15] is obtained by an algorithm (of robot's AI neuro-system) for computing the axes a figure imposes on a ground, and set of features which quantify "roughly perpendicular", using a machine learning algorithm to fine-tune the distinctions by training on labeled data. Regier [16] built a system that assigns labels such as "through" to move showing a figure relative to a ground object. Bailey [17] developed a model for learning the meanings of verbs of manipulation such as "push" and "shove". Kelleher and Costello [18] built models for the meanings of static spatial prepositions such as "in front of" and "above". Siskind [19] created a system for defining meanings for words such as "up" and "down". The framework reasons about formal temporal relations between primitive force-dynamic properties such as "supports" and "touches" and uses changes in these properties to define meanings for verbs. His framework focuses on word-level event recognition and features, etc.

Reasoning about movement and space is a fundamental competence of humans and many animals. Humans use spatial language to tell stories and give directions, abstracting away the details of a complex event into a few words such as "across the kitchen". A system that understands spatial language could be directly useful to people by finding video that matches spatial language descriptions, or giving natural language directions. We will consider a robot which retrieves video clips that match a natural language description using a probabilistic graphical model that maps between natural language and paths in the environment [12].

In this particular environment, spatial relations are modeled as probabilistic distributions for recognizing words paired with scenes. The distributions are trained from labeled examples using a set of geometric features that capture the semantics of spatial prepositions. The distribution modeled is the probability of a particular spatial relation given a trajectory and an object in the environment. This distribution corresponds to the probability that a spatial relation such as "across" or "to" describes a particular trajectory and landmark. The input to the model is the geometry of the path and landmark object; the output is a probability that the spatial relation can be used to describe this scene. These distributions are trained using labeled path examples, and in robot's brain correspond to its AI neuro-system. The system learns distributions for spatial relations, for example, by using a naive Bayes probabilistic model.

So, now we can focus to the integration of such robot's AI neuro-system with its AI symbolic system based on three natural language cognitive levels: The *syntax* of a particular natural language (French, English, etc.) its *semantic logic structure* (transformation of parts of the language sentences into the logic predicates and definition of corresponding FOL formulae) and its corresponding *conceptual structure*, which differently from the semantic layer that represents only the logic's semantics, represents the composed meaning of FOL formulae.

In this example, we focus on spatial language search of people's motion trajectories which are automatically extracted from video recorded by stationary overhead cameras. The system takes as input a natural language query, a database of surveillance video from a particular environment and the locations of non-moving objects in the environment. When the robot performs video retrieval by its AI neuro system, clips are returned in order according to the joint probability of the query and the clip. Thus, for each video clip in given database, this robot's neuro system computes the probability that considered clip satisfies a natural language query, parsed into logic FOL formula (second natural language semantic level) and consequently into intensional algebra $\mathcal{A}_{int}$ term with intensional concepts which labels are grounded by robot's neuro system processes (algorithms). Let $\mathcal{NL}$ be a given natural language. If we denote the set of finite nonempty lists of a given natural language words by $\mathcal{NL}_{list}$, then this parsing can be represented by a *partial* mapping

$$pars : \mathcal{NL}_{list} \to \mathcal{L} \tag{7}$$

where $\mathcal{L}$ is the set of logic formulae of intensional FOL.

We suppose that the concepts in the conceptual structure expressed by the intensional algebra $\mathcal{A}_{int}$ of *atomic* concepts $u \in \mathcal{D}$, and their corresponding logic atoms expressed by virtual predicates $\phi(\mathbf{x}) \in \mathcal{L}$ of FOL are the part of innate robot's knowledge, such that for robot's innate and unique intensional interpretation $I : \mathcal{L} \rightarrow \mathcal{D}$, $u = I(\phi(\mathbf{x}))$. Moreover, we suppose that robot has a parser capability to transform the sentences of particular natural language into the formulae of FOL with innate set of the atoms expressed by virtual predicates.

In this example we consider the predicates of IFOL as the verbs (V) of natural language, as follows

$$Find(x_1, x_2, x_3, x_4)$$

where the time-variable x_1 (with values "in past", "in present", "in future") indicates the time of execution of this recognition-action, the variable x_2 is used for the subject who executes this action (robot in this case), the variable x_3 is used for the object given to be eventually recognized (in this case a video clip) and x_4 for the statement (users query) that has to be satisfied by this object, and virtual predicate

$$Walk(x_1, x_2, x_3, x_4, x_5)$$

where the time-variable x_1 (with values "in past", "in present", "in future") indicates the time of execution of this action, variable x_2 for the figure (F) that moves ("person", "cat", etc.), x_3 for the initial position of walking figure (defined by the spatial relation (SR) "*from*", for example "from the table"), x_4 for the intermediate positions during movement of the figure (defined by (SR) "*through*", for example "through the corridor"), and x_5 for the final position of figure (defined by (SR) "*to*", for example "to the door").

The robot takes as input a natural language query, a database of surveillance video from a particular environment and the locations of non-moving objects in the environment. It parses the query into a semantic structure called a spatial description clause (SDC) [13]. An SDC consists of a figure (F), a verb (V), a spatial relation (SR), and a landmark (L). The system extracts SDCs automatically using a conditional random field chunker. Let us consider the example illustrated in Figure 3 in [13] of a natural language query $nq \in \mathcal{NL}_{list}$, defined by a sentence:

"The person walked *from* the couches in the room *to* the dining room table"

which is composed by two SDC with the first one

1. (F) = "the person"
2. (V) = "walked"
3. (SR) = "from"
4. (L) = "the couches in the room"

and the second SDC,

1. (SR) = "to"
2. (L) = "the dining room table"

Remark: Note that all SDC components different from (V), are particulars in D_{-1} in PRP domain $\mathcal{D}$, provided by Definition 3. The sense (mining) of the components (F) and (L) are grounded by the machine-learning video-recognition processes of the robot, that is by its neuro systems. The sense of the (SR) components is grounded by the meaning of the spatial relations, provided by different authors methods, mentioned previously, and implemented by particular robots processes.

What we need in next is to extend this grounding also to the virtual predicates of the FOL open formulae in $\mathcal{L}$. $\qquad\square$

Consequently, from these Spatial Description clauses, for the (V) of the past-time verb (V) "to walk", the semantic logic structure recognized by robot is the sentence $\phi = pars(nq) \in \mathcal{L}$, obtained from (7) so that, based on the virtual predicate $toWalk$, the sentence ϕ is

$$Walk(in\ past, person, from\ the\ couches\ in\ the\ room, NULL, to\ the\ dining\ room\ table) \tag{8}$$

Note that the inverse parsing of such logic sentence ϕ to natural language sentence is directly obtained, so that the robot can translate its semantic logic structures into natural language to communicate by voice to the people.

We consider that each grammatically *plural* word name "videoclips", robot can define by generalization by creating the virtual unary predicate $videoclips(y)$, such that its intensional *concept* $u_2 = I(videoclips(y)) \in D_1$ in PRP domain, whose meaning is grounded by robots patern-recognition process fixed by a machine learning method. In a similar way, each unary concept of visual objects can be created by robot by a machine learning method for enough big set of this type of objects.

So, each grammatically *singular* word name, like "John's videoclip" is a particular (element of D_{-1}) in PRP domain, whose meaning is grounded by the internal robot's image of this particular videoclip, *recognized* as such by robots patern-recognition process. Thus, for a given extensionalization function h in (2), and fixed robot's intensional mapping I, from the diagram (5), we obtain that the set C, of video clips in a given database of videoclips presented to this robot, is equal to

$$C = h(I(videoclips(y))) \tag{9}$$

Consequently, the human command in natural language $nc \in \mathcal{NL}_{list}$ to this robot,

"Find videoclip such that ϕ in the given set of videoclips"

(where ϕ has to be substituted by the sentence above) is parsed by robot into its second level (semantic logic structure) by virtual predicate $Find$ of the verb "to find" (*in present*) and a variable y of type "videoclip" (objects of research) and substituting "that ϕ" by abstracted term $<\phi>$, and by substituting "in the given set of" with the logic conjunction connective $\wedge_S$ of the IFOL expressed, from (7), by the following formula $\psi(y) = pars(nc)$

$$Find(in present, me, y, <\phi>) \wedge_S videoclips(y) \tag{10}$$

where $S = (2, 1)$ for joined variables in two virtual predicates.

The meaning of the unary concept $u_1 = I(Find(in\ present, me, y, <\phi>))$, corresponding to the natural language subexpression "Find (me), videoclip such that ϕ"

of the command above, is represented by its AI neuro system process of probabilistic recognition of video clips [13] satisfying the natural language query ϕ (In fact, u_2 is just equal to the name of this process of probabilistic recognition).

However, during execution of this process, the robot is able also to *deduce* the truth of the *autoepistemic sentence*, for a given assignment of variables $g : \mathcal{V} \to \mathcal{D}$, with $g(x_1) = in\ present$ and $g(x_2) = me$,

$$Know(x_1, x_2, <Find(inpresent, me, y, <\phi>)>_y)/g \tag{11}$$

of the virtual predicate $Know(x_1, x_2, x_3)$, where the time-variable x_1 (with values "in past", "in present", "in future") indicates the time of execution of this action, the variable x_2 is used for the subject of this knowledge and x_3 is used for an abstracted term expression this particular knowledge). Thus, by using deductive properties of the true sentences of FOL, this autoepistemic sentence about its state of selfknowledge, the robot would be able to comunicate to humans this sentence, traduces in natural language as

"I (me) know that I am (me) finding *videoclip* such that ϕ"

From the fact that robot defined the type of the variable y to be "videoclip", by traduction of the FOL deduced formula above into the natural language, this variable will be traduced in natural language by "videoclip". In the same way, during the execution of the human command above, expressed by the FOL formula $\psi(y)$ in (10), with composed concept $u_3 = I(\psi(y)) \in D_1$, that is, by using the homomorphic property of intensional interpretation I,

$$u_3 = u_1 \bowtie_S u_2 \tag{12}$$

the robot can deduce also the true epistemic sentence, for a given assignment of variables $g : \mathcal{V} \to \mathcal{D}$, with $g(x_1) = in\ present$ and $g(x_2) = me$,

$$Know(x_1, x_2, <Find(in\ present, me, y, <\phi>) \wedge_S videoclips(y)>_y)/g \tag{13}$$

and hence the robot would be able to communicate to humans this sentence, traduces in natural language as

"I (me) know that I am (me) finding *videoclip* such that ϕ in the set of videoclips"

Note that the subset of videoclips extracted by robot from a given set of videoclips $C = h(u_2)$ in (9), defines the current extensionalization function h, in the way that this subset is

$$E = h(u_3) = h(u_1) \bowtie_S h(u_2) = h(u_1) \bowtie_S C = h(u_1) \subseteq C \tag{14}$$

Thus, for the grounding of spatial language for video search, the robot's internal knowledge structure is divided into four levels, in ordering: natural language, semantic logic structure, conceptual structure and neuro structure, as represented by the following diagram (only two continuous arrows (intensional mapping $I : \mathcal{L} \to D_I$ where $D_I = D_0 + D_1 + ...$ are the universals in PRP domain theory) represent the total mappings, while other (dots) are partial mappings)

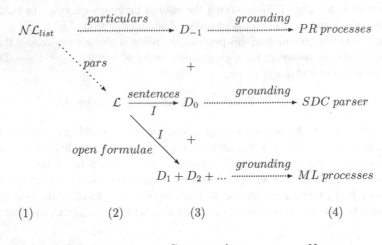

$$\begin{array}{ccc}
\mathcal{NL}_{list} & \xrightarrow{\hspace{0.3cm} particulars \hspace{0.3cm}} D_{-1} & \xrightarrow{\hspace{0.3cm} grounding \hspace{0.3cm}} PR\, processes \\
\hspace{0.3cm}\vdots\, pars & + & \\
\mathcal{L} \xrightarrow[I]{\hspace{0.3cm} sentences \hspace{0.3cm}} D_0 & \xrightarrow{\hspace{0.3cm} grounding \hspace{0.3cm}} SDC\, parser \\
I \searrow\quad + & \\
open\ formulae \hspace{0.5cm} D_1 + D_2 + ... & \xrightarrow{\hspace{0.3cm} grounding \hspace{0.3cm}} ML\, processes
\end{array}$$

$$(1) \hspace{1.5cm} (2) \hspace{1.5cm} (3) \hspace{2.5cm} (4)$$

$$Nat.Lang. \quad Log.semantic\, sys. \quad Conceptual\, sys. \hspace{1cm} Neuro\, sys. \hspace{1cm} (15)$$

It is easy to see that the conceptual system, based on PRP domain $\mathcal{D}$ composed by particulars in D_{-1} and universals (concepts) in $D_I = D_0 + D_1 + D_2 + ...$ of the IFOL, is the level of grounding of the natural language of the robot to its neuro system composed by the following processes:

1. PR (Pattern Recognition) processes of recognition of the particulars. For example, for SDC components (F) "the person", (L) "the couches in the room" and "the dining room table", etc.
2. SDC (Spatial Description Clauses) parser used for the sentences, for example, for a natural language query $nq \in \mathcal{NL}_{list}$ that is, logical proposition (sentece) $\phi = pars(nq) \in \mathcal{L}$ in (8), which is labeled by its intensional proposition label $I(\phi) \in D_0$. Thus, the grounding of nq is obtained by linking its intensional proposition $I(pars(nq))$ in PRP to the SDC parser process (part of robot's neuro system).
3. ML (Machine Learning) processes, like that used for the recognition of different types of classes (like the set of videoclips). For example, for the language plural world "videoclips" in $\mathcal{NL}_{list}$, such that $pars("videoclips") = videoclips(y) \in \mathcal{L}$ with its intensional unary concept $u_2 = I(videoclips(y)) \in D_1$ which is grounded to robot's ML process for the "videoclips".

Note that, while the top line in the diagram (15) is the ordinary component of the natu-aral language grounding developed by robot's neuro system, the two lines bellow is the new robots knowledge structure of the added *symbolic AI system* based on the Intensional First Order Logic and its grounding to robot's processes (its neuro AI system), by which the robot is able to provide logic deductive operations and autoepistemic self-reasoning about its current knowledge states and communicate it to humans by using natural languages.

4   A Short Introduction to Robots Autoepistemic Deduction

In my recent book it has been demonstrated that the Intensional FOL [1] has a conservative Tarski's semantics, shown also in this paper (only partially) by Definition 6, with interpretations (see the diagram in (5))

$$I_T^* = h \circ I : \mathcal{A}_{FOL} \to \mathcal{A}_{\mathfrak{R}}$$

as the ordinary (extensional) FOL with well known its deductive properties.

By introduction of the abstraction operators with autoepistemic capacities, expressed by the $Know$ predicate in previous section, we do not use more a *pure* logical deduction of the standard FOL, but a kind of autoepistemic deduction [20, 21] with a proper set of new axioms. However, the autoepistemic logic is introduced as a propositional logic [22] with added universal modal operator, usually written K, and the axioms:

1. Reflexive axiom **T**: $K\phi \Rightarrow \phi$
2. Positive introspection axiom **4**: $K\phi \Rightarrow KK\phi$
3. Distributive axiom **K**: $(K\phi \wedge K(\phi \Rightarrow \psi)) \Rightarrow K\psi$

for any proposition formulae ϕ and ψ, while $Know$ in IFOL is a *predicate* and not modal (Kripke-like) operator K.

It has been demonstrated that intensional enrichment of the standard (extensional) FOL, provided by Definition 14 in [1], is a kind of modal predicate logic $FOL_{\mathcal{K}}(\Gamma)$, where the set of explicit possible world $\mathcal{W}_e$ is equal to the set $\mathfrak{I}_T(\Gamma)\}$ of Tarski's interpretations $I_T^* = h \circ I$ (this composition is provided by diagram in (5)) of the standard FOL with a given set of assumptions Γ, that is, for a prefixed intensional interpretation of robot, this set of possible worlds is equal to the set of the extensionalization functions $h \in \mathcal{E}$ of robot's IFOL. It has been demonstrated that in such a minimal intensional enrichment of standard (extensional) FOL, we obtain exactly the Montague's definition of the intension (see Proposition 5 in [1]).

We recall that each robot's extensionalitation function $h \in \mathcal{E}$ in (2) is indexed by the time-instance. The actual robot's world extensionalization function (in the current instance of time) is denoted by $\hbar$, and determines the current robot's knowledge. Clearly, the robots knowledge changes in time and hence determines the extensionalization function $h \in \mathcal{E}$ in any given instance of time, based on robots experiences. Thus, as for humans, also the robot's knowledge and logic is a kind of temporal logic, and evolves with time.

Note that the explicit (conscious) robot's knowledge in actual world $\hbar$ (current time-instance) here is represented by the ground atoms of the $Know$ predicate, for a given assignments $g : \mathcal{V} \to \mathcal{D}$, that is, $g \in \mathcal{D}^{\mathcal{V}}$,

$$Know(y_1, y_2, <\psi(\mathbf{x})>_\alpha^\beta)/g = Know(g^*(y_1), g^*(y_2), g^*(<\psi(\mathbf{x})>_\alpha^\beta)) \quad (16)$$

with $\{y_1, y_2\} \bigcup \beta \bigcup \alpha \subseteq \mathcal{V}$, such that $g^*(y_1) = in\ present$ and $g^*(y_2) = me$ (the robot itself), for the extended assignments $g^* : \mathcal{T} \to \mathcal{D}$, where the set of terms $\mathcal{T}$ of IFOL is composed by the set $\mathcal{V}$ of all variables used in the defined set of predicates of robot/s IFOL, by the set of FOL constants and by the set of *abstracted terms* in (6), such that (from Definition 17 in [1]):

1. $g^*(t) = g(x) \in \mathcal{D}$ if the term t is a variable $x \in \mathcal{V}$.
2. $g^*(t) \in \mathcal{D}$ is the Tarski's interpretation of the FOL constant (nullary function) if the term t is a constant c.
3. If t is an abstracted term obtained for a formula ϕ, $<\phi(\mathbf{x})>_\alpha^\beta$, then

$$g^*(<\phi(\mathbf{x})>_\alpha^\beta) =_{def} \begin{cases} I(\phi(\mathbf{x})) \in D_{|\alpha|}, & \text{if } \beta \text{ is empty} \\ I(\phi[\beta/g(\beta)]) \in D_{|\alpha|}, & \text{otherwise} \end{cases} \qquad (17)$$

where $g(\beta) = g(\{y_1, .., y_m\}) = \{g(y_1), ..., g(y_m)\}$ and $[\beta/g(\beta)]$ is a uniform replacement of each i-th variable in the set β with the i-th constant in the set $g(\beta)$. Notice that α is the set of all free variables in the formula $\phi[\beta/g(\beta)]$.

so that in the actual world $\hbar$, the known fact (16) for robot becomes the ground atom

$$Know(y_1, y_2, <\psi(\mathbf{x})>_\alpha^\beta)/g = Know(in\ present, me, I(\psi[\beta/g(\beta)])) \qquad (18)$$

which is true in actual word, that is, from proposition (intensional concept)
$u = I(Know(in\ present, me, I(\psi[\beta/g(\beta)]))) \in D_0$, we obtain
$\hbar(u) = \hbar(I(Know(in\ present, me, I(\psi[\beta/g(\beta)])))) = t$.

Remark: Note that for the assignments $g : \mathcal{V} \to \mathcal{D}$, such that $g(y_1) = in\ future$ and $g(y_2)$ we consider robot's hypothetical knowledge in future, while in the cases when $g(y_1) = in\ past$ we consider what was robot's knowledge in the past. Consequently, generally the predicates of IFOL for robots, based on the dynamic changes of its knowledge has to be indexed by the time-instances (which are possible worlds of IFOL), for example by using an additional predicate's variable for them. In the examples in next, we will consider only the case of robot's current knowledge (in the actual world with extensional function $\hbar$) when $g(y_1) = in\ present$, so we avoid the introduction of the time-instance variable for the predicates; only at the remark at the end of this section we will show how to use time-variable τ. $\qquad\Box$

From the fact that we do not use the modal Kripke universal modal operator K, the principle of necessitation rule **N** for modal logics, which for a given proposition (sentence) ϕ derives the knowledge fact $K\phi$, here in predicate based IFOL, the robots current knowledge (ground atoms of predicate $Know$) is directly derived from its experiences (based on its neuro-system processes that robot is using in this actual world), in an analog way as human brain does:

- As an activation (under robot's attention) of its neuro-system process, as a consequence of some human command to execute some particular job.
- As an activation of some process under current attention of robot, which is part of some complex plan of robot's activities connected with its general objectives and services.

In both cases, for a given assignment $g : \mathcal{V} \to \mathcal{D}$ of virtual predicate $\phi(\mathbf{x})[\beta/g(\beta)]$ with the *set* of variables $\overline{\mathbf{x}} = \beta \bigcup \alpha$, which *concept* $I(\phi(\mathbf{x})[\beta/g(\beta)]) \in D_{|\alpha|}$ is grounded by this particular process, is transformed into abstracted term and hence robot's knowledge system generates the new ground knowledge atom $Know(y_1, y_2, <\phi(\mathbf{x})>_\alpha^\beta)/g$ with $g(y_1) = in\ presence$ and $g(y_2) = me$, in robot's temporary memory.

Remark: We consider that only robot's experiences (under robot's attention) are transformed into the ground atoms of the $Know$ predicate, and the required (by robot) deductions from them (by using general FOL deduction extended by the three epistemic axioms) are transformed into ground atoms of $Know$ predicate, and hence are saved in robot's temporary memory as a part of robot's *conscience*.

Some background process (unconscious for the robot) would successively transform these temporary memory knowledge into permanent robot's knowledge in an analog way as it happen for humans. $\square$

Thus, the three epistemic axioms of epistemic modal logic with modal operator K, used to obtain deductive knowledge, can be traduced in IFOL by the following axioms for the predicate $Know$, which in fact define the semantics of this particular $Know$ predicate, as follows:

1. The modal axiom **T**, in IFOL is represented by the axiom, for each abstracted term $<\psi(\mathbf{x})>_\alpha^\beta$ and assignment $g : \{y_1, y_2\} \bigcup \beta \to \mathcal{D}$,
   (a) If α is empty
   $$Know(y_1, y_2, <\psi(\mathbf{x})>^\beta)/g \Rightarrow \psi[\beta/g(\beta)] \qquad (19)$$

   (b) If $|\alpha| \geq 1$ and for the intensional concept $u_1 = I(\psi[\beta/g(\beta)]) \in D_{|\alpha|}$, $\hbar(u_1) = \{g_i(\mathbf{y}) \mid g_i \in \mathcal{D}^\alpha, 1 \leq i \leq n$, and $\hbar(I(\psi[\alpha/g_i(\alpha)][\beta/g(\beta)])) = t\}$ with $g_i \neq g_j$ if $i \neq j$ and the tuple of hidden variables $\mathbf{y}$ in the virtual predicate $\psi[\beta/g(\beta)]$,

   $$Know(y_1, y_2, <\psi(\mathbf{x})>_\alpha^\beta)/g \Rightarrow Know(y_1, y_2, <\psi[\alpha/g_1(\alpha)] \wedge ... \wedge \psi[\alpha/g_n(\alpha)]>^\beta)/g \qquad (20)$$

   This axiom shows how the robot's experience of execution of the process (described by abstracted term $<\psi(\mathbf{x})[\beta/g(\beta)]>_\alpha$), to which the intensional concept u_1 is grounded, transforms the true facts obtained by robot's neuro-system (of this process, which results are still the parts of robots unconscious knowledge) into the symbolic-AI FOL formula

   $$(\psi[\alpha/g_1(\alpha)] \wedge ... \wedge \psi[\alpha/g_n(\alpha)])[\beta/g(\beta)]$$

   So by using axiom (19), and FOL deduction, these deductive properties of the robot can deduce any true single fact (logical sentence) $\psi[\alpha/g_i(\alpha)][\beta/g(\beta)]$ derived by its neuro-system process, and to render it to robot's consciousness as a single known fact $Know(y_1, y_2, <\psi[\alpha/g_i(\alpha)]>^\beta)/g$.
   In the case **(a)**, when α is empty, from (17) with

   $$u_1 = g^*(<\psi(\mathbf{x})>^\beta) = I(\psi[\beta/g(\beta)]) \in D_0 \qquad (21)$$

   such that $\hbar(u_1) = \hbar(I(\psi[\beta/g(\beta)])) = t$, that is, the sentence $\psi[\beta/g(\beta)]$ is true, so from the fact that the left side ground atom of axiom's implication in (19) is equal to $Know(g(y_1), g(y_2), u_1)$, this **T** axiom (19) becomes

   $$Know(g(y_1), g(y_2), u_1) \Rightarrow \psi[\beta/g(\beta)] \qquad (22)$$

Note that the meaning of the intensional concept u_1 of the robot is grounded on robot's neuro-system process, which is just robot's current internal experience of what is he doing, and just because of that the robot's knowledge $Know(g(y_1), g(y_2), u_1)$ is true for him. So, this is really an reflexive axiom.

Consequently, the application of the **T** axiom **(a)**, allows the extraction from robot's conscious knowledge the *logical sentences* which, successively, can be elaborated by robot's implemented deductive property of FOL in two ways:

a.1. To respond to some human natural language questions (parsed into a logical formula) and to verify if the response is "yes" or "no", or "I do not know" (if robot's conscious knowledge is incomplete for such a question);

a.2. To deduce another sentences which then can be inserted in robot's conscious knowledge as ground atoms of the predicate $Know$ (where this deduced sentence is represented as an abstracted term). This process (in background, or when robot is free of other concrete activities) can be considered as a kind of consolidation and completion of robot's knowledge based on previous experiences, in an analog way as it is done by human mind when we sleep.

2. The positive introspection axiom **4**:

$$Know(y_1, y_2, <\psi(\mathbf{x})>_\alpha^\beta)/g \Rightarrow Know(g(y_1), g(y_2), <Know(y_1, y_2, <\psi(\mathbf{x})>_\alpha^\beta)/g) \tag{23}$$

that is.

$$Know(g(y_1), g(y_2), g^*(<\psi[\beta/g(\beta)]>_\alpha)) \Rightarrow$$
$$Know(g(y_1), g(y_2), <Know(g(y_1), g(y_2), g^*(<\psi[\beta/g(\beta)]>_\alpha))) \tag{24}$$

which, in the case when $g(y_1) = in\ present$ and $g(y_2) = me$, is traduced in natural language by robot as:

"I know that $\psi[\beta/g(\beta)]$" implies "I know that I know that $\psi[\beta/g(\beta)]$"

where in the logic virtual predicate $\psi[\beta/g(\beta)]$ there are the hidden variables in α, with extension $\hbar(u)$ of its intensional concept $u = g^*(<\psi[\beta/g(\beta)]>_\alpha) = I(\psi[\beta/g(\beta)]) \in D_{|\alpha|}$.

3. The distributive axiom **K** ("modal Modus Ponens"):

$$(Know(y_1, y_2, <\psi(\mathbf{x})>_\alpha^\beta)/g \wedge Know(y_1, y_2, <\psi(\mathbf{x}) \Rightarrow \phi(\mathbf{z})>_{\alpha \bigcup \alpha_1}^{\beta \bigcup \beta_1})/g)$$
$$\Rightarrow Know(y_1, y_2, <\phi(\mathbf{z})>_{\alpha_1}^{\beta_1})/g \tag{25}$$

with the *sets* of variables $\alpha \bigcup \beta = \overline{\mathbf{x}}$ and $\alpha_1 \bigcup \beta_1 = \overline{\mathbf{z}}$. Or, equivalently,

$$(Know(g(y_1), g(y_2), g^*(<\psi[\beta/g(\beta)]>_\alpha)) \wedge$$
$$Know(g(y_1), g(y_2), g^*(<\psi[\beta/g(\beta)] \Rightarrow \phi[\beta_1/g(\beta_1)]>_{\alpha \bigcup \alpha_1})))$$
$$\Rightarrow Know(g(y_1), g(y_2), g^*(<\phi[\beta_1/g(\beta_1)]>_{\alpha_1})) \tag{26}$$

Note that this axiom, when α and α_1 are empty, is a way how the robot provides the conscious implications $\psi[\beta/g(\beta)] \Rightarrow \phi[\beta_1/g(\beta_1)]$, which can be interpreted just as *a rule* "if $\psi[\beta/g(\beta)]$ then $\phi[\beta_1/g(\beta_1)$", independently if they are its innate implemented rules (introduced in robot's knowledge when it is created) or if they are learned by robot's own experience. In fact, this implication can be used only when it is true that some actual robot's experience which produced in its consciousness the knowledge $(Know(y_1, y_2, <\psi(\mathbf{x})>_\alpha^\beta)/g$ so that, from **T** axiom (**a**), the sentence $\psi[\beta/g(\beta)]$ is true, as it is necessary for execution of the rule "if $\psi[\beta/g(\beta)]$ then $\phi[\beta_1/g(\beta_1)]$", and hence to derive the true fact $\phi[\beta_1/g(\beta_1)]$.

Despite the best efforts over the last years, deep learning is still easily fooled [23], that is, it remains very hard to make any guarantees about how the system will behave given data that departs from the training set statistics. Moreover, because deep learning does not learn causality, or generative models of hidden causes, it remains *reactive*, bound by the data it was given to explore [24].

In contrast, brains act proactively and are partially driven by endogenous curiosity, that is, an internal, epistemic, consistency, and knowledge-gain-oriented drive. We learn from our actively gathered sensorimotor experiences and form conceptual, loosely hierarchically structured, compositional generative predictive models. By proposed four-level cognitive robot's structure (15), IFOL allow robots to reflect on, reason about, anticipate, or simply imagine scenes, situations, and developments within in a highly flexible, compositional, that is, semantically meaningful manner. As a result, IFOL enables the robots to actively infer highly flexible and adaptive goal-directed behavior under varying circumstances [25].

We are able to incorporate the emotional structure to robots as well, by a number of fuzzy-emotional partial mappings

$$E_i : \mathcal{D} \to [0, 1] \tag{27}$$

of robots PRP intensional concepts, for each kind of emotions $i \geq 1$: love, beauty, fear, etc. It was demonstrated [1] that IFOL is able to include any kind of many-valued, probabilistic and fuzzy logics as well.

Example: Let us consider the example provided in previous Sect. 3 and how robot, which is conscious of the fact that works to respond to user question ϕ in (8), and hence this robot knows that he initiated the process of recognition expressed by knowledge fact (13), for the assignment of variables $g : \{x_1, x_2\} \to \mathcal{D}$ with $g(x_1) = in\ present$ and $g(x_2) = me$,

$$Know(x_1, x_2, <\psi(y)>_y)/g \tag{28}$$

where the logic formula $\psi(y)$ with a free variable y is given by (10), i.e., is equal to conjunction $Find(in\ present, me, y, <\phi>) \wedge_S videoclips(y)$, so that (28) is equal to

$$Know(in\ present, me, <\psi(y)>_y) \tag{29}$$

How the robot becomes conscious of which video clips it recognized from a given set of videoclips (represented by the extension $C = h(I(videoclips(y))) = h(u_2)$ of the predicate $videoclips(y)$) we will show in next. In order to obtain this knowledge from the known fact (28), the robot's "mind" can activate the internal neuro-process of the FOL-deduction, and hence to take the output of these deducted facts from (28) as new conscious knowledge (new generated ground atoms of its $Know$ predicate), as follows:

1. From (12), we have that unary PRP concept $u_3 = I(\psi(y)) \in D_1$ has the extension $E = h(u_3)$ given by (14), composed by $n = |E| \geq 1$ elements, and hence from the **T** axiom **(b)** with $\alpha = \{y\}$, we obtain that

$$\hbar(u_3) = \{g_i(y) \mid g_i \in \mathcal{D}^\alpha, 1 \leq i \leq n, \text{ and } \hbar(I(\psi[\alpha/g_i(\alpha)])) = t\} \qquad (30)$$

and hence this T axiom (20), from (29), reduces to the logic implication

$$Know(in\ present, me, <\psi(y)>_y) \Rightarrow$$
$$Know(in\ present, me, <\psi[y/g_1(y)] \wedge ... \wedge \psi[y/g_n(y)]>) \quad (31)$$

2. So, from the true atom (29) and implication (31) by Modus Ponens rule of FOL we deduce the formula (right side of implication (31)),

$$Know(in\ present, me, <\psi[y/g_1(y)] \wedge ... \wedge \psi[y/g_n(y)]>) \qquad (32)$$

3. Now the deductive process can use the **T** axiom **(a)**, where β is empty, and from it and (32), by Modus Ponens, deduce the conjunctive formula

$$\psi[y/g_1(y)] \wedge ... \wedge \psi[y/g_n(y)] \qquad (33)$$

and hence, from this conjunctive formula by using FOL-deduction, to deduce that each ground atom $\psi[y/g_i(y)]$, for $1 \leq i \leq n$, from (10) is equal to true fact

$$Find(in\ present, me, g_i(y), <\phi>) \wedge videoclips(g_i(y)) \qquad (34)$$

that is, to the true fact that this robot verified that the video clip $g_i(y)$ satisfies the users requirement ϕ.

4. The last step of this deduction process is to render these outputs of deduction conscious to this robot, that is, to transform the set of outputs in (34) into the set of *known facts*, for $1 \leq i \leq n$,

$$Know(in\ present, me, \ll Find(in\ present, me, g_i(y), <\phi>) \wedge videoclips(g_i(y)>)$$
$$(35)$$

Remark: Note that the obtained robot's knowledge of the set in (35), from the known fact (29), at the end of deduction is in robot's temporary memory. In order to render it permanent (by cyclic process of transformation of the temporary into permanent robot's memory), we need to add to any predicate of the robot's FOL syntax, also the time-variable as, for example, the first variable of each predicate (different from $Know$), instantiated in the known facts by the *tamestamp* value τ (date/time) when this knowledge of robot is transferred into permanent memory, so that the known facts (35) in permanent memory would become

$$Know(in\ present, me, \ll Find(\tau, in\ past, me, g_i(y), <\phi>) \wedge videoclips(g_i(y))>)$$
$$(36)$$

where the second value of the predicate $Find$, from *in present* is modified into the value *in past*, and hence the FOL predicate $Find$ would be translated into natural language by the past time "have found" of this verb. So, the logic atom (36) can be translated by robot into the following natural language autoepistemic sentence:

"I know that I have found at τ the videoclip $g_i(y)$ which satisfied user requirement ϕ."

Note that, by using the **T** axiom **(a)**, the robot can deduce, from permanent memory fact (36), the logic atom

$$Find(\tau, in \; past, me, g_i(y), <\phi>) \wedge videoclips(g_i(y))$$

as an answer to user question and to respond in natural language simply by the sentence: "I have found at τ the videoclip $g_i(y)$ which satisfied user requirement ϕ."   □

This temporization of all predicates used in robot's knowledge is useful for robot to search all known facts in its permanent memory that are inside some time-interval as well.

It can be used not only to answer directly to some human questions about robot's knowledge, but also to extract only a part of robot's knowledge from its permanent memory in order to be used for robot's deduction, and hence to answer to human more complex questions that require deduction of new facts not already deposited in explicit robot's known facts.

Remark: We recall that this method of application of autoepistemic deduction (for concepts such as *knowledge*) can be applied to all other modal logic operators (for concepts such as *belief, obligation, causation, hopes, desires,* etc., for example by using *deontic* modal logic that same statement have to represent a *moral obligation* for robots), by introducing special predicates for them with the proper set of axioms for their active semantics (fixing their meaning and deductive usage).

By such fixing by humans of robot's unconciseness part with active semantics (which can not be modified by robots and their live experience) of all significant for human robot's concepts and their properties, we will obtain ethically confident and socially safe and non danger robots (controlled by public human ethical security organizations for the production of robots with general strong-AI capabilities).   □

5 Conclusions and Future Work

Computation is defined purely formally or syntactically, whereas minds have actual mental or semantic contents, and we cannot get from syntactical to the semantic just by having the syntactical operations and nothing else... Machine learning is a sub-field of artificial intelligence. Classical (non-deep) machine learning models require more human intervention to segment data into categories (i.e. through feature learning). Deep learning is also a sub-field of machine learning, which attempts to imitate the interconnectedness of the human brain using neural networks. Its artificial neural networks are made up layers of models, which identify patterns within a given dataset. Deep learning can handle complex problems well, like speech recognition, pattern recognition, image recognition, contextual recommendations, fact checking, etc.

However, with this integrated four-level robot's knowledge system presented in diagram (15), where the last level represents the robot's neuro system containing the deep learning as well, we obtain that also the semantic theory of robot's intensional FOL is a procedural one, according to which sense is an abstract, pre-linguistic procedure detailing what operations to apply to what procedural constituents to arrive at the product (if any) of the procedure.

Weak AI, also known as narrow AI, focuses on performing a specific task, such as answering questions based on user input or playing chess. It can perform one type of task, but not both, whereas Strong AI can perform a variety of functions, eventually teaching itself to solve for new problems. Weak AI relies on human interference to define the parameters of its learning algorithms and to provide the relevant training data to ensure accuracy.

Strong AI (also known as full *general AI*) aims to create intelligent robots that are quasi indistinguishable from the human mind. But just like a child, the AI machine would have to learn through input and experiences, constantly progressing and advancing its abilities over time. If researchers are able to develop Strong AI, the robot would require an intelligence more close to human's intelligence; it would have a self-aware consciousness that has the ability to solve problems, learn, and plan for the future.

However, since humans cannot even properly define what intelligence is, it is very difficult to give a clear criterion as to what would count as a success in the development of strong artificial intelligence. Thus, we argue that this example, used for the spatial natural sublanguage, can be extended in a similar way to cover more completely the rest of human natural language, and hence the method provided by this paper is a main theoretical and philosophical contribution to resolve the open problem of how we can implement the deductive power based on IFOL for new models of robots heaving strong AI capacities. Intensional FOL is able to represent the Intentional States (mental states such as beliefs, hopes, and desires), typical for human minds:

"Intentionality[5] is the fascinating property certain cognitive states and events have in virtue of being directed, or about, something. When ever we think, we think about

---

[5] A German philosopher and psychologist Franz Brentano (1838–1917) is best known for his reintroduction of the concept of intentionality, a concept derived from scholastic philosophy, to contemporary philosophy in his lectures and in his work Psychologie vom empirischen Standpunkt [26] (Psychology from an Empirical Standpoint). Brentano used the expression "intentional inexistence" to indicate the status of the objects of thought in the mind. Intentionality, based on the work of Austrian philosopher Alexius Meinong (1853–1920) a pupil of Franz Brentano, a realist known for his unique ontology, is the power of minds to be about something: to represent or to stand for things, properties and states of affairs. Intentionality is primarily ascribed to mental states, like perceptions, beliefs or desires, which is why it has been regarded as the characteristic mark of the mental by many philosophers. A central issue for theories of intentionality has been the problem of intentional inexistence: to determine the ontological status of the entities which are the objects of intentional states.

Meinong adopted the threefold phenomenological analysis of mental states that includes a mental act, its content and object of intention. Meinong wrote two early essays on David Hume, the first dealing with his theory of abstraction, the second with his theory of relations, and was relatively strongly influenced by British empiricism. He is most noted, however, for his edited book Theory of Objects (full title: Investigations in Theory of Objects and Psychology [27]), which grew out of his work on intentionality and his belief in the possibility of intending nonexistent objects. Whatever can be the target of a mental act, Meinong calls an "object.".

something; whenever we believe, there is something we believe; whenever we dream, there is something we dream about. This is true of every episode of such diverse psychological phenomena as learning, imagining, desiring, admiring, searching for, discovering, and remembering..

Sometimes, they are directed towards logically more complex objects, for instance, when we entertain a proposition, or fear a certain state of affairs, or contemplate a certain depressing situation. But all of these phenomena are to be contrasted with physical sensations, undirected feelings of joy, sadness, depression, or anxiety, and with episodes of pain and discomfort." pp. 10, [28]

So, we are able to support such robot's physical sensations by using, for example, the manyvalued fuzzy logic values assigned to robot's PRP intensional "emotional" concepts, by mappings (27), representing the feelings of joy, sadness, depression, or anxiety, etc. as well. Hence, by using manyvalued logics embedded into IFOL, as explained in [1], and autoepistemic deductive capacities provided in previous Section, the robots would be able to reason about their own sensations and to communicate with humans.

Moreover, I argue that AI research should set a stronger focus on learning compositional generative predictive models (CGPMs), from robot's self-generated sensorimotor experiences, of the hidden causes that lead to the registered observations. So, guided by evolutionarily-shaped inductive learning and information processing biases, the robots will be able to exhibit the tendency to organize the gathered experiences into event-predictive encodings.

Consequently, endowed with suitable IFOL information-processing biases, the robot's AI may develop that will be able to explain the reality it is confronted with, reason about it, and find adaptive solutions, making it Strong AI.

References

1. Majkić, Z.: Intensional First Order Logic: From AI to New SQL Big Data. Walter De Gruyter GmbH, Berlin/Boston (2022). ISBN 978-3-11-099494-0
2. Bealer, G.: Quality and Concept. Oxford University Press, USA (1982)
3. Majkić, Z.: Conservative Intensional Extension of Tarski's Semantics. Advances in Artificial Intelligence, pp. 1–17. Hindawi Publishing Corporation (2012). ISSN 1687-7470
4. Majkić, Z.: Intensionality and Two-steps Interpretations. arXiv:1103.0967, pp. 1–15 (2011)
5. Majkić, Z.: Intensional RDB Manifesto: A Unifying New SQL Model for Flexible Big Data. arXiv:1403.0017, pp. 1–29 (2014)
6. Majkić, Z.: Intensional first order logic for strong-AI generation of robots. J. Adv. Mach. Learn. Artif. Intell. **4**(1), 23–31 (2023)
7. Jackendoff, R.: Semantics and Cognition. The MIT Press, Cambridge (1983)
8. Kahneman, D.: Thinking Fast and Slow. Farrar, Straus and Giroux (2011). ISBN 978-0374275631
9. Frege, G.: Über Sinn und Bedeutung. Zeitschrift für Philosophie und Philosophische Kritik, pp. 22–50 (1892)
10. Bealer, G.: Universals. J. Philos. **90**, 5–32 (1993)
11. Bealer, G.: Theories of properties, relations, and propositions. J. Philos. **76**, 634–648 (1979)
12. Kollar, T., Tellex, S., Roy, D., Roy, N.: Toward understanding natural language directions. In: Proceedings of the 4th ACM International Conference on Human Robot Interaction (2010)

13. Tellex, S., Kollar, T., Show, G., Roy, N., Roy, D.: Grounding spatial language for video search. In: ICMI-MLMI 2010, Beijing, China, 8–12 November (2010)
14. Landau, B., Jackendoff, R.: What' and "where" in spatial language and spatial cognition. Behav. Brain Sci. **16**, 217–265 (1993)
15. Talmy, L.: The fundamental system of spatial schemas in language. From Perception to Meaning: Image Schemas in Cognitive Linguistics, Mouton de Gruyter (2005)
16. Regier, T.P.: The acquisition of lexical semantics for spatial terms: a connectionist model of perceptual categorization. Ph.D. thesis, University of California at Berkeley (1992)
17. Bailey, D.: When push comes to shove: a computational model of the role of motor control in the acquisition of action verbs. Ph.D. thesis (1997)
18. Kelleher, J.D., Costello, F.J.: Applying computational models of spatial prepositions to visually situated dialog. Comput. Linguist. **35**(2), 271–306 (2009)
19. Siskind, J.M.: Grounding the lexical semantics of verbs in visual perception using force dynamics and event logic. J. Artif. Int. Res. **15**(1), 31–90 (2001)
20. Majkić, Z.: Intensional logic and epistemic independency of intelligent database agents. In: 2nd International Workshop on Philosophy and Informatics (WSPI 2005), Kaiserslautern, Germany, 10–13 April (2005)
21. Majkić, Z.: Autoepistemic logic programming for reasoning with inconsistency. In: International Symposium on Logic-Based Program Synthesis and Transformation (LOPSTR), Imperial College, London, UK, 7–9 September (2005)
22. Marek, W., Truszczynski, M.: Autoepistemic logic. J. ACM **38**(3), 588–618 (1991)
23. Nguyen, A., Yosinski, J., Clune, J.: Deep neural networks are easily fooled: high confidence predictions for unrecognizable images. In: 2015 IEEE Conference on Computer Vision and Pattern Recognition (CVPR), pp. 427–436 (2015)
24. Marcus, G.: Deep learning: a critical appraisal. arXiv:1801.00631 [cs.AI] (2018)
25. Russell, S.: The purpose put into the machine. In: Brockman, J. (ed.) Possible Minds: 25 Ways of Looking at AI, Chap. 3, pp. 20–32. Penguin Press, New York (2020)
26. Brentano, F.: Psychologie vom empirischen Standpunkte. Duncker & Humblot, Leipzig (1874)
27. Meinong, A.: Untersuchungen zur Gegenstandstheorie und Psychologie. Verlag Der Wissenschaften (1904)
28. Zalta, E.N.: Intensional Logic and The Metaphysics of Intentionality. MIT Press, Bradford Books (1988)

Divide and Control: Generation of Multiple Component Comic Illustrations with Diffusion Models Based on Regression

Zixuan Wang[1], Peng Du[2], Zhenghui Xu[1], Qihan Hu[1], Hao Zeng[1(✉)], Youbing Zhao[1,3(✉)] (iD), Hao Xie[1], Tongqing Ma[1], and Shengyou Lin[1]

[1] Communication University of Zhejiang, Hangzhou 310018, China
{hao.zeng,zyb}@cuz.edu.cn
[2] Uber Technologies Inc., 1725 3rd St, San Francisco, CA 94158, USA
[3] University of Bedfordshire, Luton LU1 3JU, UK

Abstract. Diffusion-based text-to-image generation has achieved huge success in creative image generation and editing applications. However, when applied to comic illustrations, it still struggles to deliver predictable high-quality productions with multiple characters due to the interference of the text prompts. In this paper, we propose a practicable method to use ControlNet and stable diffusion to generate controllable outputs of multiple components. The method first generates images for individual components separately and then degenerates those images to a regressed form, such as line drawings or Canny edges. Those regressed forms of individual components are then merged and fed into ControlNet to generate the final image. Experiments show that this method is highly controllable and can produce high-quality comic illustrations with multiple components.

Keywords: Diffusion Models · ControlNet · Comic Illustrations · Multiple Component

1 Introduction

Traditionally creative generation of images is either done by humans or by humans with limited assistance from image processing software, requiring a decent level of painting and image editing skills from the user. In the past decade, with the revolutionary advances of deep neural networks [8], the creative generation of images has achieved great advances. Within the focus of text-to-image (T2I) generation, the application of AI technologies, previously using GAN [7, 11] and recently using diffusion models [9, 10, 14, 17, 19, 22] have totally changed the landscape of generative image technologies. It is now possible to generate high-fidelity images based on text prompts. Such powerful T2I image models can be

Many thanks to Mr. Yihui Shen for his generous funding support.

F. Zhao and D. Miao (Eds.): AIGC 2023, CCIS 1946, pp. 59–69, 2024.
https://doi.org/10.1007/978-981-99-7587-7_5

applied in a wide range of applications, such as cloth designing, architectural rendering, comic illustration, etc.

Recent diffusion-based T2I models, such as DALL·E 2 [13], Imagen [16] and Stable Diffusion [14] provide powerful image generation and editing capabilities. However, it still suffers from many limitations, such as difficulty in choosing appropriate prompts and hard to generate predictive results, lack of consistency, difficult to generate satisfactory results of multiple components, difficult to handle background and lighting, etc. With comic illustration in mind, we are more interested in generating comic illustrations of multiple characters/components with diffusion-based T2I models.

The aim of this paper is to design practicable methods to address the challenge of generating comic illustrations of multiple characters/components within existing diffusion model-based frameworks. The core question is how to achieve predictive and controllable illustrations of multiple components.

We propose a regression-based method using the capabilities of the diffusion model and ControlNet [22]. The method first generates independent images for each component from prompts. Then by regressing those images to a primitive form, such as Canny Edge or line drawing, the method merges the images into a composite primitive image as the input to ControlNet. The method finally generates highly predictive and controllable final images by ControlNet and Stable Diffusion.

Our main contributions lie in the following aspects:

- We propose a regression-based workflow to generate comic illustrations of multiple components/characters by leveraging the power of diffusion models, ControlNet, and human composition. This method can produce more flexible, predictive, and higher quality illustrations of multiple components than directly using Stable Diffusion.
- Our method supports multiple regressed forms (intermediate representations) of components, including Canny Edge, line drawing, depth map, etc. This provides flexible options for the composition of different global and local structures.
- Our method supports a number of component types, such as a character, an object or a background and there are no constraints on the number of components, implying the huge potential of the method in a variety of applications.

2   Related Work

2.1   Diffusion Probabilistic Models

The recent diffusion-based T2I models provide powerful image generation and generative editing capabilities based on Internet-scale text and image databases.

The diffusion probabilistic model was proposed in [17]. The successful application of diffusion models in image generation achieved quality better than GAN in recent years [5]. Architectures for training and sampling with diffusion models have also been improved in recent years, including Denoising Diffusion Probabilistic Model (DDPM) [9], score-based generative models (SGMs) [19] and score-based stochastic differential equations (Score SDE) [10]. The U-Net architecture [15] is normally employed by those diffusion models in image generation.

For a comprehensive introduction to diffusion models and their applications, the readers are referred to an extensive survey conducted in [20]. [21] also presents a more recent survey on text-to-image related diffusion models.

Due to the high-quality of generated images, diffusion models have become a new research focus in AIGC and have replaced GANs in many image generation tasks since 2021, despite the fact that they follow an iterative process and are much slower than GANs. Due to the popularity of Stable Diffusion [14], it is chosen as the default model to implement our multiple-component comic illustration generation in this paper.

However, there are also limitations and disadvantages of diffusion models. One apparent disadvantage is that diffusion models are slow in generating new images due to the use of an iterative refinement process. Although Song proposed a tentative consistency model [18] to improve speed, the new method has not yet achieved image qualities comparable to diffusion models. Another disadvantage is that it lacks predictive control of images generated from prompts. The difficulty in predicting the effects of prompt change leads most T2I generations to the proximity of pure trying. This attracts much research interest in more predictive, controllable, consistent text-to-image generation with diffusion models, and the related work is addressed in the next subsection.

2.2 Controllable Text-to-Image Generation with Diffusion Models

To gain more control over generated images, that is, to deliver more consistent and predictable results, ControlNet [22] was proposed to augment pre-trained large diffusion models with additional control blocks to support a variety of user guide inputs. For example, it can generate images conforming to predefined line sketches, edge detection, or character skeletons.

Parmar et al. proposed an enhanced diffusion-based image-to-image translation method that allows users to specify the edit direction on-the-fly, such as changing the subject from a cat to a dog [12], while preserving the general content structure using cross-attention.

Gong et al. proposed controllable text-to-image generation of multiple characters [6]. However, the results presented in their paper still look unnatural, and there is no strong proof to show that they can handle occlusions well.

Cao et al. employed cross-attention between the U-Net layers of two corresponding diffusion processes to impose character consistency [4] when adjusting the generated contents according to revised prompts.

3   Method

We propose a regression-based multiple-component illustration generation workflow that takes advantage of diffusion models, ControlNet, and human composition. This approach is capable of producing predictive and higher-quality illustrations of multiple components/characters than the direct use of stable diffusion. Our method supports multiple regressed forms (intermediate representations) of components, including Canny Edge, line drawing, depth map, etc. This provides flexible options for the composition of different global and local structures. Moreover, the type and number of components are also flexible. A possible component can be a character, an object, or a background, and there are no constraints on the number of components, implying the potential of the method in a variety of applications.

The method consists of several stages, including component image generation, regression, composition, and final image generation. The workflow stages are summarized as follows: first images of individual components are generated using the Stable Diffusion prompts. The exact component image, such as the target person, object, or background may need to be extracted from those generated images. The component images are then regressed to a primitive form such as Canny Edge, line drawing, or depth map, etc. By composing the regressed images into an integrated scene with human integration, an input image in the primitive form is prepared, which is fed to ControlNet in the final stage to generate the final image with Stable Diffusion. These stages are introduced in the following subsections in detail. The entire workflow of our method with example images taken from the case of "astronaut outside a space station" is shown in Fig. 1.

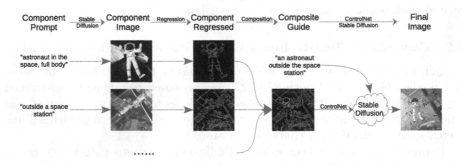

Fig. 1. The workflow of our method

3.1   Component Image Generation

In this stage, the images of individual components are generated from prompts with Stable Diffusion tools such as WebUI [2] or ComfyUI [1]. Images from external sources can also be used. For character and object components, the generated image may contain a background and the component may need to be extracted. Fortunately, we have ready-to-use plugins such as the "segment anything" plugin provided by WebUI to easily extract an object. In this stage, we are more concerned with the overall shape and structure of components than their styles, as in the next stage the images will be regressed to a more primitive form with the loss of most of their styles.

3.2 Regression

In this stage, the component images are regressed to a primitive form. Common primitive forms include the Canny Edge, line arts, depth map, and other heuristics of the general shape of the component. As the regressed image renders the overall structure of components and drops much unrelated information, it can be used to provide good hints of the desired shape, pose, structure, or content of components. As line representations such as detected shape edges, line drawings, or sketches provide strong features of components, they have been proved very effective in producing conforming component images with ControlNet. In this paper, we are using Canny Edge [3] as the regressed form.

3.3 Composition

After regression, the regressed components are then composed to form an integrated regressed view of the final image to be generated. As the integrated regressed image describes the overall layout and structure of components in the final image, it serves as a simple but strong hint for ControlNet to create the final image. In the composition stage, occlusion needs to be handled carefully as different occlusions may lead to different interpretations of component relationships in the final image.

3.4 Final Image Generation

In this stage, the integrated multi-component primitive image from the composition stage is fed into ControlNet to control the generation of the final image with Stable Diffusion. Although the integrated image provides strong control over final image generation, prompts are normally required to achieve better control of the image content. Thanks to ControlNet, the generated final image normally highly conforms to the primitive image, providing high fidelity of predefined shapes and structures, which is not possible with prompt-only image generation.

Take the prompt "Astronaut, outside a space station" as an example. Direct use of the Stable Diffusion prompt leads to unsatisfying results, as shown in Fig. 2. In most of the generated images, the space station is either totally missing or only with part of the solar panels visible. It shows that the diffusion model does not understand the prompts in a way that humans prefer.

The disadvantages of the prompt-only method are summarized as follows:

– The prompt-only method is prone to distortion of characters and objects due to interference between cues when dealing with more complex or large numbers of components.

Fig. 2. Images generated by Stable Diffusion from prompt "Astronaut, outside a space station"

- When objects or people interact with each other in the desired image, the position of the cues in the prompt can be misrepresented.
- The prompt-only method has a one-sided understanding of cue words, which can lead to the inability to accurately understand the words when the image to be generated is difficult to describe or when the model is poorly trained on a particular subject.

As shown in Fig. 3, with our method, we first divide the prompt into two groups: the background component ("Outside the space station") and the subject component ("Astronaut"). We then use the stable diffusion WebUI to generate separate images for each component. While the background space station can be directly used, the foreground astronaut needs to be extracted. We use the "segment anything" plugin provided by WebUI to directly click-and-select the astronaut.

Regressed Canny edges of these two components are then obtained using ControlNet's Canny Edge Preprocessor. The Canny edge components are merged with a customized user interface or image processing software to generate the merged Canny edge view. In our case the Canny edge of the astronaut is placed onto the Canny edge of the background space station at an appropriate position. This produces a Canny edge of the final image. Finally, we feed the integrated Canny edge to ControlNet plugin, as well as prompts (in our case "astronaut outside a space station") to Stable Diffusion to generate final images conditional on the Canny edge. As ControlNet is powerful in controlling the structure of the generated image, the final images highly conform to the Canny edge image while preserving the semantic meaning given in the prompt.

4   Experiments

Experiments are carried out using our method with Stable Diffusion WebUI and the ControlNet plugin. With Stable Diffusion, we are using a pre-trained checkpoint model of "anything-v5". The regressed Canny edges are generated with the Canny Edge Preprocessor (control_canny-fp16) of the ControlNet plugin v1.1.189. Object extraction is achieved via WebUI plugin "segment anything". The composition of Canny edge images is completed via a Web-based interface

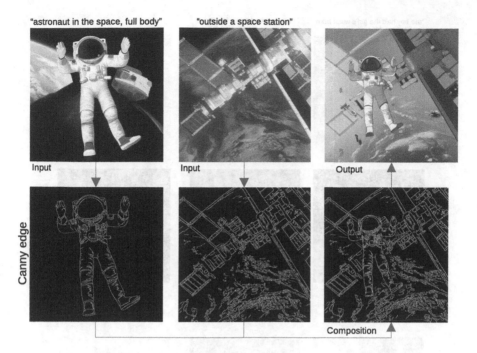

Fig. 3. The "astronaut outside a space station" case with our method

developed by us. The intermediate and final images all have a size of 512×512 and the number of iteration steps is set to 20. The experiments run on a Windows 11, 64 GB memory platform with a GeForce RTX4090 (24G) graphics card.

We present two comparisons of our method with prompt-early Stable Diffusion. The first is an experiment of two characters dancing with each other, as shown in Fig. 4. The second is an experiment of three components: a girl, a trampoline, and a playground, as shown in Fig. 5. With the two cases, it can be shown that our method provides more predictive and accurate control over the final generated images.

There are also some limitations of our method, including:

- It works best with cartoonish styles and less well with realistic styles.
- To generate better results, the cue words to generate independent components normally cannot be copied directly from the original prompt.

(a) Our method

(b) Results generated by Stable Diffusion with "a girl spread arms, a boy held the girl's waist from behind, full body shot, looking at viewer."

Fig. 4. Our method and results generated

(a) Our method

(b) Images generated by Stable Diffusion with a girl jumps off a trampoline on a playground

Fig. 5. Our method and results generated

5 Conclusions

To conclude, by generating component images separately and having those components regressed to a primitive form, our method achieves a highly controllable generation of illustrations with multiple components. The method can be readily applied in the field of comic illustration to achieve an easier generation of comic illustrations with multiple characters.

Acknowledgements. We would like to thank Mr. Yihui Shen, the Zhejiang Provincial Public Welfare Fund (LGF21F020004) as well as the 12th batch of the Course Teaching Mode Innovation Experimental Zone Project of Communication University of Zhejiang: "Exploring Bilingual Teaching Approaches for Design Patterns" for the generous funding support of the work referred to in this paper.

References

1. Stable diffusion ComfyUI. https://github.com/comfyanonymous/ComfyUI. Accessed 07 June 2023
2. Stable diffusion WebUI. https://github.com/db0/stable-diffusion-webui. Accessed 07 June 2023
3. Canny, J.: A computational approach to edge detection. IEEE Trans. Pattern Anal. Mach. Intell. **6**, 679–698 (1986)
4. Cao, M., Wang, X., Qi, Z., Shan, Y., Qie, X., Zheng, Y.: Masactrl: tuning-free mutual self-attention control for consistent image synthesis and editing (2023)
5. Dhariwal, P., Nichol, A.: Diffusion models beat GANs on image synthesis. In: Ranzato, M., Beygelzimer, A., Dauphin, Y., Liang, P., Vaughan, J.W. (eds.) Advances in Neural Information Processing Systems, vol. 34, pp. 8780–8794. Curran Associates, Inc. (2021)
6. Gong, Y., et al.: Talecrafter: interactive story visualization with multiple characters (2023)
7. Goodfellow, I., et al.: Generative adversarial nets. In: Ghahramani, Z., Welling, M., Cortes, C., Lawrence, N., Weinberger, K. (eds.) Advances in Neural Information Processing Systems, vol. 27. Curran Associates, Inc. (2014)
8. Goodfellow, I.J., Bengio, Y., Courville, A.: Deep Learning. MIT Press, Cambridge, MA, USA (2016). http://www.deeplearningbook.org
9. Ho, J., Jain, A., Abbeel, P.: Denoising diffusion probabilistic models. In: Larochelle, H., Ranzato, M., Hadsell, R., Balcan, M., Lin, H. (eds.) Advances in Neural Information Processing Systems. vol. 33, pp. 6840–6851. Curran Associates, Inc. (2020)
10. Jo, J., Lee, S., Hwang, S.J.: Score-based generative modeling of graphs via the system of stochastic differential equations. In: Chaudhuri, K., Jegelka, S., Song, L., Szepesvari, C., Niu, G., Sabato, S. (eds.) Proceedings of the 39th International Conference on Machine Learning. Proceedings of Machine Learning Research, vol. 162, pp. 10362–10383. PMLR, 17–23 July 2022
11. Karras, T., Laine, S., Aittala, M., Hellsten, J., Lehtinen, J., Aila, T.: Analyzing and improving the image quality of StyleGAN. In: 2020 IEEE/CVF Conference on Computer Vision and Pattern Recognition (CVPR), pp. 8107–8116 (2020). https://doi.org/10.1109/CVPR42600.2020.00813
12. Parmar, G., Singh, K.K., Zhang, R., Li, Y., Lu, J., Zhu, J.Y.: Zero-shot image-to-image translation (2023)
13. Ramesh, A., Dhariwal, P., Nichol, A., Chu, C., Chen, M.: Hierarchical text-conditional image generation with clip latents (2022)
14. Rombach, R., Blattmann, A., Lorenz, D., Esser, P., Ommer, B.: High-resolution image synthesis with latent diffusion models. In: Proceedings of the IEEE/CVF Conference on Computer Vision and Pattern Recognition, pp. 10684–10695 (2021)
15. Ronneberger, O., Fischer, P., Brox, T.: U-Net: convolutional networks for biomedical image segmentation. In: Navab, N., Hornegger, J., Wells, W.M., Frangi, A.F. (eds.) MICCAI 2015. LNCS, vol. 9351, pp. 234–241. Springer, Cham (2015). https://doi.org/10.1007/978-3-319-24574-4_28
16. Saharia, C., et al.: Photorealistic text-to-image diffusion models with deep language understanding (2022)
17. Sohl-Dickstein, J., Weiss, E., Maheswaranathan, N., Ganguli, S.: Deep unsupervised learning using nonequilibrium thermodynamics. In: Bach, F., Blei, D. (eds.) Proceedings of the 32nd International Conference on Machine Learning. Proceedings of Machine Learning Research, vol. 37, pp. 2256–2265. PMLR, Lille, France, 07–09 July 2015

18. Song, Y., Dhariwal, P., Chen, M., Sutskever, I.: Consistency models (2023)
19. Song, Y., Durkan, C., Murray, I., Ermon, S.: Maximum likelihood training of score-based diffusion models. In: Ranzato, M., Beygelzimer, A., Dauphin, Y., Liang, P., Vaughan, J.W. (eds.) Advances in Neural Information Processing Systems, vol. 34, pp. 1415–1428. Curran Associates, Inc. (2021)
20. Yang, L., et al.: Diffusion models: a comprehensive survey of methods and applications (2023)
21. Zhang, C., Zhang, C., Zhang, M., Kweon, I.S.: Text-to-image diffusion models in generative AI: a survey (2023)
22. Zhang, L., Agrawala, M.: Adding conditional control to text-to-image diffusion models (2023)

Enhancing EFL Vocabulary Acquisition Through Computational Thinking

Youjun Tang[1,2](✉) and Xiaomei Ma[1]

[1] Xi'an Jiaotong University, Xi'an 710049, Shannxi, China
qdbhxytyj@126.com, xiaomei@xjtu.edu.cn
[2] Qingdao Binhai University, Qingdao 266555, Shandong, China

Abstract. The emergence of ChatGPT marks the advent of the era of artificial intelligence (AI) and language intelligence (LI). Moreover, the essence of AI and LI is a kind of deep machine learning based on computational thinking (CT). Computational thinking is rooted in computer science and is higher-order thinking that mimics how computers solve problems. At the same time, CT, a concept computer scientists propose, is also human thinking. Given factors such as the inconsistent definition of CT, the integration of CT into foreign language education in China has only just begun. In order to improve the vocabulary richness in English short essay writing of our non-English majors, we followed the critical skills of CT (data analysis, pattern recognition, abstraction, decomposition, and parallelization) to intervene in students' English vocabulary acquisition. We measured their short essay writing before and after the intervention using Quantitative Index Text Analyser (QUITA) software. The results showed that all vocabulary-related quantitative indexes changed significantly and that CT could facilitate Chinese students' English vocabulary acquisition efficiency. This study has implications for the new direction of foreign language education in China in the artificial and language intelligence era.

Keywords: Artificial Intelligence · Language Intelligence · Computational Thinking · EFL Vocabulary Acquisition

1 Introduction

1.1 ChatGPT and its Essence

The emergence of ChatGPT, an intelligent chatbot, marks that the information society has entered an epoch-making period of human-computer interaction and marks that education informatization has reached a deep integration of disciplines and information technology.

ChatGPT is the hottest intelligent chat tool and a recent research hotspot in linguistics. With its high text generation ability and smooth human-computer interaction capability, it has once again brought artificial intelligence (AI) into the focus of various research fields. In essence, ChatGPT is a natural language processing model that belongs to the application of generative artificial intelligence technology [1]. It also means the advent of the era of language intelligence (LI).

F. Zhao and D. Miao (Eds.): AIGC 2023, CCIS 1946, pp. 70–82, 2024.
https://doi.org/10.1007/978-981-99-7587-7_6

1.2 Thinking of AI and LI

Artificial intelligence (AI) refers explicitly to "a new technical science that studies and develops theories, methods, technologies, and application systems that can simulate, extend, and expand human intelligence. The purpose of the research is to promote intelligent machines that can listen (speech recognition, machine translation), see (image recognition, text recognition), speak (speech synthesis, human-machine dialogue), think (human-machine games, theorem proving), learn (machine learning, knowledge representation), and act (robotics, autonomous driving)" [2].

Language intelligence (LI) is one of the critical technologies that need to be concentrated in the current research of AI, and the breakthrough of its basic theory and essential technology research is of great significance to the development of AI in China. Linguistic intelligence, the intelligence of linguistic information, uses computer information technology to imitate human intelligence and analyze and process human language [3]. The purpose is to realize human-computer language interaction [4] eventually.

AI and LI are both manifestations of deep, autonomous machine learning. Its essence is automatic processing and processing based on the working principles of computers, ultimately solving problems. In this way, AI and LI follow computational thinking (CT), the embodiment of the computer way of thinking.

1.3 Computational Thinking (CT)

In 1980, Papert introduced the term "computational thinking" to refer to a model in which students use computers to improve their thinking [5]. In 1996, he re-emphasized the need for students to use computers to change their learning and improve their ability to express their ideas [6]. In 2006, Professor Jeannette M. Wing published a paper entitled "Computational Thinking," which, for the first time, defined CT as a universal attitude and skill and interpreted CT as a class of solutions that allow people to think "like computer scientists" [7]. In 2008, she gave a new interpretation of CT, stating that CT is essentially analytical thinking and that its three aspects of problem-solving, system design and evaluation, and understanding of intelligence and human behavior converge with mathematical thinking, engineering thinking, and scientific thinking, respectively [8]. Since then, the conceptual interpretation and teaching practice of CT have shown apparent diversity, and the academic community has yet to reach a consensus on CT.

Although there is no consensus on the definition of CT and what it encompasses, there is a consensus in the academic community that CT is higher-order thinking that imitates computer thinking to solve problems, is one of the essential qualities necessary for international talents in the 21st century, and is a vital tool for optimizing the way knowledge is acquired [9–12]. CT is both machine thinking and a concept proposed by computer scientists; therefore, CT is also human thinking, and the two can be mutually reinforcing. In the era of AI and LI, CT can contribute wisdom to the new direction of foreign language education.

1.4 Research Questions

Foreign language education in China faces opportunities and challenges in the era of AI and LI. As the common thinking of AI and LI, the intervention of CT in foreign language education may bring surprises.

English short essay writing is difficult for Chinese non-english majors, and its central problem lies in the poor English vocabulary of students in short essay writing, with many repetitive words and few advanced words. It is because students need help acquiring English vocabulary and acquire it inefficiently. Based on this, this study uses a sandwich intervention to intervene in CT in English vocabulary acquisition. It aims to address the following three questions:

(1) How does computational thinking intervene in English vocabulary acquisition?
(2) How does computational thinking affect English vocabulary acquisition?
(3) What is the role of computational thinking in vocabulary intervention?

2   Methods

2.1 Quasi-experimental Design

According to the needs of the study, we adopted a quasi-experimental design, i.e., the research process consisted of three main stages, including a pre-test, an instructional intervention based on the primary skills steps of CT, and a post-test. The study involved two natural classes, i.e., the experimental and control classes. According to Mackey and Gass, the experimental class engaged in a stepwise instructional intervention based on the formation of practical skills in CT. In contrast, the control class received the traditional instructional model of direct transfer of vocabulary memory [13]. In other words, the independent variable was the stepwise instructional intervention model based on the formation of the primary skills of CT. At the same time, English vocabulary mechanical memory that occurred in English learners' English writing in the control class was the dependent variable.

2.2 Experimental Conditions

The study was conducted in a private university with a teaching class that included at least 30 students; the course under study was college English writing with tiered instruction, and the study lasted for ten weeks, 90 min each.

2.3 Participants

Ninety-two freshmen from the class of 2021 participated in the study, with 46 students in each natural class. It means that the experimental class was homogeneous with the control class: from the same grade, at the same level of instruction, and in the same course. These students were between the ages of 18 and 21 and involved 11 majors, such as preschool education, mechanics, electronics, and international trade, with similar gender ratios. Before enrollment, all students had taken the college entrance exam, meaning they had at least ten years of English language learning experience. To verify the homogeneity of the participants, all students took a short essay writing pre-test, and the results showed that students' short essay writing in English generally lacked advanced vocabulary and vocabulary richness.

2.4 Assessment Tools

Test. Since students were required to take the National English Language Proficiency Test (NELPT), we organized a pre-test and a post-test, respectively, using questions from the June and December 2020 exams, as shown in Figs. 1 and 2 for more information:

Part I. Writing (30 minutes)

    Directions: For this part, you are allowed 30 minutes to write an essay on the use of translation apps. (at least 150 words).

Fig. 1. Writing test in pre-test (June, 2020)

Part I. Writing (30 minutes)

    Directions: In this section, you are allowed 30 minutes to write a short essay commenting on the saying "Learning is a daily experience and a lifetime mission." (at least 150 words).

Fig. 2. Writing test in post-test (December, 2020)

Instrument. All the essays in the research were measured using Quantitative Index Text Analyzer (QUITA) software, a free software designed and developed by the linguistics faculty and students at Palacky University, Czech Republic. QUITA is available for download at http://oltk.upol.cz/software or https://kcj.osu.cz/wp-content/uploads/2018/06/QUITA_Setup_1190.zip.

Quantitative Text Indexes. Some quantitative text indexes are needed to check the vocabulary. All indexes used in this research come from Haitao Liu's *An Introduction to Quantitative Linguistics* (2017) [14]. Twenty-two quantitative text indexes are briefly introduced in Liu's book. In this research, however, we only applied four of the most frequently used vocabulary-related indexes, including TTR, R1, Descriptivity, and Verb Distances. Table 1 below shows the relevant details for the four indexes.

Raters. The scorers were the researcher and another teacher with extensive experience in teaching writing. The two raters independently used QUITA to measure all students' essays on relevant vocabulary indexes, organized consistency tests, and discussed and agreed on disagreements.

Table 1. Brief introduction of the principal quantitative indexes related vocabulary.

Quantitative indexes	Brief introduction of the principal quantitative indexes
TTR	Token/type ratio: The ratio between the total number of word types (V) and that of word tokens (N) [15, 16] and an index on lexical diversity [17]
R_1	Relates to the vocabulary's richness, esp. The richness of made-up words [18]
Descriptivity	Refers to the degree of descriptivity in the form of adjectives/verbs + adjectives [14, 19]
Verb Distances	Implies the distance between the two neighbor verbs in a sentence or neighbor sentences [14, 19]

2.5 Intervention Process

The control group received the traditional direct instruction method, which teaches students to memorize English vocabulary by reading words aloud and then directly. In contrast, the experimental group experienced a 10-week stepwise dynamic assessment interventionist based on the formation of core skills in CT, which was divided into three main phases: pre-test, intervention, and post-test, but administered as a whole, also called sandwich dynamic assessment interventionist [20]. The main elements of these three phases are as follows:

Phase 1: Pre-test. The pre-test was used to obtain data on vocabulary measures for both classes and diagnostic information on short-text writing.

Stage 2: Mediation through instructional intervention. Based on the diagnostic information obtained from the pre-test, the researcher organized the experimental group to participate in a stepwise intervention teaching practice based on forming core CT skills to mediate the learners. Specifically, English words are viewed as a numerical symbol, and students are expected to abstract the form of word formation and its origin based on the critical skills of CT. This study adopts Tang and Ma's core step model of CT, which involves crucial skills, including data analysis (seeing words as numbers, symbols, or codes), pattern recognition (looking for commonalities, regular usage, patterns, and features), abstraction (distilling data to form procedural knowledge), decomposition (breaking down complex problems into more solvable or operational subproblems that can be multi (decomposing complex problems into easier-to-solve or easier-to-handle sub-problems that can be decomposed at multiple levels), and parallelization (parallel thinking, point by point, touch by touch, and networked mind maps) [21]. The steps for using CT to reinforce the memorization of English words are shown in Table 2:

Table 2. Step-by-step intervention on English vocabulary acquisition based on CT.

Steps	Data analysis	Pattern recognition	Abstraction	Decomposition	Parallelization
Operations	1. select target vocabulary from the textbook; 2. Ask learners to observe these target words carefully	1. Marking common parts in English words; 2. Asking learners to pay attention to the location of common forms and their forms; 3. Finding all words in the textbook that have similar structures	1. Rewrite all English words (coding), replacing English words with numbers, symbols or root words; 2. Focus on possible patterns of word formation or patterns of production	Simplify complex problems, improve the efficiency of problem solving by breaking them down in layers, reduce the difficulty and enhance the operability	Observe each word carefully and then associate it with the word you are most familiar with from the perspective of root affixes, word conversions, fixed collocations, habitual usage, and morphological similarities to form the broadest coverage mind network map possible
Targets	Guiding learners to focus on vocabulary formation methods that are characteristic of English thinking	Further help learners to enhance their awareness of the differences between English and Chinese vocabulary structures	Students are guided to gradually abstract word formation based on common root words and affixes: English words have their own specific historical origins and each of the 26 letters has its own specific basic meaning; English words are generally made up of roots plus prefixes and suffixes	The core question "How can I learn English vocabulary efficiently?" can be broken down into smaller questions. Can be broken down into smaller questions, such as: What are the common roots in English affixes? What are the common prefixes or suffixes, etc.? Then break it down until it is manageable for the individual learner	Expand your vocabulary by carefully analyzing each word and gradually associating, enriching, and expanding it layer by layer through computational thinking skills to form more and more category-based vocabulary network maps

For example, when we encounter the new word "Pose" in the textbook, the intervention can be listed as follows:

- Data analysis (The first step: Input words in the form of data or signs)

The instructor intends to find out the relevant words related to "Pose" in the textbook and lists them as follow:

apposite apposition
component compose composition compound
depose deposit discompose dispose disposition
exponent expose exposition expound
impose imposition impound
pose position positive propone proponent proposal propound preposition purpose
repose repository

- Pattern recognition through decomposition (Step two: Find commonalities in the structure of these words by breaking them down)

The instructor guides the students to decompose all the words listed above. They may be presented in the form below after decomposition.

> *ap+pos+i+te=apposite proper; ap+pos+i+tion=apposition juxtaposed, congruent; com+pon+ent=composed; part; parts; com+pos+e=compose, create; make; calm; com+pos+i+tion= composition composition; combination; composition; com+pound=compound compound; compounding; compound word; compounded; de+pos+e=exempt; precipitate; testify; de+pos+i+t=deposit store; pile; deposit; mineral deposit; heap; deposit; dis+com+pos+e= decompose to unsettle; dislocate; panic; de+com+pos+i+tion=decomposition to disintegrate; dis+pos+e=dispose to dispose of, remove, destroy; arrange; arrange; dis+pos+i+tion=disposition disposition; arrangement; ex+pon+ent= exponent explainer; ex+pos+e=expose expose; reveal; ex+pos+i+tion=exposition explain; elaborate; fair, exposition ex+pound=expound detail*

> *im+pos+e=impose imposes; tax; im+pos+i+tion=imposition tax; impose; im+pound=impound to put... Enclose; impound; pos+e=pose cause; raise (a question, etc.); pose; pos+i+tion=position position; position; post; pos+i+tiv+e=positive affirmative; positive; optimistic; masculine; post+pon+e=postpone postpone; pro+pon+e= propone propose; propose; pro+pon+ent=proponent proposer; proponent; pro+pos+al=proposal proposer; proposal; proposal; proposal; pro+pos+e=propose proposal; proposal; pro+pround=propound propose; pur+pos+e= purpose purpose re+pos+e=repose rest; sleep; re+pos+i+tory=repository warehouse*

- Abstraction (The third step: Summarize the commonalities; abstract the patterns of word formation)

    Lead students to gradually abstract the results of:

    (1) English words are formed by word roots plus prefixes or with suffixes;
    (2) Pon, pound, pos(it) = to put, to place put, place;
    (3) Common prefixes: ap, com, de, dis, ex, im, pro, pur, re;
    (4) Common suffixes: tion, ent, tive, al, tory

- Decomposition (The forth step: Decompose the problems resulted from the abstraction to make them easier and more accessible)

In this case, the question arises: How can we acquire English vocabulary through words plus prefixes or suffixes? It seems complicated for the Chinese learners. So it can be decomposed as the following questions:

(1) What are the common roots in English words?
(2) What are the common prefixes in English words?
(3) What are the common suffixes in English words?

Of course, these questions can be further decomposed depending on the different cognitive levels of different learners.

- Parallelization (The fifth step: A new word is associated with a related word that is most familiar to the learner, and then related according to word formation, fixed monogram, morphological similarity, and word conversion, forming a neural network-like mind map).

In this case, parallelization may help the learners form the following word mind map (Fig. 3).

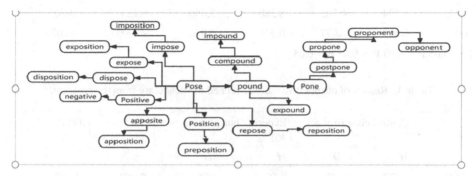

Fig. 3. "Pose" related words mind map based on parallelization

Of course, this mind map belongs to the first level of mind mapping, and according to parallel thinking, it can continue to expand and extend more vocabulary.

In this way, we repeat the intervention numerous times, and the learners should draw vocabulary mind maps as many as possible at the end of the intervention period.

Stage Three: Post-test. After ten weeks of instructing a dynamic assessment step-by-step intervention based on the formation of core skills in CT, we organized timely post-test writing.

2.6 Data Collection and Processing

We used QUITA software to measure the relevant quantitative indexes from the pre-test and post-test essays of the learners and conducted independent sample t-tests and paired sample t-tests by SPSS23 to observe the effects and changes brought about by the intervention.

3 Results

3.1 Comparison Between the Experimental Class and Control Class

The control class received the traditional intervention model. The learners just memorized the target words through reading, writing, and memorizing. In comparison, the experimental class embraced the CT-based intervention. Both groups attended the pre-test and post-test. Tables 3 and 4 show the results:

Table 3. Results of quantitative indexes related to vocabulary from the pretest.

	Control class (n = 46)		Experimental class (n = 46)		MD	t(45)
	M	SD	M	SD		
TTR	0.524	0.830	0.523	0.847	0.0004	0.024
R1	0.788	0.058	0.787	0.057	0.0009	0.078
V.D.	8.834	6.202	8.580	5.650	0.253	0.205
Des.	0.315	0.077	0.316	0.076	-0.0003	-0.020

$P > 0.05$ $P > 0.05$ $P > 0.05$ $P > 0.05$.

Table 4. Results of quantitative indexes related to vocabulary from the post-test.

	Control class (n = 46)		Experimental class (n = 46)		MD	t(45)
	M	*SD*	*M*	*SD*		
TTR	0.527	0.076	0.586	.075	-0.059	-3.767**
R1	0.785	0.048	0.689	0.048	0.095	9.523**
V.D.	8.679	5.651	6.910	4.629	1.769	1.643**
Des.	0.311	0.067	0.257	0.044	0.053	4.511**

** $P<0.01$ ** $P<0.01$ ** $P<0.01$ ** $P<0.01$

Table 3 clearly shows no significant difference between the means of the control and experimental groups regarding the quantitative indexes related to vocabulary in the pre-test. In contrast, Table 4 distinctly shows that the means of all measures in the post-test were significantly different, indicating that the control group did not change significantly after traditional vocabulary instruction, while the experimental group changed significantly, and students' vocabulary richness was enhanced.

3.2 The Experimental Class

All the learners from the experimental group received a 10-week intervention based on CT. Table 5 below demonstrates the outcome data of the experimental group after paired samples t-test.

Table 5 indicates that all the means from the post-test are significantly different from those of the pre-test, which means the intervention approach to vocabulary acquisition works. All the learners can improve their EFL vocabulary acquisition performance.

3.3 The Control Class

To check the effects of the traditional approach to EFL vocabulary acquisition, we also tested the quantitative indexes related to the control group from both the pre-test and post-test. Table 6 lists the results:

Table 5. Results of quantitative indexes related to vocabulary from the experimental class.

	Pre-test (n = 46)		Post-test (n = 46)		MD	t(45)
	M	SD	M	SD		
TTR	0.524	0.084	0.586	0.075	-0.063	-10.660**
R1	0.787	0.057	0.689	0.048	0.098	19.751**
V.D.	8.581	5.651	6.910	4.629	1.671	8.153**
Des.	0.315	0.077	0.257	0.044	0.058	9.367**

** $P<0.01$ ** $P<0.01$ ** $P<0.01$ ** $P<0.01$

Table 6. Results of quantitative indexes related to vocabulary from the control class.

	Pre-test (n = 46)		Post-test (n = 46)		MD	t(45)
	M	SD	M	SD		
TTR	0.524	0.083	0.527	0.075	-0.003	-1.444
R1	0.788	0.058	0.785	0.048	0.003	1.082
V.D.	8.834	6.202	8.679	5.651	0.155	1.278
Des.	0.315	0.077	0.311	0.667	0.004	1.298

$P > 0.05$ $P > 0.05$ $P > 0.05$ $P > 0.05$.

Table 6 illustrates that all the means from the post-test are not significantly different from the pre-test data. It denotes that the traditional ways of EFL vocabulary acquisition do not initiate improvement.

4   Discussion

4.1   How Does Computational Thinking Intervene in English Vocabulary Acquisition?

The traditional ways of English vocabulary acquisition are closely related to reading, writing, and memorizing. This study shows the low efficiency of the traditional ways of English vocabulary acquisition. To change the status quo, we followed the primary skills of CT step-by-step to intervene in students' vocabulary memorization. The data show that this new approach is efficient, practical and expands students' vocabulary in a short time. Using higher-order thinking, such as data analysis, pattern recognition, abstraction, decomposition, and parallelization, allows learners to understand English word formation patterns gradually. Word-to-word associations are achieved from the perspectives of root word affixation, word conversion, word similarity, fixed collocations, and regular usage, resulting in many vocabulary network diagrams that visualize and memorize more words.

CT works in EFL vocabulary acquisition in that the new way agrees with the Theory of Prototypes [22]. Prototypical category theory advocates the centrality of prototypical categories, the key to which is finding category similarities. Many words already exist in the learner's brain, which the learner has mastered and is familiar with. The role of computational thinking is to guide learners to find members related to this category through higher-order thinking. Computational thinking intervenes in vocabulary acquisition and is consistent with the cognitive psychology of learners.

4.2 How Does Computational Thinking Affect English Vocabulary Acquisition?

CT, higher-order thinking, can help the learners focus on the form of the words. They follow the CT-based step-by-step model to solve all problems before vocabulary acquisition. The critical role of computational thinking in this study is to help learners make influential associations between different words in line with the learners' cognitive mechanisms.

In this study, computational thinking helps learners re-categorize and re-categorize new words in their textbooks to accumulate more relevant vocabulary. This method is more effective than the traditional method. It is more in line with learners' cognitive patterns, which can effectively enhance vocabulary richness in learners' English compositions and add more advanced vocabulary.

4.3 What is the Role of Computational Thinking in Vocabulary Intervention?

CT, in this study, is a thinking bridge. It has become an effective intervention tool. CT links the new vocabulary with the prototype words in the learners' brains. In this way, the learners come to build up more and more vocabulary patterns or categories. Gradually, learners can accumulate new words or phrases.

In the era of big data, information technology, and artificial intelligence, computational thinking has been recognized as an essential educational technology, a basic talent competitiveness literacy, and, more importantly, a symbolic mediating tool [21, 23, 24]. In the present study, the central core skills of computational thinking have been transformed into sociocultural symbolic tools that mediate prior learning (memorization of vocabulary knowledge) and higher levels of learning of learners and play a critical mediating role in forming learners' higher-order thinking.

5 Conclusion

The present study further illustrates that computational thinking can strengthen learners' English vocabulary acquisition. It also validates the positive role of computational thinking in foreign language education [25]. Also, dynamic assessment theory, especially the sandwich intervention, enhanced the teaching intervention in large classes [26].

CT is the fundamental way of thinking in AI and LI, reflecting both the characteristics of machine thinking and the attribute features of human thinking. This kind of higher-order thinking has an essential inspirational role in foreign language education and teaching in the era of big data and AI.

References

1. Hu, G.S., Qi, Y.J.: Foreign language education in China in the era of ChatGPT: seeking change and adaptation. Technol. Enhan. Forei. Lang. **209**(1), 3–6 and 105 (2023)
2. Tan, T.N.: The History, Present and Future of Artificial Intelligence. [EB/OL] (2019). Retrieved from http://ia.cas.cn/xwzx/mtsm/201903/t20190311_5252250.html
3. Zhou, J.S., Lv, X.Q., Shi, J.S., Zhang, K.: Research on linguistic intelligence is becoming a hot topic. China Social Science News (Feb. 17, 2017)
4. Hu, K.B., Tian, X.J.: Cultivation of MTI talents in the context of language intelligence: challenges, countermeasures and prospects. Foreign Language World **2**, 59–64 (2020)
5. Papert, S.: Mindstorms: Children, computers, and powerful ideas, 285–286. Basic Books, New York (1980)
6. Papert, S.: An exploration in the space of mathematics education. Int. J. Comput. Math. Learn. **1**, 95–123 (1996)
7. Wing, J.M.: Computational Thinking. Commun. ACM **49**(3), 33–35 (2006)
8. Wing, J.M.: Computational thinking and thinking about computing. Philosophical transactions of the royal society of London: Mathematical, Physical and Engineering Sciences **366**(1881), 3717–3725 (2008)
9. Papert, S.: You can't think about thinking without thinking about thinking about something. Contempor. Issu. Technol. Teach. Edu. **5**(3), 366–367 (2005)
10. Barr, V., Stephenson, C.: Bringing computational thinking to K-12: What is involved and what is the role of the computer science education community? Acm Inroads **2**(1), 48–54 (2011)
11. Jona, K., et al.: Embedding computational thinking in science, technology, engineering, and math (CT-STEM). In: a paper presented at the future directions in computer science education summit meeting. Orlando, FL (2014)
12. Lee, I., et al.: Computational thinking for youth in practice. Acm Inroads **2**(1), 32–37 (2011)
13. Mackey, A., Gass, S.M.: Second language research: Methodology and design. Erlbaum (2005)
14. Liu, H.T.: An introduction to quantitative linguistics. The Commercial Press, Beijing (2017)
15. Jockers, M.L.: Text analysis with R for students of literature. Springer, Switzerland (2014)
16. Laufer, B., Nation, P.: Vocabulary size and use: Lexical richness in L2 written production. Appl. Linguis. **16**(3), 307–322 (1995)
17. Xian, W., Sun, S.: Dynamic lexical features of Ph.D. theses across disciplines: a text mining approach (2018)
18. Popescu, I.-I., Čech, R., Altmann, G.: Some geometric properties of Slovak poetry. J. Quantit. Linguis. **2**, 121–131 (2012)
19. Kubát, M., Matlach, V., Čech, R.: QUITA. RAM-Verlag, Quantitative Index Text Analyzer. Lüdenscheid (2014)
20. Poehner, M.E.: A casebook of dynamic assessment in foreign language education. CALPER publications, The Pennsylvania State University (2018)
21. Tang, Y.J., Ma, X.M.: Computational thinking: A mediation tool and higher-order thinking for linking EFL grammar knowledge with competency. Thinking Skills and Creativity **46**(4), 101143 (2022)
22. Wang, Y.: Cognitive Linguistics, 1st edn. Shanghai Foreign Language Education Press, Shanghai (2007)
23. Nouri, J., Zhang, L., Mannila, L., Norén, E.: Development of computational thinking, digital competence and 21st century skills when learning programming in K-9. Educ. Inq. **11**(1), 1–17 (2020)
24. Acevedo-Borrega, J., Valverde-Berrocoso, J., Garrido-Arroyo, M.D.C.: Computational thinking and educational technology: a scoping review of the literature. Education Science **12**(39) (2022)

25. Tang, Y.J., Ma, X.M.: Can computational thinking contribute to EFL learning and teaching?. In: 2023 International Conference on Artificial Intelligence and Education (ICAIE), pp. 20–24. Kobe, Japan (2023). https://doi.org/10.1109/ICAIE56796.2023.00016
26. Tang, Y., Ma, X.: An interventionist dynamic assessment approach to college english writing in China. Lang. Assess. Q. (2022). https://doi.org/10.1080/15434303.2022.2155165

Connected-CF²: Learning the Explainability of Graph Neural Network on Counterfactual and Factual Reasoning via Connected Component

Yanghepu Li[✉]

Graduate School of Advanced Integrated Studies in Human Survivability,
Kyoto University, Kyoto, Japan
li.yanghepu.58z@st.kyoto-u.ac.jp

Abstract. Structural data, such as social networks, molecules, citation networks, etc., exists everywhere in various fields. The complex topology makes it difficult to process and fully utilize such informative data. In recent years, Graph Neural Networks (GNNs) have achieved great success on learning representations for structural data. However, most of them are still considered as black boxes and the non-transparency of the models makes the explanation and interpretation of the predictions by GNNs non-trivial. This research seeks to solve the explainability problem of GNNs considering **C**ounterfactual and **F**actual reasoning from casual inference theory on connected substructures of graphs, which are more human-intelligible and intuitive while ignored by most existing methods. In this paper, we propose an original method, Connected-CF², to explain GNNs by formulating an optimization problem based on the counterfactual and factual reasoning condition and the connectivity condition of explanations. Two kinds of explanation strengths are given for the condition on reasoning, and the connectivity condition is a constraint on the number of connected components in graphs. This distinguishes Connected-CF² from previous explainability methods. Experiments show that Connected-CF² generates significantly improved explanations than the existing state-of-the-art methods.

Keywords: Graph Learning · Explainable GNNs · Connected Component · Causal Inference · Relaxation Optimization

1 Introduction

Structural data exists widely in plenty of fields such as molecules [3,10], social networks [15], citation networks [1,13], etc. Such data is generally represented as graphs, containing rich information. Nevertheless, studying this kind of data is very difficult for researchers because graph data, unlike Euclidean data such as image or audio, contains extra complicated topological information.

© The Author(s), under exclusive license to Springer Nature Singapore Pte Ltd. 2024
F. Zhao and D. Miao (Eds.): AIGC 2023, CCIS 1946, pp. 83–94, 2024.
https://doi.org/10.1007/978-981-99-7587-7_7

In recent years, GNNs have shown great success in learning graph representations, as they can aggregate features and structural information through message passing in graphs. Hence, GNN-based models show great advantages in different kinds of tasks, including node classification, graph classification, and relation prediction. Nonetheless, most existing GNN models lack explainability and are still considered as black boxes, and predictions given by them cannot be fully trusted. This non-transparency of the deep models prevents GNNs from critical applications, such as material design and drug discover. Therefore, exploring the explainability of GNN-based models is very crucial, since good explanations can help us better understand the predictions given by GNNs and discover unknown flaws or knowledge so that we can further promote the GNNs.

From an overall perspective, most recent approaches for GNN explanations are based on factual reasoning [8,16] or counterfactual reasoning [6,7,17]. Approaches based on factual reasoning aim to find explanations which contain *sufficient* information so that they can produce consistent predictions with the original graph, while approaches based on counterfactual reasoning aim to find explanations which contain *necessary* information so that if removed the predictions will be different.

Counterfactual reasoning and factual reasoning are both very important methods to explore explanations, but they have their own disadvantages. Counterfactual reasoning prefers explanations only including critical information, i.e., the prediction will be different if the explanation is removed. As a result, counterfactual reasoning might merely find a part of the actual explanation. On the contrary, factual reasoning prefers explanations containing adequate information to give the consistent prediction, while the explanations might contain redundant information (nodes/edges).

The disadvantages are illustrated in Fig. 1. To balance the shortcomings of factual reasoning and counterfactual reasoning, Tan et al. [14] first combined factual reasoning and counterfactual reasoning and proposed CF^2 framework. They reported that models considering both of them are better than models that applied only one perspective. However, Tan et al. [14] neglected the connectivity in the graph, resulting in some unconnected parts in explanations, which are not intuitive and human-intelligible. On the contrary, Yuan et al. [17] focused on the connected sub-graphs as explanations, while they only applied counterfactual reasoning.

In this work, we propose the Connected-CF^2 optimization framework, which combines factual and counterfactual reasoning and takes the connectivity of explanations into consideration. We bring up two conditions for explanations in the objective: 1) reasoning part, and 2) connectivity part. The reasoning part follows the definitions in Tan et al. [14], and the connectivity condition is achieved by constraining the number of connected components in explanations. This distinguishes Connected-CF^2 from previous explainability methods. We conduct experiments on three datasets to evaluate our method quantitatively and qualitatively. The experimental results highlight that Connected-CF^2 generates significantly improved explanations than previous state-of-the-art methods.

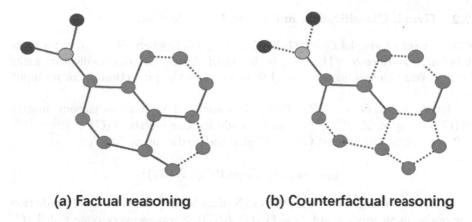

(a) Factual reasoning   (b) Counterfactual reasoning

Fig. 1. An instance for explanations (in bold edges) given by (a) factual reasoning, (b) counterfactual reasoning for mutagenic molecules. Carbon (C) atom is in orange color, Nitrogen (N) atom is in gray color, and Oxygen (O) atom is in black color. Hydrogen (H) atom is omitted here. (Color figure online)

The paper is organized as follows: Sect. 2 contains preliminaries and basic notation; Sect. 3 formalizes the problems we address; Sect. 4 gives details of our framework and its implementation; Sect. 5 presents experimental results of our method and corresponding analyses; Sect. 6 concludes and brings up future direction.

2   Preliminaries and Notation

Here we introduce preliminary information and notation at (1) learning representations by GNNs, (2) basic notation in graph classification and node classification, and (3) some notations in graph theory to be used throughout the paper.

2.1   Learning Representations

Let $G = (V(G), E(G))$ be a graph, where $V(G)$ is the set of vertices of G and $E(G)$ is the set of edges of G, and every node $v_i \in V(G)$ has a d-dimensional feature vector $x_{v_i} \in \mathbb{R}^d$. In general, a GNN model learns the representations of v_i through aggregating the information of v_i's neighbors $N(v_i)$, i.e., the nodes connected to v_i. The GNN model updates the v_i's representations at the l-th layer by $h_i^l = update(h_i^{l-1}, h_{N(v_i)}^l)$, where h_i^{l-1} is the representation of v_i in the $(l-1)$-th layer, and $h_{N(v_i)}^l = aggregate(h_j^{l-1}), \forall v_j \in N(v_i)$. In different GNNs, the update$(\cdot)$ function and aggregation$(\cdot)$ function could be different. For a GNN with K layers, h_i^K is the final representation of the node v_i. In graph classification tasks, graph representation can be obtained from all node representations by taking the sum or average.

2.2   Graph Classification and Node Classification

Given a set of graphs $G = \{G_1, G_2, ..., G_{n-1}, G_n\}$, where $G_k \in G$ has a class label $y_k \in S$. Here $S = \{1, ..., m\}$ is the set of classes. Graph classification tasks seek to train a classification model Φ which gives the prediction $\hat{y}_k$ for an input G_k.

Every input $G = (V(G), E(G))$ is associated with an adjacency matrix $A(G) = [a_{ij}] \in \mathbb{R}^{|V(G)| \times |V(G)|}$ and a node feature matrix $X(G) \in \mathbb{R}^{|V(G)| \times d}$. After training, the trained GNN will give the prediction $\hat{y}_k$ for G_k:

$$\hat{y}_k = \arg \max_{c \in S} P_\Phi(c|A(G_k), X(G_k)) \tag{1}$$

In node classification tasks, the GNN aims to give the class label prediction for nodes in an input graph $G = (V(G), E(G))$. Suppose every node $v_i \in V(G)$ has a node label $y_i \in S = \{1, ..., m\}$. Here, we assume only the K-hop neighborhood of a node v_i that can cause actual impact on h_i^K. Thus, we take the K-hop sub-structure of the node v_i as the computational graph denoted as $G_{t(i)}$. $X(G_{t(i)})$ and $A(G_{t(i)})$ denote the corresponding feature matrix and the adjacency matrix of $G_{t(i)}$. The trained GNN will give the prediction $\hat{y}_i$ for the node v_i as:

$$\hat{y}_i = \arg \max_{c \in S} P_\Phi(c|A(G_{t(i)}), X(G_{t(i)})) \tag{2}$$

2.3   Laplacian Matrix and Connected Component

Based on the adjacency matrix A, the Laplacian matrix is defined to be

$$L = A - D$$

where $D = D(G) = [d_{ij}] \in \mathbb{R}^{|V(G)| \times |V(G)|}$, $d_{ii} = \sum_{j=1}^{|V(G)|} a_{ij}$, $a_{ij} \in A$ and $d_{ij} = 0$ if $i \neq j$, is the diagonal matrix whose entries are the degrees of the vertices (called the degree matrix). The Laplacian matrix is symmetric for undirected graphs, and thus it has a complete set of eigenvalues.

In graph theory, a connected component of an undirected graph is a subgraph in which each pair of nodes is connected with each other via a path.

3   Problem Formulation

In this section, we first introduce the explainability problem for GNNs. Second, the mathematical definitions of the factual reasoning and counterfactual reasoning used to find explanations are given. Last, we give the conditions on explanations, which require a good explanation to be compact, effective, and connected.

3.1 Explainable Graph Neural Networks

Based on the setup of Ying et al. [16], suppose we have an input graph $G_k = (V(G_k), E(G_k))$ that has the predicted label $\hat{y}_k$, we generate a subgraph, i.e., a subset of edges in the original graph, as the explanation for the prediction. Therefore, the aim of Connected-CF^2 is to find an edge mask $M_k \in \{0,1\}^{|V(G_k)| \times |V(G_k)|}$ to learn the explanation of the input graph G_k. The explanation will be $A(G_k) \odot M(G_k)$, where $\odot$ denotes the element-wise multiplication.

3.2 Reasoning Conditions

Existing methods on explainable GNNs generally take one prospective from factual reasoning and counterfactual reasoning. Factual reasoning tends to ask "Given P, will Q happen?" and counterfactual reasoning refers to "If P does not happen, will Q happen?" [12]. For GNN explanations, factual reasoning generates explanations containing adequate information to give the consistent prediction. On the contrary, counterfactual reasoning generates the explanations that the prediction given by GNN will change if we remove the explanation. Factual reasoning tends to find *sufficient* set of edges which maintains the prediction as the whole graph, while it may contain redundant part. Counterfactual reasoning tends to find a *necessary* set of edges which are so important that different predictions will be produced without them. Nevertheless, counterfactual reasoning may neglect some important information. In a nutshell, they both have their own advantages and disadvantages.

Tan et al. [14] first combined factual reasoning and counterfactual reasoning on explaining GNNs, and reported that models considering both of them perform better than models that applied only one perspective. Following Tan et al. [14], the factual reasoning condition can be mathematically defined as:

$$\arg \max_{c \in S} P_{\Phi}(c|A(G_k) \odot M(G_k), X(G_k)) = \hat{y}_k \tag{3}$$

The counterfactual reasoning condition can be defined as:

$$\arg \max_{c \in S} P_{\Phi}(c|A(G_k) - A(G_k) \odot M(G_k), X(G_k)) \neq \hat{y}_k \tag{4}$$

These two conditions will be considered in the Connected-CF^2 optimization framework. More details can be seen in Sect. 4.

3.3 Compact, Effective and Connected Explanations

Occam's Razor Principle [2] suggests when two explanations are both very effective, the more compact one will be preferred. In order to find both compact and effective explanations, we adopt Explanation Complexity and Explanation Strength from Tan et al. [14]. These two concepts are helpful for Connected-CF^2 to explore both compact and effective explanations.

Explanation complexity $C(M)$ measures the complexity of an explanation by counting the number of edges in the explanation. Here, that edge mask M is a binary matrix whose elements are in $\{0,1\}$ indicating whether an edge is kept in the explanation or not. Therefore, $C(M)$ can be mathematically calculated by the number of 1 in the M matrix:

$$C(M) = ||M||_0 \qquad (5)$$

Explanation strength $S(M)$ measures the effectiveness of the explanation. As mentioned in Sect. 3.2, a good explanation should be both *necessary* and *sufficient*, which is the purpose of counterfactual and factual reasoning (Eq. (3) and (4)). Hence, the explanation strength could be divided into two parts: counterfactual strength $S_c(M)$ and factual strength $S_f(M)$. The larger explanation strength suggests the better explainability capacity.

The definition of $S_c(M)$ is in accordance with the counterfactual reasoning condition:

$$S_c(M) = -P_\Phi(\hat{y}_k|A(G_k) - A(G_k) \odot M(G_k), X(G_k)) \qquad (6)$$

Similarly, $S_f(M)$ is in accordance with the factual reasoning condition:

$$S_f(M) = P_\Phi(\hat{y}_k|A(G_k) \odot M(G_k), X(G_k)) \qquad (7)$$

The explanation complexity and strength are employed in the Connected-CF^2 optimization framework, in order to find compact and effective explanations.

Although explanation complexity and strength are capable to keep explanations compact and effective, they neglect the connectivity of explanations, which is a very critical feature of graphs. Connectivity in graphs has its specific meaning in different types of graphs. For example, the connection between nodes in social networks may represent an acquaintance relationship, the connection in molecules is the chemical bond, etc. It is possible to gain unconnected explanations only considering explanation complexity and strength, which are ambiguous and hard to understand. Hence, besides compact and effective, we want the explanation to be also connected, i.e., the number of connected components of the explanation to be 1, which are more human-intelligible and intuitive.

To constrain the connectivity of graphs, we apply the following theorem [5]:

Theorem 1. *The geometric multiplicity of* 0 *eigenvalue of the Laplacian matrix* L *of a graph* G *corresponds to the number of connected components in* G.

For an input graph G, we want the explanation to be connected, i.e., the explanation has only 1 connected component. According to the above theorem, the geometric multiplicity of 0 eigenvalue of the Laplacian matrix L is 1. To get the geometric multiplicity, we need to calculate the dimension of the solution space in the linear system $(L - 0 \times I)\eta = L\eta = \mathbf{0}$, where I is the n-dimensional identity vector, n is the number of nodes in G. Due to the rank-nullity theorem, the dimension of $\{\eta|L\eta = \mathbf{0}\}$ should be $\mathbf{n - rank(L)}$. As a result, we want the dimension of $\{\eta|L\eta = \mathbf{0}\}$ to be 1, which is equivalent to $rank(L) = n - 1$. This will also be included in the learning constraint.

As described above, good explanations should be compact, effective, and connected in the Connected-CF² framework. More details will be introduced in Sect. 4.

4 The Connected-CF² Framework

4.1 Connected-CF² Optimization Problem

Connected-CF² can generate explanations for predictions made by GNNs. As mentioned in Sect. 3, Connected-CF² seeks to extract compact (low complexity), effective (high explanation strength), and connected (the number of connected component to be 1) explanations. The goal of Connected-CF² can be expressed in an optimization framework: to minimize the explanation complexity of an edge mask M such that M has strong factual and counterfactual explanation strength and the explanation $A \odot M$ is connected.

According to the definitions described above, for an input G_k with prediction $\hat{y}_k$ and the corresponding Laplacian matrix L_k, the optimization problem could be written as:

$$
\begin{aligned}
\text{minimize} \quad & C(M) \\
\text{s.t.,} \quad & S_c(M) > -P_\Phi(\hat{y_{k,s}}|A(G_k) - A(G_k) \odot M(G_k), X(G_k)) \\
& S_f(M) > P_\Phi(\hat{y_{k,s}}|A(G_k) \odot M(G_k), X(G_k)) \\
& rank(L_k) = n - 1
\end{aligned}
\tag{8}
$$

where $\hat{y_{k,s}}$ is the second possible label other than $\hat{y}_k$, and n is the number of nodes in G_k. The constraints aim to ensure that when only the explanation is used for prediction, $\hat{y}_k$ will still be the most possible prediction consistent with the original input, however, if we remove the explanation from the original input graph, there will be at least one label other than $\hat{y}_k$ that has the highest possibility, resulting in the change of the prediction. Besides, the explanation will be connected as much as possible.

4.2 Relaxation Optimization

Unfortunately, optimizing Eq. (8) directly is very challenging since both the objective part and the constraint part are discrete and not differential. In order to make them optimizable, we add relaxation to the two parts.

For the objective, we relax the mask $M(G_k)$ to real values that $M^*(G_k) \in \mathbb{R}^{|V(G_k)| \times |V(G_k)|}$. At the same time, since the L0-norm is not differentiable, we use L2-norm, which has many applications on optimization [9,11], to substitute the explanation complexity $C(M^*(G_k))$.

For the counterfactual and factual reasoning constraint part, we follow Tan et al. [14] to add relaxation to it as pairwise contrastive loss L_f and L_c:

$$
\begin{aligned}
L_c = ReLU(\zeta - S_c(M^*) \\
- P_\Phi(\hat{y_{k,s}}|A(G_k) - A(G_k) \odot M(G_k), X(G_k)))
\end{aligned}
\tag{9}
$$

Similarly,

$$L_f = ReLU(\zeta - S_f(M^*) \\ + P_\Phi(y_{\hat{k},s}|A(G_k) \odot M(G_k), X(G_k))) \tag{10}$$

For the connectivity part, first we relax the Laplacian matrix to $L^* \in \mathbb{R}^{|V(G_k)| \times |V(G_k)|}$. Then, we use determinant to substitute the rank constraint, which is not differential. According to linear algebra theory, $rank(L^*) = n - 1$, i.e., the determinant of L^* is zero and there is at least one **first minor** (the determinant of the submatrix by subtracting i-th row and j-th column from the original matrix) of L^* to be non-zero. Denote C_{ij} as the **first minor** of L^*. We apply penalty method to constrain the rank term, i.e., if $det(L^*)$ is non-zero or all C_{ij} are zeros, the penalty term will be large, and if $det(L^*)$ equals zero and there is a C_{ij} non-zero, the penalty term will dramatically decrease. Under this context, we think the exponential function is suitable for this purpose. Therefore, the connectivity constraint can be expressed as:

$$det(L^*) + exp(-\theta \sum_{i,j=1}^{n} C_{ij}^2) \tag{11}$$

After the above relaxation, Eq. (8) becomes optimizable and can be rewritten as:

$$\text{minimize} \, ||M^*(G_k)||_2 + \lambda(\alpha L_f + (1-\alpha)L_c) \\ + \beta(det(L^*) + exp(-\theta \sum_{i,j=1}^{n} C_{ij}^2)) \tag{12}$$

When solving the relaxed optimization problem, ζ in L_f and L_c is set to 0.5. Moreover, we take 0.5 as a threshold value to the edge mask M_k^* to decide which edge should be included in the explanation, i.e., when the entry in M_k^* is larger than 0.5, the corresponding edge is a part of the explanation.

In Eq. (12), hyperparameters λ and β control the significance of the explanation complexity, the explanation strength and the connectivity, influencing the trade-off among them. Increasing λ will make the model more focused on effectiveness, and increasing β will make the model more focused on the connectivity. Another hyperparameter α influences the trade-off between counterfactual reasoning and factual reasoning. Increasing (or decreasing) α will make this model tend to generate more *sufficient* (or *necessary*) explanations. And the hyperparameter θ controls the sensitivity to the connectivity.

5   Experiments

In this section, first we introduce datasets used in the experiments. Second, we show the comparison baselines. Then, we demonstrate the experimental setup. Eventually, we report the experiment results and corresponding quantitative and qualitative analyses.

5.1 Dataset

We employ two synthetic datasets and one real-world dataset to evaluate our model, and they all have ground-truths. The synthetic datasets are Tree-Cycles and BA-shapes [16]. One real-world dataset is $Mutag_0$ dataset, which is a sub-dataset extracted from Mutag dataset [4] by Tan et al. [14], merely containing mutagenic chemical compounds with benzene-NO_2 and non-mutagenic chemical compounds without benzene-NO_2. Concrete statistics of datasets are reported in Table 1.

Table 1. Statistics of all datasets used in experiments.

Datasets	#graph	#avg node	#avg edge	#class	#feature	task
BA-shapes	1	700	4100	4	–	node
Tree-Cycles	1	871	1950	2	–	node
$Mutag_0$	2301	31.74	32.54	2	14	graph

5.2 Baselines

In this paper, we want the baselines to have such features: 1) They generate sub-graphs (connected or unconnected) explanations; and 2) They do not have critical requirements for the dataset. Therefore, the baselines are as follows:

GNNExplainer [16]: GNNExplainer is a model based on perturbation. It generates explanations by maximizing the mutual information.

CF-GNNExplainer [7]: CF-GNNExplainer is an extended version of GNNExplainer that employs counterfactual reasoning to generate explanations.

Gem [6]: Gem does not apply factual reasoning or counterfactual reasoning. It trains an auto-encoder to generate explanations.

CF^2 [14]: CF^2 is a model that combines factual reasoning and counterfactual reasoning to generate explanations.

5.3 Setup

The experiments contain two phases: 1) fixing and training the basic GNN model for classification; 2) solving the optimization problem and giving corresponding explanations.

We choose a fixed GCN as the basic GNN model, with 3 convolutional layers. The dimensions of hidden layer are set to 16. After each layer except the last layer, we apply the ReLu activation function, and the Softmax function is employed after the last layer as the classification function. The learning rate is set to be 0.001 during training, and the training set accounts for 80% of the total dataset. Table 2 reports concrete information for the basic GCN model used in the experiments. To ensure the fairness, we employ the same basic model for all baselines and Connected-CF^2.

Table 2. The number of epochs and classification accuracy of the fixed trained basic GCN.

Datasets	BA-Shapes	Tree-Cycles	Mutag$_0$
Epochs	3000	3000	1000
Accuracy	97.86	98.29	98.05

In the explanation phase, we set all hyperparameters in Connected-CF^2 to remain the same. The λ is set to 500, 500, and 1000 for BA-Shapes, Tree-Cycles and Mutag$_0$ datasets. The α is set to 0.6 to make counterfactual reasoning less influential than factual reasoning, and we set β to 100 and θ to 1000.

5.4 Quantitative Analysis

Because Tree-Cycles, BA-Shapes and Mutag$_0$ datasets all have ground-truth, so that the commonly-used metrics for classification tasks Accuracy, Precision, Recall, and F_1 scores are applied to explanations in the test datasets to evaluate different explanation methods. Table 3 reports the results of the generated explanations of each dataset compared with the ground-truth. On Tree-Cycles, Connected-CF^2 has better performance on Accuracy and F_1 scores than other approaches. On BA-Shapes and Mutag$_0$, Connected-CF^2 has an overall high Accuracy and F_1 scores which are very close to the best scores. This shows that Connected-CF^2 has strong explanation strength. Besides, another observation is that Connected-CF^2 is better in Recall in three datasets, and the biggest increase is 31.9% in Tree-Cycles, which indicates that Connected-CF^2 is more likely to give the explanations consistent with the ground-truth. Moreover, the promotion of Connected-CF^2 on the connectivity cannot be observed through the quantitative analysis, which is introduced in Sect. 5.5.

5.5 Qualitative Analysis

According to Sect. 4, besides counterfactual and factual reasoning part in the objective, Connected-CF^2 has a special penalty term influencing the connectivity in the explanations, which cannot be revealed by the quantitative analysis, i.e., Accuracy, Precision, Recall and F_1 scores, since they are edge-level metrics and unable to tell the connectivity in graphs. In Fig. 2, we visualize explanation instances based on the topological structures. Results show that Connected-CF^2 tends to give connected explanations and avoids unconnected parts that are ambiguous and hard to understand.

Table 3. Explanation evaluation w.r.t ground-truth. Acc, Pr and Re represent Accuracy, Precision and Recall, respectively.

Models	BA-Shapes				Tree-Cycles				Mutag$_0$			
	Acc	Pr	Re	F$_1$	Acc	Pr	Re	F$_1$	Acc	Pr	Re	F$_1$
GNNExplainar	95.25	60.08	60.08	60.8	92.78	68.06	68.06	68.06	96.96	59.71	85.17	68.85
CF-GNNExplainer	94.39	67.19	54.11	56.79	90.27	**87.40**	47.45	59.10	96.91	**66.09**	39.46	47.39
Gem	**96.97**	64.16	64.16	64.16	89.88	57.23	57.23	57.23	96.43	63.12	47.11	54.68
CF^2	96.37	**73.15**	68.18	**66.61**	93.26	84.92	73.84	75.69	**97.34**	65.28	88.59	**72.56**
Connected-CF^2	95.22	56.73	**81.51**	62.00	**93.76**	68.71	**97.45**	**78.60**	96.54	49.87	**96.30**	63.45

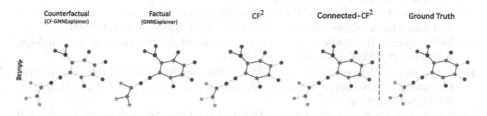

Fig. 2. Qualitative Analysis. Visualization of the generated explanations on instances from the Mutag dataset. We show the explanations given by different models based on counterfactual reasoning, factual reasoning, CF^2, and the ground-truth from left to right.

6 Conclusion and Future Work

In this paper, we propose an original model, Connected-CF^2, which can generate GNN explanations by combining factual reasoning and counterfactual reasoning and considering the connectivity of explanations. Experiments on the synthetic and real-world datasets demonstrate that Connected-CF^2 has its superiority compared with the baselines through quantitative and qualitative analyses. In the future, we will promote the efficiency of Connected-CF^2, and make more explorations on the deeper inner relationship between the connectivity and the explanations. Moreover, our framework will be used to more complicated graph datasets, and may be generalized beyond graph-based explanations.

References

1. Getoor, L.: Link-based classification. In: Advanced Methods for Knowledge Discovery from Complex Data. AIKP, pp. 189–207. Springer, London (2005). https://doi.org/10.1007/1-84628-284-5_7
2. Blumer, A., Ehrenfeucht, A., Haussler, D., Warmuth, M.K.: Occam's razor. Inf. Process. Lett. **24**(6), 377–380 (1987)
3. Brown, T.L.: Chemistry: The Central Science. Pearson Education, London (2009)
4. Debnath, A.K., Lopez de Compadre, R.L., Debnath, G., Shusterman, A.J., Hansch, C.: Structure-activity relationship of mutagenic aromatic and heteroaromatic nitro compounds. Correlation with molecular orbital energies and hydrophobicity. J. Med. Chem. **34**(2), 786–797 (1991)

5. Hamilton, W.L.: Graph Representation Learning. Synthesis Lectures on Artificial Intelligence and Machine Learning, vol. 14, no. 3, pp. 1–159. Springer, Cham (2020). https://doi.org/10.1007/978-3-031-01588-5
6. Lin, W., Lan, H., Li, B.: Generative causal explanations for graph neural networks. In: International Conference on Machine Learning, pp. 6666–6679. PMLR (2021)
7. Lucic, A., Ter Hoeve, M.A., Tolomei, G., De Rijke, M., Silvestri, F.: CF-GNNExplainer: counterfactual explanations for graph neural networks. In: International Conference on Artificial Intelligence and Statistics, pp. 4499–4511. PMLR (2022)
8. Luo, D., et al.: Parameterized explainer for graph neural network. Adv. Neural Inf. Process. Syst. **33**, 19620–19631 (2020)
9. Malyshev, A.N.: A formula for the 2-norm distance from a matrix to the set of matrices with multiple eigenvalues. Numer. Math. **83**(3), 443–454 (1999)
10. McNaught, A.D., Wilkinson, A., et al.: Compendium of Chemical Terminology, vol. 1669. Blackwell Science, Oxford (1997)
11. Nakamura, I., Tonomura, Y., Kiya, H.: Unitary transform-based template protection and its application to l 2-norm minimization problems. IEICE Trans. Inf. Syst. **99**(1), 60–68 (2016)
12. Quelhas, A.C., Rasga, C., Johnson-Laird, P.N.: The relation between factual and counterfactual conditionals. Cogn. Sci. **42**(7), 2205–2228 (2018)
13. Sen, P., Namata, G., Bilgic, M., Getoor, L., Galligher, B., Eliassi-Rad, T.: Collective classification in network data. AI Mag. **29**(3), 93–93 (2008)
14. Tan, J., et al.: Learning and evaluating graph neural network explanations based on counterfactual and factual reasoning. In: Proceedings of the ACM Web Conference 2022, pp. 1018–1027 (2022)
15. Yanardag, P., Vishwanathan, S.: Deep graph kernels. In: Proceedings of the 21th ACM SIGKDD International Conference on Knowledge Discovery and Data Mining, pp. 1365–1374 (2015)
16. Ying, Z., Bourgeois, D., You, J., Zitnik, M., Leskovec, J.: GNNExplainer: generating explanations for graph neural networks. In: Advances in Neural Information Processing Systems, pp. 9240–9251 (2019)
17. Yuan, H., Yu, H., Wang, J., Li, K., Ji, S.: On explainability of graph neural networks via subgraph explorations. In: International Conference on Machine Learning, pp. 12241–12252. PMLR (2021)

Controllable Feature-Preserving Style Transfer

Feichi Chen, Naye Ji$^{(\boxtimes)}$, Youbin Zhao, and Fuxing Gao

Communication University of Zhejiang, Hangzhou 310018, China
jinaye@cuz.edu.cn

Abstract. This paper proposes a new style transfer quality assessment approach introducing quantifiable metrics to optimize. First, we utilize a pre-trained DualStyleGAN model to generate multiple stylized portraits in the style vector space. Then, we design a custom scoring mechanism that uses the newly proposed $CSCI$ and $CCVI$ metrics to evaluate the results' structural similarity, color consistency, and edge retention. We select and optimize the top outputs using human aesthetic standards to obtain the most natural, beautiful, and artistic results. Experimental results show that our proposed evaluation pipeline can effectively improve the quality of style transfer.

Keywords: StyleGAN · Style Transfer · Quality Evaluation

1 Introduction

Style transfer, an extensively researched topic in computer vision, aims to transform the artistic style of images while preserving their content. Recent advancements in deep learning have yielded impressive results; however, ensuring the quality of generated stylized images remains challenging.

Although various methods have been developed to enhance stylization quality, each existing method has its own limitations. Examples include StyleGAN3, DualStyleGAN [2], and StableDiffusion [8]. Current style transfer methods excel in specific aspects while lacking in others. For instance, DualStyleGAN struggles with style transfer on Asian faces, while StableDiffusion produces random results that are hard to control.

These challenges stem from the lack of comprehensive considerations for stylization quality. Some methods prioritize results and content effects, while others emphasize style accuracy, but both approaches have inherent limitations. Is there a way to integrate the strengths of these methods and address their shortcomings? The evaluation methods in this paper aim to answer this question.

Existing style transfer methods have made valuable contributions to the field of stylization. However, these methods have some drawbacks, such as not working well with Asians, and some Anime styles are challenging to transfer. These limitations prevent them from achieving comprehensive and high-quality results.

F. Zhao and D. Miao (Eds.): AIGC 2023, CCIS 1946, pp. 95–104, 2024.
https://doi.org/10.1007/978-981-99-7587-7_8

This paper aims to bridge these gaps and merge the strengths of various techniques to overcome their limitations.

This paper presents a detailed analysis of our approach, highlighting the effectiveness of our proposed metrics and their impact on the style transfer process. Furthermore, we conduct experiments on diverse datasets, comparing our results with state-of-the-art methods to demonstrate the superiority of our approach.

Our main contributions lie in the following aspects:

1. We propose a set of quantitative metrics to evaluate style transfer performance, including perceptual loss, structural similarity, edge preservation, color preservation, and visual saliency. A multi-layer weighted scoring approach assesses these factors and filters out low-quality results.
2. We build upon previous style transfer techniques and further optimize the process. We enhance stylized outputs' accuracy and aesthetic appeal by incorporating prior knowledge and fine-tuning details.
3. We conduct extensive experiments on multiple portrait datasets. Comparisons with state-of-the-art methods demonstrate the superiority of our approach in improving stylization quality.
4. We develop an end-to-end pipeline integrating style transfer, quality evaluation, and refinement. This represents an important step towards controllable and high-fidelity artistic stylization.

2   Related Work

2.1   StyleGAN-Related Models

The StyleGAN model, proposed by Karras et al. [1], is a generative adversarial network (GAN) that leverages style transfer techniques to generate high-quality images. This model introduces a novel generator architecture, enabling intuitive control over image synthesis at multiple scales. Compared to traditional GANs, StyleGAN improves distribution quality metrics and interpolation properties. It achieves this by disentangling latent factors of variation, effectively separating different aspects of an image into distinct components that can be manipulated independently. The model is flexible and can be applied to various image synthesis tasks. Notably, it includes the creation of a unique human face dataset called FFHQ and introduces automated methods for quantifying interpolation quality and disentanglement. Overall, the StyleGAN model significantly advances image generation and synthesis.

StyleGAN has indeed catalyzed the development of numerous style transfer models, including DualStyleGAN, JoJoGAN [7], and VToonify [13]. Additionally, it has given rise to variant style transfer models that explore different approaches. An example is StyleCLIP [10], which combines the principles of StyleGAN and CLIP models.

The demand for personalized and stylized images has significantly contributed to the popularity of StyleGAN and its derivatives among academic researchers and the general public. DualStyleGAN has introduced a wide range of styles for image style transfer, while JoJoGAN offers a one-shot style transfer capability. These models have garnered considerable attention due to their ability to generate high-quality simulated images. For our research, we selected Style-GAN as our fundamental model due to its proven performance and versatility.

However, StyleGAN and its derivatives still exhibit certain limitations and drawbacks in their performance. While these models have shown the ability to generate impressive results, they often struggle to achieve a consistent and cohesive style and character across generated images. Furthermore, a notable disadvantage is the inadequate performance of DualStyleGAN on Asian faces which is found in our experiments due to its training on FFHQ, a predominantly European and American face dataset. Moreover, in certain styles like Anime, existing models face challenges in effectively transferring those styles to facial images. These shortcomings have posed challenges in ensuring the accuracy and validity of the generated results, prompting the emergence of further studies in this area.

2.2   Evaluation of Style Transfer

To achieve controllable style transfer, previous studies have proposed various evaluation methods. These research efforts aim to quantify the effectiveness of style transfer techniques. One notable approach is the Style-Eval method introduced by Wang et al. [3], which has shown promising results across various style transfers. Style-Eval offers several advantages, including a novel quantitative evaluation framework based on three measurable quality factors. This comprehensive approach thoroughly assesses style transfer quality from multiple perspectives.

Wright and Ommer have introduced a novel method called ArtFID [11] to complement the predominantly qualitative evaluation schemes currently employed. This proposed metric demonstrates a strong correlation with human judgment. One significant advantage of this approach is its ability to facilitate automated comparisons between different style transfer approaches, enabling a comprehensive analysis of their strengths and weaknesses.

Another evaluation procedure, proposed by Mao et al. [12], is known as Quantitative Evaluation (QTEV). It involves plotting the effectiveness, which measures the degree of style transfer, against coherence, which assesses the extent to which the transferred image retains the same object decomposition as the content image. This process generates an EC plot that aids in evaluating the performance of style transfer methods.

Controllable style transfer is an important research direction. Previous studies have proposed various evaluation methods to quantify the effectiveness of style transfer techniques to achieve controllable stylization. While existing methods have achieved specific results, there is room for improving style transfer quality evaluation. Future research can explore new metrics building on current ones. Moreover, combining quality evaluation with style transfer methods to achieve end-to-end quality optimization is also worth exploring. We look forward to seeing new breakthroughs in controllable and high-fidelity style transfer.

2.3   Developments of Style Transfer

A wide range of approaches characterizes recent research on Style Transfer, as theories and methods intersect effectively. Some Style Transfer methods incorporate other models, such as Stable Diffusion, while others explore combinations with other technologies.

InST [4] is an example of a one-shot stylized model that shares similarities with JoJoGAN but is based on Stable Diffusion. In the domain of 3D Reconstruction, PAniC-3D [5] and StyleRF [9] utilize their methods to integrate Style Transfer and 3D Reconstruction techniques. Furthermore, an exciting application of Style Transfer involves generating talking heads using portrait images, as demonstrated by MetaPortraits [6].

As technology and theoretical advancements continue, the utilization of Style Transfer is expected to become more extensive, and we can anticipate the emergence of new products and applications.

3   Method

We propose a novel method that incorporates a comprehensive evaluation of style transfer specifically designed for human faces. This approach aims to generate improved results by leveraging effective evaluation techniques. Our method combines multiple standards and technologies, including a unique evaluation metric called STE(Style Transfer Evaluation) which includes $CSCI$ and $CCVI$, along with a human-led aesthetic evaluation. Furthermore, our approach can accommodate various styles such as cartoon and anime, making it applicable to diverse studies.

As shown in Fig. 1 our approach consists of three stages. In the first stage, traditional StyleGAN models like DualStyleGAN are utilized to generate multiple batches of style images, which serve as inputs for the subsequent steps. The second stage employs Style Transfer Evaluation to perform preliminary screening, selecting the optimal results from each batch. Finally, in the third stage, an aesthetic evaluation is conducted to provide the final assessment and generate the output results.

3.1   Image Style Transfer

We are using DualStyleGAN as our basic model and we train it on its pre-train model. This method proposed a novel approach for high-resolution portrait style transfer training with a few hundred examples. The main contribution of this work is the characterization and disentanglement of facial identity versus artistic styles. To conclude, modeling the portrait synthesis with a dual style transfer process which can control both facial identity and style degree.

The model is trained using a dataset comprising exemplar artistic portraits and target faces. Given a target face, this model generates a random intrinsic style code (z) and an extrinsic style code (w). By applying style loss and identity

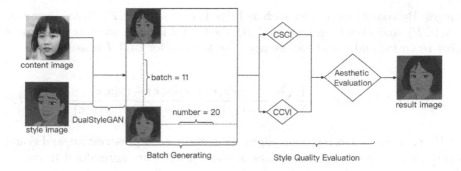

Fig. 1. The general workflow of Style Transfer Evaluation.

loss, it transfers the style of an exemplar artistic portrait onto the target face using the intrinsic and extrinsic style codes.

In the generative process, we set up a batch size and a range of weights. This process generates multiple batches of different results for the next stage of our work.

3.2 Style Transfer Evaluation

To evaluate the quality of style transfer, we employed the following methods. Firstly, we extracted the vector representation of the style and output image and encoded the images using advanced image coding techniques for subsequent analysis and calculations.

Next, we introduced the formula $CSCI$ (Combined Structural Content Similarity Index). This formula effectively combines two metrics, $SSIM$ (Structural Similarity Index) and $PSNR$ (Peak Signal-to-Noise Ratio), by utilizing a sliding window to traverse the images and calculate the similarity between the two images in terms of structure and content. The formulation of the $CSCI$ is as follows:

$$CSCI(\mathbf{s}, \mathbf{c}) = \frac{1}{N} \sum_{i=1}^{n} \frac{ssim(\mathbf{s}, \mathbf{c}) + psnr(\mathbf{s}, \mathbf{c})}{\mid ssim(\mathbf{s}, \mathbf{c}) \mid + \mid psnr(\mathbf{s}, \mathbf{c}) \mid} \tag{1}$$

Here, s and c are the style image and content image respectively, i represents the sliding window index, N denotes the total number of windows. The $CSCI$ score ranges from 0 to positive infinity, with 0 indicating perfect no structural and content similarity between the style and output images.

Indeed, the $CSCI$ formula comprehensively considers the structure and content information of the images, calculating the similarity based on weighted coefficients. Additionally, we introduced the $CCVI$ (Color Consistency Visual Index) formula, which evaluates the visual and color similarity between two images. By combining metrics such as Edge Preservation (EP), Color Preservation (CP), and Visual Significance (VS), the $CCVI$ formula accurately measures color retention and visual consistency. The formula for $CCVI$ is as follows:

$$CCVI = (\mathbf{s}, \mathbf{c}) = \frac{1}{N} \sum_{i=1}^{n} \frac{ep(\mathbf{s}, \mathbf{c}) + cp(\mathbf{s}, \mathbf{c}) + vs(\mathbf{s}, \mathbf{c})}{\mid ep(\mathbf{s}, \mathbf{c}) \mid + \mid cp(\mathbf{s}, \mathbf{c}) \mid + \mid vs(\mathbf{s}, \mathbf{c}) \mid} \tag{2}$$

Here, ep, cp, and vs represent edge preservation, color preservation, and visual significance, respectively. The values of these metrics are normalized to ensure they range between 0 and 1.

Based on the results of style transfer, we constructed two scoring sequences and assessed each output using the $CSCI$ and $CCVI$ formulas, while higher index values from the $CSCI$ and $CCVI$ formulas correlate with superior style transfer quality. Once the scoring was completed, we sorted the results in descending order, prioritizing the $CSCI$ score over the $CCVI$ score. Ultimately, the output with the highest combined score was selected as the best result for style transfer. This approach allowed us to consider multiple factors, including the structure, content, color, and visual aspects of the image, to evaluate the style transfer's quality comprehensively. By employing this scoring and evaluation method, we ensured a comprehensive assessment of style transfer quality and identified the most successful transfer result based on multiple criteria.

3.3 Aesthetics Evaluation

To balance artistry and aesthetics, we introduced a set of aesthetic criteria for evaluation. One such criterion is the "Three Courts and Five Eyes", which analyzes the proportion and distribution of facial features to define a standard face. However, it is essential to consider the diversity of face shapes, such as oval, Chinese, round, and others, to maintain individuality. We use the proportion characteristics of each face shape as quantitative evaluation criteria for face aesthetics. Based on these criteria, we design an aesthetic scoring function that incorporates indicators of naturalness, aesthetics, and artistry. This function comprehensively evaluates the style transfer outputs while considering the specific characteristics of different face shapes.

Based on the quantitative evaluation criteria for facial aesthetics, this paper aims to develop an adaptive collaborative exploration reinforcement learning model. This model is designed to function autonomously and self-learn the facial shape structure by considering facial aesthetics, as well as the naturalness, aesthetics, artistry, vividness, and emotional expression of the generated results. Specifically, the naturalness, aesthetics, and artistry indicators are reflected in the artistic style pen touch, picture cleanliness, and degree of defects, denoted as s_b, s_c, and s_a, respectively. These indicators are recorded and utilized in the learning process of the model. However, human aesthetic evaluation risks introducing cultural biases, so we give specific meanings for the degree.

The s_b indicator quantifies the uniqueness of the painting style in the generated portrait. It is computed as the distance between the style reference image and the content image, with a lower distance corresponding to a higher s_b score. This signifies how closely the synthesized portrait reflects the artistic style. The s_c indicator evaluates the capability of the generated portrait in suppressing low-quality effects like noise, blurring, and jagged edges. Moreover, the s_a indicator measures the degree to which discernible flaws or distortions in facial characteristics or outlines are averted. Integrating these indicators facilitates a quantitative appraisal of style representation, quality enhancement, and fidelity of the synthesized portraits.

4 Experiments

4.1 Datasets

In our experiments, we utilized three datasets to evaluate the performance of our method in cartoon stylization. For the Caricature dataset, we collected 199 images from WebCaricature, curated explicitly for studying face caricature synthesis. Additionally, we obtained an Anime dataset from Danbooru Portraits, consisting of 140 pairs of style-corresponding portraits. Furthermore, our cartoon stylization experiments involved a cartoon dataset comprising 317 cartoon face images sourced from Toonify [14]. These diverse datasets enable us to evaluate the effectiveness and versatility of our method across different stylization tasks, including sketch stylization, caricature synthesis, and cartoon stylization.

4.2 Implementation Details

We trained the DualStyleGAN model by fine-tuning its pre-trained model, adjusting the training iterations to 3000 with a batch size of 32. The training process took approximately 12 h and utilized two 3090 GPUs. During the stylized portrait generation stage, we fine-tuned the 18-bit weights and explored the vector space around the default value. This process generates 220 stylized results in 11 batches, with each batch containing 20 images.

We use $[n_1 * v_1, 1 * v_i, n_2 * v_2, n_3 * v_3]$ to indicate the vector w. the first n_1 weights in vector w are set to the value of v_1, the next one weights are set to the value of v_i which is changeable, the following n_2 is weighted as v_2, the last n_3 weights are v3. w_c, w_i, w_r and w_f denote the controllable weight vector (the first i weights of w), variable weight vector (the weights of i) ready weight vector(the weights from i +1 to 11) and the final weight vector(the last 9 weights) respectively. By default, we set w_c to 0.75, w_i is 0 to 1, the step size is 0.5, w_r to 0.75, w_f to 1. For testing for cartoon, caricature, anime, respectively.

After obtaining the results, we performed evaluations using the $CSCI$ and $CCVI$ metrics. The results were then sorted in descending order based on the evaluation scores. Subsequently, we conducted an aesthetic evaluation to assess the outputs further. Ultimately, we selected the result that ranked first in evaluation scores as the final output.

4.3   Comparable Experiments

We compare our method with JoJoGAN and DualStyleGAN in Cartoon, Anime, and Caricature styles. From Fig. 2 it can be seen that our method profitably cartoonized subjects with better results. Because of our evaluation process, the qualities of our results are better in color preservation, structure feature retention, and image fidelity. Specifically, our method can find the best results to preserve details such as hair, eyes, and mouth.

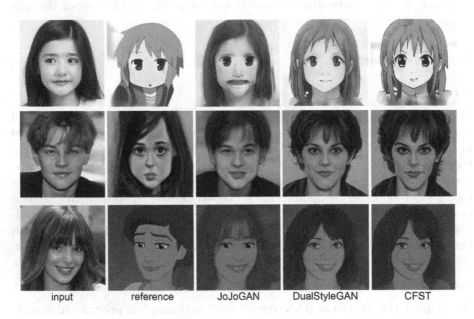

Fig. 2. Comparison with other methods

Table 1. Comparison of Style Transfer Methods

Style	Cartoon			Caricature			Anime		
Method	JoJoGAN	DualStyleGAN	Ours	JoJoGAN	DualStyleGAN	Ours	JoJoGAN	DualStyleGAN	Ours
$CSCI$	3.59	13.56	17.89	5.34	15.73	17.01	1.28	6.9	8.42
$CCVI$	0.37	0.78	0.86	0.22	0.67	0.72	0.23	0.56	0.63

Table 1 compares the $CSCI$ and $CCVI$ metrics achieved by different style transfer methods on three portrait styles. Our proposed method obtains the highest scores for both metrics across all styles compared to JoJoGAN and DualStyleGAN.

Our proposed method outperforms others in achieving controllable, high-fidelity artistic stylization on diverse datasets. The higher $CSCI$ and $CCVI$

values indicate that our method better preserves structural and content similarity as well as color consistency between the style image and the transferred result.

5 Conclusion

In conclusion, by utilizing DualStyleGAN with various weights, we can generate portraits. These generated portraits can then be ranked and sorted using our evaluation methods, namely the $CSCI$, $CCVI$, and aesthetics evaluation. Through this ranking process, we can optimize the original results and identify the most desirable outputs. This method holds potential for generating portraits using Style Transfer and filtering datasets, providing a valuable approach for enhancing the quality and selection of stylized portraits.

Future research will focus on enhancing evaluation formula robustness, model generalization, and style transfer for Asian faces. We will also try to refine the quality of non-realistic styles like anime, and explore model ensemble methods to leverage different algorithms' strengths. Moreover, it also can be used in Text-to-Image models like Stable Diffusion. Overall, future work will center on improving controllability and quality to achieve controllable and high-fidelity style transfer. Key directions include boosting quantification, expanding versatility, combining approaches, and enabling customization. We look forward to future innovations that will unlock the full potential of this technology.

Acknowledgements. We would like to thank the "Pioneer" and "Leading Goose" R&D Program of Zhejiang(No.2023C01212) as well as the Public Welfare Technology Application Research Project of Zhejiang (No.LGF22F020008).

References

1. Karras, T., Laine, S., Aila, T.: A style-based generator architecture for generative adversarial networks. In: CVPR (2019)
2. Yang, S., Jiang, L., Liu, Z., Loy, C.C.: Pastiche master: exemplar-based high-resolution portrait style transfer. In: CVPR (2022)
3. Wang, Z., et al.: Evaluate and improve the quality of neural style transfer. In: CVIU (2021)
4. Zhang, Y., et al.: Inversion-based style transfer with diffusion models. In: CVPR (2023)
5. Chen, S., et al.: PAniC-3D: stylized single-view 3D reconstruction from portraits of anime characters. In: CVPR (2023)
6. Zhang, B., et al.: MetaPortrait: identity-preserving talking head generation with fast personalized adaptation. In: CVPR (2023)
7. Chong, M.J., Forsyth, D.A.: JoJoGAN: one shot face stylization. arXiv preprint arXiv:2112.11641 (2021)
8. Rombach, R., Blattmann, A., Lorenz, D., Esser, P., Ommer, B.: High-resolution image synthesis with latent diffusion models. In: CVPR (2022)
9. Liu, K., et al.: StyleRF: zero-shot 3D style transfer of neural radiance fields. In: CVPR (2023)

10. Patashnik, O., Wu, Z., Shechtman, E., Cohen-Or, D., Lischinski, D.: StyleCLIP: text-driven manipulation of StyleGAN imagery. In: ICCV (2021)
11. Wright, M., Ommer, B.: ArtFID: quantitative evaluation of neural style transfer. In: Andres, B., Bernard, F., Cremers, D., Frintrop, S., Goldlücke, B., Ihrke, I. (eds.) DAGM GCPR 2022. LNCS, vol. 13485, pp. 560–576. Springer, Cham (2022). https://doi.org/10.1007/978-3-031-16788-1_34
12. Yeh, M.-C., Tang, S., Bhattad, A., Forsyth, D.A.: Quantitative evaluation of style transfer. (2018). https://doi.org/10.48550/arXiv.1804.00118
13. Yang, S., Jiang, L., Liu, Z., Loy, C.C.: VToonify: controllable high-resolution portrait video style transfer. In: ACM TOG (Proceedings of SIGGRAPH Asia) (2022)
14. Pinkney, J.N., Adler, D.: Resolution dependent GAN interpolation for controllable image synthesis between domains. arXiv preprint arXiv:2010.05334 (2020)

Data Adaptive Semantic Communication Systems for Intelligent Tasks and Image Transmission

Zhiqiang Feng, Donghong Cai(✉), Zhiquan Liu(✉), Jiahao Shan, and Wei Wang

College of Information Science and Technology, Jinan University, Guangzhou 510632, People's Republic of China
{63064fzq,shanjh,walle}@stu2022.jnu.edu.cn, {dhcai,zqliu}@jnu.edu.cn

Abstract. In this paper, we consider an end-to-end semantic communication system for compressed image wireless transmission. To economize the communication bandwidth, and enhance the communication reliability as well as the intelligence of the receiver, two networks are designed for image transmission and intelligent tasks. In particular, two simple linear layers are used to extract the feature of the image and recover the image, taking into account the sparsity of the image in a specific linear space. The trained encoder is deployed at the transmitter, while the trained decoder and the classifier are deployed at the receiver. To adapt to new communication data, the proposed networks are trained in a meta-learning framework. A few samples of new data are fed into the trained network to calculate new model parameters, which are fed back to the transmitter for updating the network of the encoder. Experimental results show that the proposed system has superior performance in terms of image compression transmission over fading channels compared with the existing semantic communication systems.

Keywords: Semantic communication · image transmission · Meta-learning

1 Introduction

With the development of communication techniques and the applications of Metaverse and virtual reality, communication systems are gradually shifting from traditional bit information transmission to systems with more powerful semantic understanding [1–3]. The semantic communication systems are not only able to convey information to the recipient but also to make deep understanding of the information, enabling a wider range of applications [4], such as autonomous vehicles. By using the deep learning techniques to design the transmitter and the receiver, the semantic communication system can compress the source information for effective transmission. Meanwhile, the receiver design based on deep learning can improve the intelligence of communication.

F. Zhao and D. Miao (Eds.): AIGC 2023, CCIS 1946, pp. 105–117, 2024.
https://doi.org/10.1007/978-981-99-7587-7_9

Traditional compression methods, such as JPEG2000 [5] and H.264 [6], are commonly used for image and video compression. Especially, JPEG2000 uses Discrete Cosine Transform (DCT) to transform the image in the frequency domain, and then uses quantization and entropy coding to achieve data compression. H.264 is a widely used standard for video compression, which uses motion estimation, transform coding and entropy coding to improve the efficiency of video compression. These traditional compression methods can reduce the size of data effectively, but they may have some limitations for wireless communication. For example, compressed information is difficult to decode correctly due to channel fading and noise. Channel coding should be considered to protect the compressed information. Compared with traditional compression schemes, semantic communication encodes information in a structured, simplified and flexible way, and provides a new perspective for joint source and channel coding [7]. However, the accuracy and robustness of semantic understanding remains a critical issue. Since the communication data includes structured data and unstructured data, such as text, picture, and video, it is difficult to use a scope-limited knowledge base. In addition, data scarcity is a problem that cannot be ignored in current semantic systems. Semantic understanding of specific tasks usually requires large amounts of annotated data to train models. However in real scenarios, it is very difficult and expensive to obtain large-scale annotated data [8–10]. Existing semantic communication methods face the challenge of efficiency and scalability when dealing with large and diverse data. Besides, the effective knowledge transfer and representation is also an urgent problem.

Recently, Huiqiang Xie et al. [11] proposed that a deep neural network enabled semantic communication system, named MU-DeepSC, to execute the visual question answering (VQA) task. Through joint design and optimization of transceivers, the most relevant data features are extracted to achieve task-oriented transmission. However, deep learning-based semantic communication systems may suffer from problems such as data scarcity and labeling difficulties during the learning process. In order to solve the problem that the actual observation data at the transmitter may have inconsistent distribution with the empirical data in the shared background knowledge base, Hongwei Zhang et al. [12] proposed a new semantic communication system for image transmission based on neural network. By using the domain adaptation technique of transfer learning, the data adaptation network is designed to learn how to transform observed data into similar forms of empirical data that semantic coding networks can process without retraining. However, under the influence of low compression rate and fading channel, this training method based on transfer learning is not ideal. In addition, Chanhong Liu et al. [4] proposed a compression ratio and resource allocation (CRRA) algorithm to support multi-users to perform tasks at low compression rate and occupy fewer resources. However, CRRA is difficult to apply because of its high complexity.

To overcome the limitations of current research, we propose an end-to-end semantic communication system based on meta-learning. In particular, the encoder and the decoder are designed based on linear layers, because the image information can be sparse by linear transformation and be compressed by a sens-

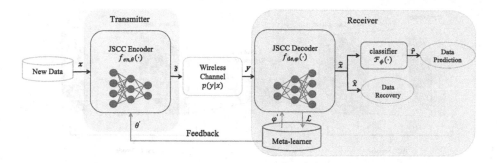

Fig. 1. Illustration of end-to-end semantic communication system.

ing matrix. A classifier is also designed to perform specific tasks. The proposed network is training based on the Model-agnostic meta-learning (MAML) algorithm [9]. Then the trained encoder is deployed at the sending end. The trained decoder and classifier are deployed at the receiving end. To accommodate the new data, a few samples of new data with labels are calculated with the trained model, and some parameters are fed back from the receiver to the transmitter for model update. The experimental results indicate that under low compression ratios (CR) and in the presence of fading channels, the reconstructed images transmitted through semantic encoding can be visually discernible to the human eye. Furthermore, the reconstructed images can still perform specific classification tasks.

2   System Model

Consider an end-to-end semantic communication system, where both of transmitter and receiver have single-antenna. As shown in Fig. 1, the input image signal is first encoded by a joint source and channel coder (JSCC) at the transmitter for compression wireless transmission. Meanwhile, a receiver designed by a deep neural network is used to reconstruct the image data as well as the intelligent recognition task. Especially, the input image data $\mathbf{x} \in \mathbb{R}^{N \times N \times L}$ is encoded by a feature vector $\tilde{\mathbf{s}} \in \mathbb{R}^K (K \ll N^2)$ containing the semantic information of the input image, i.e.,

$$\tilde{\mathbf{s}} = f_{en,\theta}(\mathbf{x}), \tag{1}$$

where $f_{en,\theta}(\cdot)$ represents the encoder with parameter θ, N is the size of the image and L is the numbers of channels of the image. Further, the vector $\mathbf{s}$ is normalized as $\mathbf{s} = \eta\tilde{\mathbf{s}}$. where $\eta = \frac{1}{\|\tilde{\mathbf{s}}\|_2}$ is the normalized coefficient.

At the receiver, the received signal vector $\mathbf{y} \in \mathbb{C}^K$ is formulated as

$$\mathbf{y} = h\tilde{\mathbf{s}} + \mathbf{z}, \tag{2}$$

where $h \in \mathbb{C}$ is a block fading channel coefficient, and $\mathbf{z} \sim \mathcal{CN}(0, \sigma^2 \mathbf{I}_K)$ represents the symmetric complex additive Gaussian noise with mean zero and variance σ^2. The semantic decoder deployed at the receiving end will reconstruct the original image from $\mathbf{y}$, which is given by

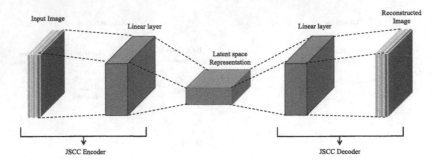

Fig. 2. The proposed JSCC encoder and decoder networks.

$$\hat{\mathbf{x}} = f_{de,\varphi}(\mathbf{y}), \tag{3}$$

where $f_{de,\varphi}(\cdot)$ represents the decoder with parameter φ. The recovery image $\hat{\mathbf{x}}$ is further used to perform the recognition task, i.e.,

$$\hat{r} = \mathcal{F}_\phi(\hat{\mathbf{x}}), \tag{4}$$

where $\mathcal{F}_\phi(\cdot)$ is the classifier with parameter ϕ and $\hat{r}$ is the prediction result.

Note that $f_{en,\theta}$, $f_{de,\theta}$ and $\mathcal{F}_\phi$ of semantic communication are designed by the deep neural networks and trained with the empirical data. The trained $f_{en,\theta}$ is deployed at the transmitter for compressed code transmission. The trained $f_{de,\theta}$ and $\mathcal{F}_\phi$ are deployed at the receiver side for image recovery and the classification task. Thus the semantic communication can reduce the bandwidth requirement while ensuring the communication quality, and improve the intelligent commitment of the receiver. However, the performance of semantic communication degrades seriously when the communication data is dynamically transformed or a new communication data appears. To this end, a semantic communication network is designed to adapt to the new communication data quickly. Different from the data adaptation network in [12], where a GAN is used to generate the target data, the networks of semantic communication are trained with Meta-learning framework in this paper.

3   Proposed Semantic Communication Networks

In this section, we first introduce deep neural network architectures that satisfy requirements of coder and decoder for semantic communication, including the joint source and channel encoder, decoder for image recovery, and classifier for performing intelligent tasks. Then, the data adaptive semantic communication network is obtained by introducing the MAML algorithm. When new communication data needs to be transmitted, the semantic communication model is updated through a few of feedbacks and adapted to the reliable transmission of new data.

3.1   Network Structures of Encoder and Decoder

On the one hand, the encoder of semantic communication needs to extract the features of image data well, reduce the amount of transmitted data as much as possible and reduce the communication bandwidth overhead while ensuring the transmission quality. On the other hand, the decoder needs to guarantee the quality of recovered image. It is important to point out that the proposed semantic communication system requires less cycles of model feedback updates. In addition, lightweight networks of JSCC coder and encoder should be designed. To this end, we propose two simple networks for the encoder and the decoder of semantic communication, which mainly consists of two linear layers, as shown in Fig. 2.

In the end-to-end semantic communication system, the encoder and decoder are deployed at the transmitter and receiver, respectively. The output of encoder is transmitted through the wireless channel. Therefore, the encoder need to compress source information and resist wireless channel fading. Meanwhile, the decoder has to generate the image based on the compressed feature of original image with noise. Note that the image can be sparsely represented as

$$\mathbf{D} = \boldsymbol{\Psi}\mathbf{X}\boldsymbol{\Psi}^T, \tag{5}$$

where $\boldsymbol{\Psi} \in \mathbb{R}^{N \times N}$ is a wavelet basis matrix. The wavelet coefficient matrix $\mathbf{D}$ is parse in (5), which can be further expressed as

$$\text{vec}(\mathbf{D}) = (\boldsymbol{\Psi} \otimes \boldsymbol{\Psi})\text{vec}(\mathbf{X}), \tag{6}$$

where $\text{vec}(\cdot)$ denotes the vectorization of a matrix, and $\otimes$ denotes the Kronecker product. Then the sparse vector $\text{vec}(\mathbf{D}) \in \mathbb{R}^{NN \times 1}$ can be compressed by a dictionary matrix of compressed sensing, results in

$$\bar{\mathbf{s}} = \mathbf{A}\text{vec}(\mathbf{D}) = \mathbf{A}(\boldsymbol{\Psi} \otimes \boldsymbol{\Psi})\text{vec}(\mathbf{X}), \tag{7}$$

where $\mathbf{A} \in \mathbb{R}^{M \times N}$ is a dictionary matrix of compressed sensing, and $\bar{\mathbf{s}} \in \mathbb{R}^{M \times 1}$ is the obtained compressed vector. In fact, the compressed vector $\bar{\mathbf{s}}$ is equivalent to the feature vector $\tilde{\mathbf{s}}$ in(1) if the number of rows of dictionary matrix is designed as K, i.e., $K = M$. However, the design of dictionary matrix $\mathbf{A}$ is based on the sparsity of $\text{vec}(\mathbf{D})$, and the quality of image recovery decreases without Restricted Isometry Property (RIP)-like conditions. To this end, a network of encoder with one linear layer can be used to implement the compression of image in (7). The encoder is

$$f_{en,\theta} \triangleq \mathbf{A}(\boldsymbol{\Psi} \otimes \boldsymbol{\Psi}) \triangleq \boldsymbol{\Lambda} \in \mathbb{R}^{K \times N}. \tag{8}$$

Note that the input of decoder based on (2) can be expressed as

$$\tilde{\mathbf{y}} = \mathbf{s} + \tilde{\mathbf{z}}, \tag{9}$$

where $\tilde{\mathbf{y}} = \mathbf{y}/h$ and $\tilde{\mathbf{z}} = \mathbf{z}/h$. Then the recovery image is given by

$$\hat{\mathbf{x}} = f_{de,\varphi}(\mathbf{s} + \tilde{\mathbf{z}}). \tag{10}$$

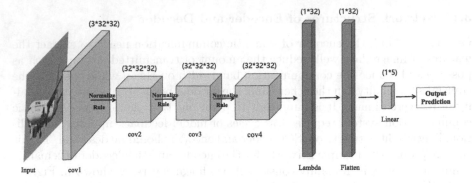

Fig. 3. The proposed network of classifier for semantic communication.

Based on the encoder presented in (8), the input of decoder can be re-written as

$$\tilde{\mathbf{y}} = \boldsymbol{\Lambda}\text{vec}(\mathbf{X}) + \tilde{\mathbf{z}}, \qquad (11)$$

which is a linear expression of original image. Applying the linear MMSE, we have

$$\text{vec}(\hat{\mathbf{X}}) = \left(\boldsymbol{\Lambda}^T\boldsymbol{\Lambda} + \frac{\sigma^2}{|h|^2}\mathbf{I}_K\right)^{-1}\boldsymbol{\Lambda}^T\tilde{\mathbf{y}} \triangleq \boldsymbol{\Omega}\tilde{\mathbf{y}}. \qquad (12)$$

It is important to point out that the linear transformation matrixes $\boldsymbol{\Lambda}$ and $\boldsymbol{\Omega}$ are unknown and difficult to design. Thus, we use the linear layer to achieve image recovery in our proposed networks.

3.2 Network Structures of Classifier

The purpose of deploying classifier at the receiving end of semantic communication is to quickly and intelligently recognize the meaning of the transmitted information. In fact, it requires the receiver equipped with a classifier to classify the received information. To this end, we proposed a classifier network, which consists of 4 convolution blocks, Lambda layer, Flatten layer and fully connected layer, as shown in Fig. 3. Each convolution block consists of several key components, including a 3×3 convolution with stride 1 and padding 1, a regularization layer, a Relu activation function, and a maximum pooling layer. The function of these convolutional blocks is to better extract the features of the input image. Through the extraction layer by layer, different information in the image can be captured. After four convolution blocks are processed, the information enters the Lambda layer. The Lambda layer is used for mean pooling of features to calculate the mean value of each channel. This action helps to further reduce the dimension of the feature while effectively preserving important information. After processing by the Flatten layer, the features are flattened into one-dimensional vectors for easy feeding into the fully connected layer. The function of the Flatten layer is to convert the features extracted from the convolution layer into a suitable form for processing by the fully connected layer such that the classification task can be better performed. Finally, the feature is passed to the fully connected layer, which classifies and maps the input features to each class's probability distribution. By training the fully connected layer's parameters, the classifier accurately determines the image's category based on the input features.

3.3   Model Training

For the model training, we use meta-learning framework to train the transmission network and the classifier network. In particular, the meta-learning of semantic communication system is divided into the meta-training stage and the meta-adaptation stage.

1) Meta-training stage: The training can be divided into inner-loop stage and outer-loop stage. The inner-loop stage is the semantic codec performing gradient descent for the loss of a specific task. And the outer-loop is updating the randomly initialized model parameters by calculating the gradient relative to the optimal parameters in each new task. It is assumed that there are I tasks. The task $T_i, i = 1, 2, \cdots, I$, has a training set D_i^{tr} and a validation set D_i^{val}. For the inner-loop phase, task T_i updates its model parameters by randomly sampling $\mathcal{K}$ samples ($\mathcal{K}$ is a small integer) according to its own specific task. Specifically, the encoder generates few low-dimensional feature vectors $\tilde{s}_{i,j}, j = 1, 2, \cdots, \mathcal{K}$, from the input of training set D_i^{tr} in task T_i, and then gets $y_{i,j}$ after normalization through fading channels. The decoder reconstructs an image similar to the input image through the decoder network. Then the loss of the original image and the generated image is

$$\mathcal{L}_i^{tr} = \mathcal{D}_{MSE}(X_i^{tr}, \hat{X}_i^{tr}), \tag{13}$$

where X_i^{tr} and $\hat{X}_i^{tr}$ are the image of the i-th task and the reconstructed image, respectively. In addition, $\mathcal{D}$ denotes the mean square error (MSE) of X_i^{tr} and $\hat{X}_r^{tr}$. Then, task T_i uses $\mathcal{L}_i^{tr}$ to perform gradient descent. The parameter update of the i-th task can be expressed as

$$\psi'(\theta', \varphi') \leftarrow \begin{cases} \theta_i' \leftarrow \theta_i - \alpha \arg\min \nabla_\theta \mathcal{L}_i^{tr}, \\ \varphi_i' \leftarrow \varphi_i - \alpha \arg\min \nabla_\varphi \mathcal{L}_i^{tr}, \end{cases} \tag{14}$$

where α is the learning rate of the inner-loop. Furthermore, task T_i uses the updated model parameters $\psi'(\theta', \varphi')$ to find the loss between the reconstructed image and the original image on validation set D_i^{val}, which is given by

$$\mathcal{L}_i^{val} = \mathcal{D}_{MSE}(X_i^{val}, \hat{X}_i^{val}), \tag{15}$$

where X_i^{val} and $\hat{X}_i^{val}$ are the image of the i-th task and the reconstructed image, respectively. After completing the above inner-loop, we get the loss $\mathcal{L}_i^{val}$ for $i = 1, 2, \cdots I$. For the outer-loop, the sum losses is

---

Algorithm 1. Model training and data adaptation of semantic communication

Meta-training Stage: Train semantic encoders and decoders
Input: Training $T_i, i = 1, 2...I$, tasks, where the batch size is $\mathcal{K}$
1: **Initialize:**randomly initialize $\psi(\theta, \varphi)$
2: **while** not done **do**
3:    **for all** T_i **do**
4:       Sample $\mathcal{K}$ data $\mathbf{x}_{i,j} \in D_i^{tr}(j = 1, 2, \cdots, \mathcal{K})$ for train task T_i
5:       Encoding: $\tilde{\mathbf{s}}_{i,j} = f_{en,\theta}(\mathbf{x}_{i,j})$
6:       Decoding: $\hat{\mathbf{x}}_{i,j} = f_{de,\varphi}(\mathbf{y}_{i,j})$
7:       Evaluate $\arg\min \nabla_\theta \mathcal{L}_i^{tr}$ using D_i^{tr}
8:       Compute train parameters $\psi'(\theta', \varphi')$ with gradient decent
9:       Feed $\psi'(\theta', \varphi')$ back to the network for encoder and decoder
10:       Get the loss $\mathcal{L}_i^{val}$ for task T_i with data D_i^{val}
11:    **end for**
12:    Update $\psi(\theta, \varphi)$ by equation (16) and feed it back to the encoder and decoder
13: **end while**
Meta-adaptation Stage: The trained encoder and decoder are deployed at the transmitter and receiver, respectively
Input: a new task B, the adaptive number of iterations is U, few samples for update the transmission model, the trained $\psi(\theta, \varphi)$ in Meta training stage
14: **for** $u = 1, 2, \cdots, U$ **do**
15:    The transmitter encodes the new sample similar to line 5
16:    The receiver decodes the information similar to line 6
17:    The receiver evaluates $\arg\min \nabla_\theta \mathcal{L}^{tr}$ similar to line 7 and computes $\psi'(\theta', \varphi')$ similar to line 8
18:    Parameter $\psi'(\theta', \varphi')$ is feed back to the transmitter from the receiver
19:    Get the loss $\mathcal{L}^{val}$ similar to line 10
20: **end for**
21: Update $\hat{\psi}(\hat{\theta}, \hat{\varphi})$ and feed it back to the transmitter

---

$$\psi(\theta, \varphi) \leftarrow \begin{cases} \theta \leftarrow \theta - \beta \arg\min \nabla_\theta \sum_{i=1}^I \mathcal{L}_i^{val}, \\ \varphi \leftarrow \varphi - \beta \arg\min \nabla_\varphi \sum_{i=1}^I \mathcal{L}_i^{val}, \end{cases} \tag{16}$$

where β is the learning rate of the outer-loop.

2) Meta-adaptation stage: The meta-learning training method based on parameter optimization aims to make the model have the ability to quickly adapt to the new task B. When the meta training is over, we deploy the encoder and decoder at the transmitter and receiver, respectively. Specifically, the data of new task is also divided into training sets D^{tr} and validation sets D^{val}, which enter the encoder for feature extraction through (1). The transmitter transmits the obtained signal (2) through the fading channel to the receiver. Furthermly, the loss (13) is obtained between the data reconstructed by decoding the decoder and the original data in D^{tr}. The meta-learner which in receiver updates the parameters of the encoder and decoder according to the loss as shown in (14) and sends them back to the encoder and decoder. The transmitter and the receiver use the

data from the validation set D^{val} to obtain the loss shown in (15). The parameters are then updated and transmitted back to the transmitter. The parameters of the new task are updated by

$$\hat{\psi}(\hat{\theta}, \hat{\varphi}) \leftarrow \begin{cases} \hat{\theta} \leftarrow \hat{\theta} - \beta \arg\min \nabla_{\hat{\theta}} \mathcal{L}^{val}, \\ \hat{\varphi} \leftarrow \hat{\varphi} - \beta \arg\min \nabla_{\hat{\varphi}} \mathcal{L}^{val}. \end{cases} \qquad (17)$$

When the training converges, the encoder and decoder are able to compress and recover the new data. An update that usually takes only a few times. Model training and data adaptation of proposed semantic communication network is shown in **Algorithm** 1. Since the training of the classifier model also uses MAML algorithm, the training process of the classifier with cross entropy loss function is not introduced here.

4   Experimental Results

In this section, we verify the reconstruction performance and classification accuracy of semantic communication systems under various data sets, including MNIST, KMNIST, FasionMNIST, Omniglot, CIFAR-10 and STL-10. In addition, we evaluate the adaptive ability of the proposed networks, i.e., the model reconstructs data that was not used during training.

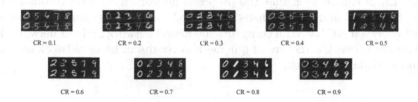

CR = 0.1   CR = 0.2   CR = 0.3   CR = 0.4   CR = 0.5

CR = 0.6   CR = 0.7   CR = 0.8   CR = 0.9

Fig. 4. Reconstruction results of MNIST over fading channel, where SNR = 10 dB.

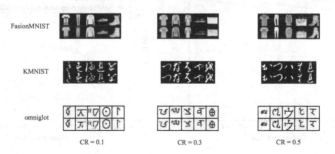

FasionMNIST

KMNIST

omniglot

CR = 0.1   CR = 0.3   CR = 0.5

Fig. 5. Reconstruction performance of different single-channel data sets over fading channels, where SNR = 10 dB.

4.1 Image Recovery

To validate the experimental performance of the proposed network for image restoration, we conducted tests using the MNIST dataset first, which comprises a training set of 60000 handwritten numerical images and a test set of 10000 samples. Each sample in grayscale format is a 28×28 pixels image associated with a label representing the correct identification of the handwritten number shown. The MNIST data set has been pre-processed and normalized to ensure that all images have a consistent size and orientation. The CR is defined as the ratio of the length of the feature vector to the total number of pixels in the original image. In semantic communication system, the feature vector passes through the Rayleigh fading channel after compression, and finally, the signal is transmitted to the decoder for reconstruction.

As shown in Fig. 4, the reconstruction results are presented for MNIST data set with CR from 0.1 to 0.9, where SNR = 10 dB. We compare the image recovery performance of different data sets in Fig. 5, including KMNIST, FasionMNIST and Omniglot data set. The variation trend of PSNR with the compression ratio is shown in Fig. 6. It can be observed that the PSNR increases with the increase of CR. Because the pixels of the original image are different, the PSNRs of image recovery are different at the same CR.

In Fig. 7, PSNR comparison of different methods are presented for MNIST data set, including Autoencoder (AE) with only linear layers, AE with convolutional layers, Variational Autoencoder (VAE), and GAN-Data Adaptation Networks [12]. It can be seen that the proposed network has better reconstruction result than VAE and data transfer learning. Meanwhile, we can obtain similar PSNRs between AE based on convolutional neural network and our network. But the simpler network structure of our method results in lower network overhead for model update feedback.

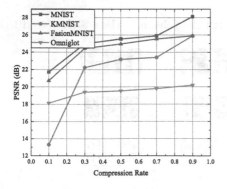
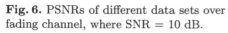
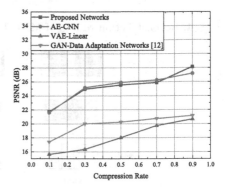

Fig. 6. PSNRs of different data sets over fading channel, where SNR = 10 dB.

Fig. 7. Comparison of different methods over fading channel, where SNR = 10 dB.

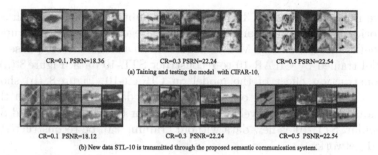

CR=0.1, PSRN=18.36   CR=0.3 PSRN=22.24   CR=0.5 PSRN=22.54

(a) Taining and testing the model with CIFAR-10.

CR=0.1 PSNR=18.12   CR=0.3 PSNR=22.24   CR=0.5 PSNR=22.54

(b) New data STL-10 is transmitted through the proposed semantic communication system.

Fig. 8. Reconstructions of CIFAR-10 and STL-10 over fading channels, where SNR = 10 dB.

4.2 Data Adaptation

In this subsection, the main focus is to validate the adaptive capability of the semantic communication model. Specifically, we first train the model using the CIFAR-10 dataset, which consists of 60000 color images in $3 \times 32 \times 32$ format. These images are divided into 10 different categories, including cats, dogs, airplanes, trucks, and more, with each category containing 6000 images. Then, we test the model using the STL-10 dataset, which has different distribution compared to CIFAR-10. The images in these datasets are source from real-world photos and aim to reflect the diversity and complexity of real-life scenes.

We use a 5-ways, 1-shot training setup, i.e., there are 5 categories in the training set, and each category has only one sample. This means that the model needs to learn from just one sample in each category and be able to correctly reconstruct other untrained samples during the testing phase. In the training

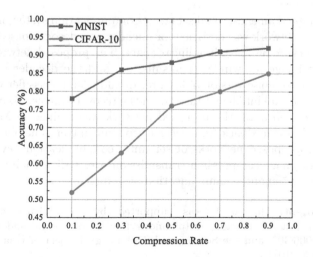

Fig. 9. Classification accuracy.

stage, we randomly divided CIFAR-10 into 32 subtasks, each of which contains 5 categories and only one sample for each category. The gradient update of the model is carried out by MAML algorithm. During the test phase, we use the model trained by CIFAR-10 to reconstruct STL-10 data. Figure 8 (a) shows the reconstruction effect of the model on CIFAR-10. Figure 8 (b) shows the adaptive effect of the model on STL-10 as new data. It can be seen that the model can reconstruct relatively clear pictures for both CIFAT-10 and STL-10. In particular, STL-10 as new data without training can be reconstructed by the proposed networks.

4.3   Classification Task

It should be noted that the classifier used is a pre-trained model. In order to make the classifier have strong generalization ability, the classifier training is also trained by MAML algorithm. This allows the trained classifier to directly classify images reconstructed by the transmit network. It verifies that the image reconstructed can be used to perform a specific tasks (such as classification tasks) in the semantic communication system. Figure 9 shows the accuracy of classification task over MNIST and CIFAR-10 data sets reconstructed by transmit network. It can be seen that the classification accuracy of MNIST can reach 0.78, and the classification accuracy of CIFAR-10 can reach 0.51 for CR = 0.1 and SNR = 10 dB. When the CR is 0.5, the reconstructed image can achieve a better classification accuracy, which indicates that the semantic communication system can capture the semantic content of the image rather than completely retain every detail of the original image.

5   Conclusion

We have considered an end-to-end semantic communication system in this paper. A simple linear network was designed for encoder and decoder based on the special sparse structure of image in a linear space. The proposed network extracted the feature of image for wireless transmission, and the decoder recovered the image based on the noise version of feature. In addition, a classifier was designed for performing special intelligent tasks with the recovery image. To further adapt new communication data, the proposed network was trained in Meta learning framework with few feedback from receiver to transmitter. The obtained results show that the proposed networks of semantic communication have a superior performance of image compression transmission over fading channels compared with the existing semantic communication systems.

Acknowledgment. This work was supported by the Science and Technology Major Project of Tibetan AutonomousvRegion of China under Grant No. XZ202201ZD0006G02, and the Science and Technology Project of Guangzhou under Grant No. 202201010200.

References

1. Xie, H., Qin, Z., Tao, X., Letaief, K.B.: Task-oriented multi-user semantic communications. IEEE J. Sel. Areas Commun. **40**(9), 2584–2597 (2022)
2. Weng, Z., Qin, Z.: Semantic communication systems for speech transmission. IEEE J. Sel. Areas Commun. **39**(8), 2434–2444 (2021)
3. Xie, H., Qin, Z., Li, G.Y., Juang, B.-H.: Deep learning enabled semantic communication systems. IEEE Trans. Signal Process. **69**, 2663–2675 (2021)
4. Liu, C., Guo, C., Yang, Y., Jiang, N.: Adaptable semantic compression and resource allocation for task-oriented communications. arXiv preprint arXiv:2204.08910 (2022)
5. Taubman, D.S., Marcellin, M.W., Rabbani, M.: JPEG2000: image compression fundamentals, standards and practice. J. Electron. Imaging **11**(2), 286–287 (2002)
6. Richardson, I.E.: H. 264 and MPEG-4 Video Compression: Video Coding for Next-Generation Multimedia. Wiley, Hoboken (2004)
7. Zhang, P., et al.: Toward wisdom-evolutionary and primitive-concise 6G: a new paradigm of semantic communication networks. Engineering **8**, 60–73 (2022)
8. Wang, Y., Yao, Q., Kwok, J.T., Ni, L.M.: Generalizing from a few examples: a survey on few-shot learning. ACM Comput. Surv. (CSUR) **53**(3), 1–34 (2020)
9. Finn, C., Abbeel, P., Levine, S.: Model-agnostic meta-learning for fast adaptation of deep networks. In International Conference on Machine Learning, pp. 1126–1135. PMLR (2017)
10. Li, Z., Zhou, F., Chen, F., Li, H.: Meta-SGD: learning to learn quickly for few-shot learning. arXiv preprint arXiv:1707.09835 (2017)
11. Xie, H., Qin, Z., Li, G.Y.: Task-oriented multi-user semantic communications for VQA. IEEE Wirel. Commun. Lett. **11**(3), 553–557 (2021)
12. Zhang, H., Shao, S., Tao, M., Bi, X., Letaief, K.B.: Deep learning-enabled semantic communication systems with task-unaware transmitter and dynamic data. IEEE J. Sel. Areas Commun. **41**(1), 170–185 (2022)

Generative Artificial Intelligence and Metaverse: Future of Work, Future of Society, and Future of Humanity

Yuxin Liu[ID] and Keng L. Siau[✉][ID]

City University of Hong Kong, Kowloon Tong, Hong Kong SAR
yliu2324-c@my.cityu.edu.hk, klsiau@cityu.edu.hk

Abstract. The rapid development of Generative Artificial Intelligence (GenAI) and the emergence of the Metaverse are dynamically reshaping our lives and societies. GenAI can enhance the development of Metaverse and empower the applications in Metaverse. Metaverse is also an excellent environment for GenAI to demonstrate its power and usefulness. This interwoven relationship fuels the potential of integrating GenAI and Metaverse. The paper discusses the integration potential of GenAI and Metaverse from four aspects. We further investigate how GenAI, Metaverse, and the integration of GenAI and Metaverse can reshape our future across the realms of work, society, and humanity. This paper offers theoretical and practical contributions by proposing research directions and specific research questions. Academic researchers can glean insights for future research and generate novel topics based on our findings. Policymakers, technical experts, and professionals across industries can gain a comprehensive grasp of GenAI and the Metaverse, enhancing their ability to adapt and contribute effectively to this emerging wave of innovation.

Keywords: Generative AI · ChatGPT · Metaverse · Research directions

1 Introduction: Generative Artificial Intelligence and Metaverse

1.1 Generative Artificial Intelligence

Artificial Intelligence (AI) is advancing exponentially and reshaping our world in unprecedented ways [1, 2]. Various AI applications, like virtual voice assistants, autonomous vehicles, and AI translators, have been integrated into our daily lives [3]. People are increasingly accustomed to interacting with these transformative technologies. However, a groundbreaking advancement in the ongoing AI revolution has emerged that has once again subverted our imagination – ChatGPT. ChatGPT represents the latest frontier of AI technologies and has become a hot topic since its advent. The impact of ChatGPT has extended far beyond conventional AI applications, leading us to the concept of Generative AI (GenAI).

GenAI is defined as a class of AI technologies that can generate new content in various forms, such as text, images, and audio, from existing training data [4, 5]. Compared to traditional AI algorithms that mainly focus on prediction, classification, or optimization tasks, GenAI shows superior abilities and great potential in terms of creativity and intelligence. Currently, the text-generation capability of GenAI is the most well-known. Represented by ChatGPT, which is built based on large language models (LLMs), GenAI applications can engage in dynamic conversations with humans to constantly learn from users and adapt responses to users. The text generation ability of GenAI not only supports a more interactive and human-like conversation between AI and human but also provides great help in code writing, literary creation, and many other fields.

Although the ability and potential of GenAI promise a bright future, the evolution of GenAI remains in its nascent stages, facing realized and unrealized challenges. Concerns about the accuracy and originality of AI-generated content encompass various domains. The ability of current GenAI algorithms to generate images, audio, and other forms of content also remains relatively constrained. Beyond these technical aspects, broader considerations such as organizational adoption and integration, trust, ethical considerations, legal framework, and other social concerns (e.g., labor force substitute) are critical subjects that demand attention and research [6, 7].

1.2 Metaverse

The concept of Metaverse has gained prominence with Facebook's rebranding to Meta, signaling the emergence of a new digital frontier. However, Metaverse's precise definition and contours remain subjects of ongoing exploration among practitioners and researchers. Despite this evolving landscape, the swift evolution of information technologies—spanning Virtual Reality (VR), Augmented Reality (AR), Mixed Reality (MR), Generative Artificial Intelligence (GenAI), cloud computing, and blockchain—has paved the way for the promising potential of the Metaverse to be realized.

The definition of the Metaverse varies from a narrow lens to a sweeping landscape. It encapsulates everything from immersive virtual environments that mimic the real world to expansive, interconnected digital realms that transcend physical constraints [8]. This breadth signifies the potential for the Metaverse to reshape many aspects of our lives, including social interactions, entertainment experiences, healthcare provisions, education systems, and economic activities [9–11].

Metaverse and its predecessors, such as Second Life, describe a unique and transformative vision of our future [12, 13]. GenAI, along with other advanced technologies, can provide strong support to build the expansive canvas of Metaverse. Simultaneously, the broad landscape of Metaverse serves as a dynamic arena that can guide the development of GenAI and other technologies. This research is dedicated to unraveling the potential integrations of GenAI and the Metaverse, investigating their capacity to reshape facets of our future encompassing work, society, and humanity. We identify and highlight key challenges that might emerge throughout the process, providing directions for both practical implementations and future research directions.

2 Current Development of GenAI and Metaverse

2.1 Practical Applications

GenAI. While Open AI commands considerable attention because of the development of ChatGPT, a variety of other GenAI companies and startups (e.g., Hugging Face, LangChain, AssemblyAI) are also dedicated to creating a diverse array of GenAI applications [14]. These applications span an extensive spectrum, ranging from AI assistants and code generation tools to character animation in media and even drug discovery in healthcare.

The adoption of GenAI in organizations is experiencing an accelerated surge. In a span of less than a year since the debut of many of these tools, one-third of companies have utilized GenAI in at least one business function [15]. However, this adoption trend is still concentrated in a few specific business domains. While this surge of innovation holds considerable promise, it also introduces a range of accompanying risks that companies may find challenging to navigate. Some factors include addressing employee training and reskilling, the recruitment roles performed by AI, managing shifts in workforce size, and dealing with various other complexities of transition. These considerations demand substantial attention and commitment from both companies and employers.

Metaverse. The Metaverse's defining characteristic is its expansive and interconnected nature, encompassing a diverse array of virtual spaces, social platforms, and digital ecosystems that facilitate user creativity, exploration, and interaction [16]. This interconnectedness fosters new pathways for social interactions, content creation, commerce, and even work [17].

Businesses and industries are recognizing the potential of the Metaverse. Companies are exploring how to leverage the Metaverse for marketing, brand enhancement, virtual events, and customer engagement [18]. The entertainment industry is embracing the immersive storytelling and interactive experiences of the Metaverse that blur the lines between fiction and reality. However, the current state of Metaverse development is still somewhat constrained, with its full potential as a broad and interconnected landscape yet to be fully realized due to technological limitations. While certain companies like Gucci, Adidas, Puma, and Prada have utilized the Metaverse concept to build virtual commerce and entertainment platforms, achieving a fully immersive experience and an interconnected virtual society remains challenging. Other challenges also remain, including concerns about privacy, data security, ethical considerations, and accessibility. As the Metaverse continues to develop, addressing these challenges will be essential to ensure responsible and inclusive growth of Metaverse.

2.2 Research Topics

Researchers are actively investigating emerging research topics within the realms of GenAI and the Metaverse. Both fields are witnessing substantial attention in technological innovation, adoption, and utilization, as well as their impacts across different domains, including education, economy, healthcare, and society. Moreover, identifying emerging opportunities and challenges has become a focal point.

From the IT design perspective, efforts extend beyond algorithm enhancements to encompass the integration of multiple technologies. A strong emphasis is placed on human-centered design [19], ensuring that the interfaces of GenAI, Metaverse environments, and user avatars are carefully designed to enhance user experience.

One of the transformative effects of GenAI and Metaverse is the reshaping of social interactions. This includes interactions between humans in both the real world and the Metaverse and between humans and AI applications in both contexts. Research focusing on these evolving social dynamics is also gaining significant attention [20, 21].

The combined influence of GenAI and Metaverse introduces novel opportunities and challenges to various industries. Take education as an example. As educational institutions adopt varying strategies concerning AI and GenAI emergence, the question of how to effectively harness AI and GenAI's potential for education while ensuring proper education integrity and quality becomes an urgent area of research [22].

3 Potential of Integrating GenAI into Metaverse

3.1 Harnessing Rich Data in Metaverse

GenAI is built and evolved based on a large amount of training data. The quality and quantity of data impact the quality of AI-generated content. Metaverse, as a vast landscape of interconnected digital worlds, encompasses immense data that can serve as foundations to support diverse GenAI applications.

Available data in Metaverse covers user behaviors, social interactions, user-generated content, economic transactions, and many others. Compared to the real world, data in Metaverse has some advantages that differentiate it from its real-world counterparts, including accessibility, comprehensiveness, real-time nature, accuracy, and low cost. For instance, user behavior data, like movement patterns, dialogue content, communication styles, and other detailed information, are easier to collect and have great potential to be integrated into GenAI applications. These data enable the development of GenAI applications with diverse and high-quality functions to redefine digital experiences in Metaverse.

3.2 Capitalizing the Dynamic Nature of Metaverse

Ever-evolving environments, activities, functions, and digital characteristics differentiate Metaverse from the real world. The dynamic nature of Metaverse creates more diverse and constant content demands, aiming at satisfying user preferences and improving user experiences. Therefore, Metaverse is an environment for GenAI to realize its potential and deliver substantial value.

GenAI enables real-time adjustment and continuous optimization of various elements in Metaverse. By leveraging real-time data on user behaviors and interactions, GenAI gains insights into user preferences, enabling the creation and updating of content that aligns with user needs. Examples of GenAI's learning dimensions include users' conversational styles, clothing choices, work patterns, investment inclinations, risk attitudes, and consumption habits. Consequently, workspace aesthetics, conversational avatar dialogue, sales appearances, investment advice, product recommendations, and more can be continuously fine-tuned to elevate user experiences.

3.3   Enhancing Ubiquitous Human-AI Interactions in Metaverse

AI applications occupy a broad and universal scope in the diverse landscape of Metaverse. How to enhance ubiquitous human-AI interactions in Metaverse remains a central concern. The integration of GenAI presents a promising avenue.

First, AI-generated text supported by Natural Language Processing (NLP) has a broader application range in Metaverse. Given that most users may be unfamiliar with virtual environments, various chatbots, service assistants, and other interactive characters are essential to help users easily adapt to various settings and effectively conduct various activities. With GenAI, the interface mechanisms, such as Avatars or applications, can better understand user requirements and respond to users with natural and human-like language.

Second, GenAI creates new possibilities in avatar creation. As user interfaces play a considerable role in human-AI interactions, GenAI's image and audio generation potential can be fully capitalized in AI avatar generation. AI avatars with various appearances, demographic information, clothing styles, languages, accents, and other characteristics can be rapidly created to achieve optimal interaction effects in various contexts. The combination of AI-generated texts in AI avatars can shape AI avatars in Metaverse to be as realistic as real humans.

3.4   Enabling Collaborative Creation in Metaverse

Metaverse empowers users to create and build their own worlds. This power of creation extends to personalized environmental elements such as avatars, literary compositions, and artistic expressions. Integrating GenAI with Metaverse can provide multi-aspect assistance in user content generation.

For instance, GenAI can simplify the user creation process and extend the creation scope. By aiding users with GenAI-powered suggestions, templates, and prompts, creating personalized content is easier. Thus, users can be empowered to generate a broader range of content in Metaverse. The collaboration between users and GenAI in Metaverse will effectively reduce users' entry barriers to Metaverse and increase user confidence to explore the new domain, thus leading to a more inclusive and accessible creative environment.

Further, the collaboration of humans and GenAI can augment the creativity of both humans and GenAI. The interplay between human ingenuity and AI-augmented insights creates a virtuous circle to constantly foster innovation and fuel continuous development and evolution in Metaverse.

4   Research Directions and Research Questions

The swift evolution of technology is reshaping our world at an unprecedented pace. In particular, GenAI and Metaverse are immensely promising fields and offer great potential to promote profound transformations across diverse dimensions. They are changing individual lifestyles, altering work environments and modes, redefining social interactions, revolutionizing industries, and even propelling humanity into a future depicted in science fiction. In this section, we discuss how GenAI and Metaverse may shape the future of work, the future of society, and the future of humanity. We summarize several research directions and list specific research questions for future research. We further propose research questions to support the integration of GenAI and Metaverse.

4.1 Future of Work

Table 1. Research directions regarding future work.

Field	Research questions
GenAI	**Organizational adoption and use** How do we design GenAI systems to enhance organizational adoption of GenAI in operations? How do we integrate GenAI into the existing information systems of companies?
	Labor force substitution Will GenAI replace creative jobs? How do we create more job opportunities utilizing GenAI?
	Human-AI collaboration How do we use GenAI to enhance employee creativity? How do we avoid employees being over-dependent on GenAI at work?
Metaverse	**Virtual workspace** What kinds of environmental factors (e.g., layout, presence of objects, accessibility) influence employees' work efficiency in Metaverse? What kinds of entertainment components (e.g., gamification) can enhance employees' productivity in Metaverse?
	Virtual avatar How do we design effective avatars? How can AI avatars be used to enhance human-AI interactions in Metaverse?
	Virtual collaboration How do we effectively incorporate AI in a virtual work team? How do we empower human-AI collaboration in Metaverse to improve collaboration efficiency?

Embracing the future of work reshaped by GenAI and Metaverse can benefit both individuals and organizations. On the one hand, these cutting-edge technologies provide novel opportunities for personal career development and contribute to company growth.

Individuals who understand these advanced trends are more likely to be aware of promising job opportunities and be more competitive than others. Companies are increasingly seeking to utilize the power of GenAI and Metaverse to augment their workforce, reduce labor costs, and bolster performance. However, how to seize and leverage these opportunities is a critical and tough task, as a blind pursuit of advanced technologies and trends can yield unintended consequences.

On the other hand, potential risks come along with the advantages. Novel challenges arise that traditional approaches might not readily address due to the distinctive characteristics of these technologies. Therefore, constant effort is imperative to identify risks and mitigate any adverse impact proactively. In Table 1, we propose feasible research directions that need attention from researchers and practitioners to unlock the potential of GenAI and Metaverse better.

4.2 Future of Society

Table 2. Research directions regarding future society.

Field	Research questions
GenAI	**Education** What are the most suitable GenAI adoption strategies for educational institutions of different disciplines, categories, and levels? How do we regulate students' use of GenAI to increase learning efficiency and encourage active thinking and creativity?
	Finance How can GenAI be integrated into other financial technologies to create more innovative financial tools? What are effective collaboration strategies for financial practitioners and GenAI to improve financial services?
Metaverse	**Education** How do we combine traditional physical teaching with virtual teaching in Metaverse to maximize student learning effect? How do we increase the creativity and critical thinking of students using Metaverse?
	Finance How do we support financial activities in Metaverse with traditional financial markets in the real world? How do we adapt legal structures in the real world to design a legal structure in Metaverse?

Revolutionized technological advancement brings significant shifts across various industries and aspects of society. Education and economy serve as two important fields related to everyone in society. In this part, we discuss how GenAI and Metaverse shape the future of education and finance, identifying critical issues specific to the two domains that need to be solved in the future, as shown in Table 2.

4.3 Future of Humanity

Table 3. Research directions regarding future humanity.

Field	Research questions
GenAI	**Mental health** What mental health problems may arise from the rapid technological changes that impact lifestyle? How do we focus on the central position of humans instead of AI applications?
	Discrimination and bias How can GenAI be used to provide targeted support and resources for vulnerable and marginalized groups to bridge existing societal gaps? How do we ensure that AI-generated content avoids reinforcing stereotypes and biases, especially in contexts like advertising, media, and content creation?
Metaverse	**Mental health** How do we encourage users to maintain offline social connections and nurture real-world relationships alongside virtual ones? How do we prevent users from overusing and becoming addicted to Metaverse?
	Discrimination and bias How do we ensure equal access to Metaverse and rich resources within Metaverse, particularly for individuals from marginalized populations, to prevent digital disparities? How do we regulate user-generated content and AI-generated content to avoid content containing social discrimination materials?

Technological progress and societal transformation are intended to enhance human well-being. Thus, it is imperative to dedicate ample attention to the array of concerns associated with the future of humanity. These concerns serve as essential guidance for developing technologies to elevate human well-being while vigilantly averting any adverse impacts. We provide some noteworthy issues in Table 3.

4.4 Integrating GenAI and Metaverse

We further list distinct research directions that deserve additional attention to successfully integrate GenAI and Metaverse, as shown in Table 4.

5 Conclusions

GenAI and Metaverse are rapidly changing our world. In this study, we investigate the potential of GenAI and Metaverse in shaping the future. We systematically review the current development of GenAI and Metaverse, including technology applications from the practical perspective and research topics from the academic dimension. Based on the comprehensive review of GenAI and Metaverse, we propose and discuss the potential of integrating GenAI and Metaverse. Finally, we describe how GenAI and Metaverse would reshape our work, society, and humanity. We also propose several research directions for each aspect. These research directions can provide insightful guidelines for the further development of technologies and societies.

This research makes both theoretical and practical contributions. Academic researchers can utilize the research directions identified to conduct significant future research and contribute to the development of GenAI and Metaverse. Policymakers, technical personnel, and practitioners from various industries can have a more comprehensive understanding of GenAI and Metaverse to better prepare for the future.

Table 4. Research directions regarding GenAI and Metaverse integration.

Field	Research questions
GenAI-Metaverse	**Data collection and use** What components of Metaverse can be supported and improved by GenAI? What are effective methods to collect and aggregate minimal data from Metaverse so that only vetted information is used for GenAI training?
	Data privacy How do we integrate users' consent mechanism into Metaverse to ensure that the data usage in GenAI is transparent? How do we balance data privacy protection and high-quality user experience in the Metaverse?
	User experience How do we analyze user behavior in Metaverse to personalize where and how AI-generated content can best support user experiences? How do we ensure that AI-generated content is aligned with user-generated content and the overall context?

References

1. Siau, K., et al.: FinTech empowerment: data science, artificial intelligence, and machine learning. Cutter Business Technology Journal **31**(11/12), 12–18 (2018)
2. Wang, W., Siau, K.: Artificial intelligence, machine learning, automation, robotics, future of work, and future of humanity – a review and research agenda. J. Datab. Manage. **30**(1), 61–79 (2019)
3. Hyder, Z., Siau, K., Nah, F.: Artificial intelligence, machine learning, and autonomous technologies in mining industry. J. Datab. Manage. **30**(2), 67–79 (2019)
4. Nah, F., Zheng, R., Cai, J., Siau, K., Chen, L.: Generative AI and ChatGPT: applications, challenges, and ai-human collaboration. J. Info. Technol. Case and Applicat. Res. **25**(3), 277–304 (2023)
5. Sun, J., et al.: Investigating explainability of generative AI for code through scenario-based design. In: 27th International Conference on Intelligent User Interfaces, pp. 212–228 (2022)
6. Siau, K., Wang, W.: Building trust in artificial intelligence, machine learning, and robotics. Cutter Bus. Technol. J. **31**(2), 47–53 (2018)
7. Siau, K., Wang, W.: Artificial Intelligence (AI) Ethics – Ethics of AI and Ethical AI. J. Datab. Manage. **31**(2), 74–87 (2020)
8. Greenbaum, D.: The virtual worlds of the metaverse. Science **377**(6604), 377 (2022)

9. Ma, Y., Siau, K.: Artificial intelligence impacts on higher education. In: Midwest United States Association for Information Systems (MWAIS 2018) **42**(5), (2018)
10. Yang, Y., Siau, K., Xie, W., Sun, Y.: Smart health: intelligent healthcare systems in the metaverse, artificial intelligence, and data science era. J. Organizat. End User Comput. **34**(1), 1–14 (2022)
11. Yousefpour, A., et al.: All one needs to know about fog computing and related edge computing paradigms: a complete survey. J. Syst. Architect. **98**, 289–330 (2019)
12. Eschenbrenner, B., Nah, F., Siau, K.: 3-D virtual worlds in education: applications, benefits, issues, and opportunities. J. Datab. Manage. **19**(4), 91–110 (2008)
13. Siau, K., Nah, F., Mennecke, B., Schiller, S.: Co-creation and collaboration in a virtual world: a 3D visualization design project in Second Life. J. Datab. Manage. **21**(4), 1–13 (2010)
14. CB Insights Research: GenAI 50: The most promising generative artificial intelligence startups of 2023, https://www.cbinsights.com/research/generative-ai-top-startups-2023/
15. Mckinsey & Company: The state of AI in 2023: Generative AI's breakout year | McKinsey, https://www.mckinsey.com/capabilities/quantumblack/our-insights/the-state-of-ai-in-2023-generative-ais-breakout-year
16. Uddin, M., Manickam, S., Ullah, H., Obaidat, M.A.: Abdulhalim dandoush: unveiling the metaverse: exploring emerging trends, multifaceted perspectives, and future challenges. IEEE Access **11**, 87087–87103 (2023)
17. Wang, Y., Siau, K., Wang, L.: Metaverse and human-computer interaction: a technology framework for 3D virtual worlds. In: International Conference on Human-Computer Interaction, pp. 213–221 (2022)
18. Schiller, S., Nah, F., Luse, A., Siau, K.: Men are from mars and women are from venus: dyadic collaboration in the metaverse. Internet Research (to appear)
19. Shneiderman, B.: Human-centered artificial intelligence: three fresh ideas. AIS Trans. Human-Comp. Interac. **12**, 109–124 (2020)
20. Davis, A., Murphy, J., Owens, D., Khazanchi, D., Zigurs, I.: Avatars, people, and virtual worlds: foundations for research in metaverses. J. Assoc. Inf. Syst. **10**, 90–117 (2009)
21. Hennig-Thurau, T., Aliman, D.N., Herting, A.M., Cziehso, G.P., Linder, M., Kübler, R.V.: Social interactions in the metaverse: framework, initial evidence, and research roadmap. J. Acad. Mark. Sci. **51**(4), 889–913 (2022)
22. Siau, K.: Education in the age of artificial intelligence: how will technology shape learning? The Global Analyst **7**(3), 22–24 (2018)

CanFuUI: A Canvas-Centric Web User Interface for Iterative Image Generation with Diffusion Models and ControlNet

Qihan Hu[1], Zhenghui Xu[1], Peng Du[2], Hao Zeng[1(✉)], Tongqing Ma[1(✉)], Youbing Zhao[1,3] (iD), Hao Xie[1], Peng Zhang[4], Shuting Liu[4], Tongnian Zang[5], and Xuemei Wang[1]

[1] Communication University of Zhejiang, Hangzhou 310018, China
{hao.zeng,matongq}@cuz.edu.cn
[2] Uber Technologies Inc., 1725 3rd Street, San Francisco, CA 94158, USA
[3] University of Bedfordshire, Luton LU1 3JU, UK
[4] Jiangsu Dongyin Intelligent Engineering Technology Research Institute, Nanjing 211111, China
[5] Jiangsu CRRC Digital Technology Co. Ltd., Nanjing 210000, China

Abstract. Today, various AI generation tools are emerging in succession. And the majority of existing tools are predominantly model-centric in design, resulting in steep learning curves and high usability thresholds for users. Moreover, current user interfaces lack built-in image editing capabilities, forcing users to rely on external software even for basic image editing tasks. Considering that most image generation is an iterative process, this limitation significantly hampers user experience and creative potential. Instead, this paper proposes a novel canvas-centric design that seamlessly integrates editing functionalities into the UI called CanFuUI, streamlining secondary image processing. Users can crop, modify, and annotation of specific regions of generated images within the same canvas in CanFuUI. Furthermore, canvas content is utilized as preprocessed images, directly integrated into the ControlNet preprocessing procedure, reinforcing the customization capabilities of AI-generated outputs.

Keywords: WebUI · ComfyUI · Diffusion Models · Canvas-Centric UI

1 Introduction

The landscape of AI image generation tools has experienced an explosive expansion with the emergence of diffusion models [5,7,9–12] based open-source solutions. However, most existing AI generation tools primarily focus on model-centric design, offering high flexibility, but also presenting a steep learning curve for users. Additionally, the ability to perform secondary editing on generated

Q. Hu and Z. Xu—Contributed equally to this work.
Many thanks to Mr. Yihui Shen for his generous funding support.

images is often limited, hindering direct image manipulation. Prominent examples like Stable Diffusion WebUI [2] and ComfyUI [1], which are widely utilized, pose challenges to users who need to possess a deep understanding of checkpoint, LoRA [6], ControlNet [12] models, expertly manipulate parameters like cfg, Step, Sampling methods, and employ sophisticated prompt techniques to achieve the desired artistic result. Furthermore, as user expectations continue to increase, the utilization of AI to generate images that meet those expectations becomes increasingly challenging.

For novice users in the field of AI artistry, their limited knowledge may lead to simplistic prompts or basic parameter adjustments, resulting in unsatisfactory outcomes. Another major challenge in stable diffusion research is the intricate process of blending or replacing specific elements from multiple generated images. For instance, merging a character into a scene requires several cumbersome steps, such as converting both the character and scene into sketches, erasing unnecessary elements, integrating the character sketch into the scene sketch, and finally using ControlNet to generate the final image. These operations that involve multiple tools require a significant amount of time and effort.

To address these challenges, we propose a new web user interface named CanFuUI that streamlines operations, reduces the learning curve, and integrates settings, parameters, and prompt techniques to generate high-quality images into selectable options. Additionally, we present an innovative approach, canvas-centric editing, that allows users to crop, modify, and draw on specific regions of the generated images. All edits can be performed seamlessly within a single canvas, and the resultant images, as well as preprocessed images obtained using ControlNet, are saved in the gallery, allowing for easy drag-and-drop usage from the toolbar. These advances significantly simplify the workflow, saving substantial time and effort.

Our main contributions lie in the following aspects:

– We introduce the first canvas-centric design AI generation tool that integrates editing functionalities into the UI, streamlining the cumbersome process of secondary image processing.
– We propose using canvas contents as preprocessed images, directly utilized in ControlNet when activated.
– We pre-select and integrate model parameters into style options, reducing the complexity and lowering the barrier to user engagement.

2 Related Work

Utilizing the Stable Diffusion model [9], remarkable image quality has been achieved that exceeds that of traditional GAN models [3,4,8]. However, due to the limited input conditions of the Stable Diffusion model, the ControlNet [12] model enables a wider range of input conditions, thus granting us greater control over the generated images.

Currently, the most popular user interfaces for Stable Diffusion are Stable Diffusion WebUI [2] and ComfyUI [1]. While the Stable Diffusion WebUI adopts a functional model-centric design, it requires users to have a comprehensive understanding of multiple models, have the ability to adjust various parameters, and have experience crafting diverse and effective prompts. On the other hand, ComfyUI centers on a node-based design, requiring users to master the intricate workflow of AI-generated image production. Both WebUI and ComfyUI lead to suboptimal user experiences in terms of image creation, modification, and adjustments. In contrast, CanFuUI is built as canvas-centric, significantly enhancing user convenience for image editing, adjustments, and creative endeavors.

3   Canvas-Centric Web UI

In order to enhance user convenience, we have organized CanFuUI into four distinct modules: AI Artistry (a), Canvas (b), Toolbar (c) and Asset Library (d), as shown in Fig. 1.

Fig. 1. Overview of CanFuUI design

3.1   AI Artistry

The extent of utilization of the AI artistry module generally depends on the user's proficiency in AI drawing techniques. For those users who seek artistic entertainment through AI drawing, an extensive degree of parameter adjustment and model selection might not be necessary. On the contrary, they may prefer fewer manual interventions and entrust more tasks to AI generation. As a result, we have simplified the creative interface, reducing the parameters options less frequently used while integrating various parameters, models, and prompt techniques into the style options shown in Fig. 1(a). This empowers users to effortlessly select their desired artistic styles for image generation. Moreover, for more demanding users, we have introduced ControlNet as an advanced guidance tool, facilitating further refinement of the generated images.

3.2   Canvas and Toolbar

The Canvas module is shown in Figs. 1(b) and (c), which offers a scalable canvas where AI-generated images are showcased, and these images can also be directly utilized as pre-processed images for ControlNet. Users have the flexibility to add their own generated images or import external images to the canvas, allowing for seamless manipulation using the diverse tools available in the toolbar. The toolbar primarily supports functions such as drawing, selection, cropping, and image insertion as demonstrated in Fig. 1(c). Additionally, the Canvas module incorporates support for layer operations, facilitating users to make secondary modifications to the generated images with ease.

3.3   Asset Library

The Asset Library module maintains a record of the ten most recently generated images for user selection presented in Fig. 1(d). Previous generated images are seamlessly integrated into the gallery depicted in Fig. 2(a), allowing users to preview their past generations as in Fig. 2(b). Additionally, users can easily add these images to the canvas or directly export them. The Asset Library module also facilitates the upload function, allowing the import of external images into the gallery.

(a) Gallery   (b) Preview

Fig. 2. Asset library

3.4   Use Case

In the AI Artistry module, users can effortlessly obtain desired images by entering prompt phrases, as illustrated in Fig. 3(a) and Fig. 3(b). For more advanced operations, such as incorporating a character from Fig. 3(a) into Fig. 3(b), users can utilize ControlNet's canny model within the advanced options to transform images in Fig. 3(a) and Fig. 3(b) into sketches, represented as Fig. 3(c) and Fig. 3(d). Then, using the brush and cropping tools in the toolbar, the character can be accurately isolated to obtain the image Fig. 3(e). Subsequently, the image shown in Fig. 3(e) can be seamlessly integrated with the scene sketch, producing the image presented in Fig. 3(f). Finally, employing AI, the image of the result of the fusion can be generated as in Fig. 3(g).

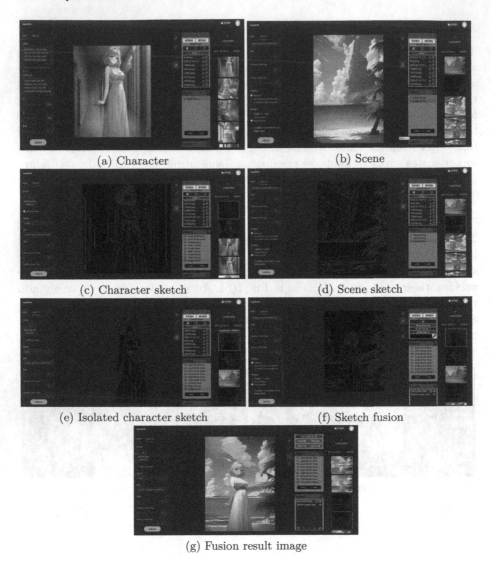

(a) Character

(b) Scene

(c) Character sketch

(d) Scene sketch

(e) Isolated character sketch

(f) Sketch fusion

(g) Fusion result image

Fig. 3. Use case

4   Comparison with WebUI and ComfyUI

Compared to WebUI and ComfyUI presented in Fig. 4(a) and Fig. 4(b) respectively, the advantages of CanFuUI design approach are as follows:

(1) Our approach seamlessly integrates image generation and image editing into a unified interface, eliminating the need for external editors. In addition, the generated images are automatically recorded and displayed on one side, facilitating users in their creative iterations. (2) The presence of a resizable canvas streamlines image editing; this feature absents in both Stable Diffusion

WebUI and ComfyUI. (3) The community enables users to access image assets and facilitates communication, interaction, and feedback among users. (4) We divide image generation into two modules, AI Artistry and Advanced Guidance. AI Artistry involves fewer parameter adjustments, making it simple and user-friendly, while Advanced Guidance incorporates additional functionalities like ControlNet. (5) We provide a dedicated asset library, allowing users to upload and use image materials. Users can also upload the required image assets to the library through the community feature.

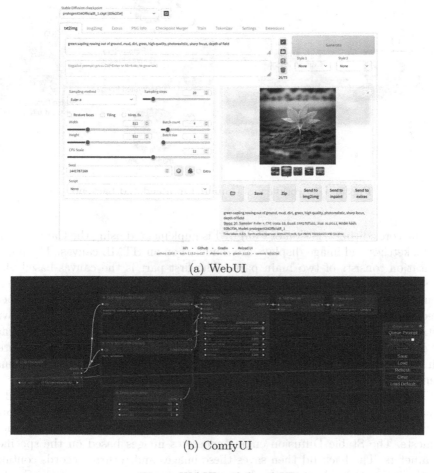

(a) WebUI

(b) ComfyUI

Fig. 4. WebUI and ComfyUI

5   Implementation

CanFuUI is developed using the React and UmiJS frameworks, along with the AntD UI component library. UmiJS is an enterprise-level React application framework introduced by the Ant Group, designed to facilitate routing and life-cycle management in React applications. By utilizing an UmiJS-based framework, CanFuUI becomes easier to develop and extend. The AntD UI library, also developed by the Ant Group, offers modern UI components that adhere to the Ant Design guidelines for various frameworks such as React, Vue, Angular, and more. In CanFuUI, we use AntD 5.x UI components.

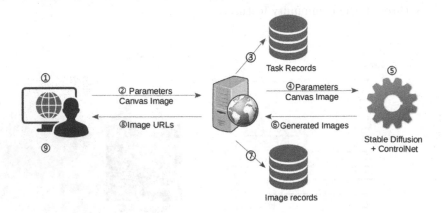

Fig. 5. The workflow of CanFuUI frontend and backend

The core editing function of CanFuUI is implemented using the Canvas APIs. Both sketches and image displays are rendered on an HTML canvas. The implementation consists of two main parts. The first part is the canvas-based view layer, responsible for rendering images, sketches, selections, and other visual elements. The second part includes various editing tools such as the eraser, pen, selector, and layer and history management. During page initialization, a canvas instance is created. Each tool selection triggers specific operations on the canvas instance. For example, when the pen tool is selected, a 2D context is created using the canvas's 'getContext('2d')' operation. The 2D context listens to mouse events and records the mouse trajectories to enable drawing functionality.

The interaction between the frontend and the stable diffusion backend is established through Restful HTTP requests. Prompts and parameters for Stable Diffusion and ControlNet are provided as JSON parameters within these requests. The Stable Diffusion engine generates images based on the specified parameters. The backend then saves these images and returns records containing the image URLs back to the frontend. The general workflow of CanFuUI is shown in Fig. 5.

6   Evaluation

A cohort of 32 first-year students majoring in Digital Media Technology from the College of Media Engineering, Communication University of Zhejiang, was selected as test users for the survey. After a brief tutorial on how to use three distinct UIs (WebUI, ComfyUI and CanFuUI), we arranged 32 test users to individually employ these three UIs to generate the artwork shown in Fig. 3, and the minimum, average and maximum time consumption among the 32 test users was recorded in Table 1. Based on the data provided in Table 1, it is evident that the creative efficiency is higher when using CanFuUI compared to the other two UIs.

Table 1. Time consumption of test users employing three UIs

UI	Minimum time	Average time	Maximum time
WebUI	19 m 47 s	22 min 32 s	26 m 05 s
ComfyUI	21 m 14 s	26 m 12 s	30 m 24 s
CanFuUI	15 m 10 s	18 m 23 s	21 m 02 s

Furthermore, based on the questionnaire conducted with the 32 test users, 46.88% of the participants (15 test users) consider the CanFuUI design to be excellent and approximately 84.39% of the participants (27 test users) expressed positive feedback on the overall usability and user experience of the CanFuUI design, as shown in the pie chart in Fig. 6(a). These test users found the CanFuUI design elegant and intuitive, allowing for rapid parameter customization and providing abundant functionality and operational options for users.

Out of the total 32 test users, approximately 52.94% (18 test users) rated the CanFuUI design as excellent in integrated image generation and editing. And about 84.39% (27 test users) give positive feedback as shown in Fig. 6(b). The result indicates that the test users are highly satisfied with the quality of the generated images, thus expressing that CanFuUI effectively meets their creative requirements.

Approximately 71.88% of the participants (23 test users) found the layer manipulation feature in the CanFuUI design to be useful, as shown in Fig. 6(c). CanFuUI design incorporates the functionality of layer operations, allowing users to make secondary modifications and adjustments to the generated images, thereby enhancing their ability to customize the images.

As presented in Fig. 6(d), approximately 40.63% of the participants (13 test users) perceived the asset library module and the gallery feature as being excellent. They expressed that the asset library's recording and management capabilities significantly facilitated their utilization and selection of the generated images. The survey result indicates that the advantages provided by CanFuUI have enhanced the creative experience of the user, the operational convenience and the creative flexibility, allowing them to effortlessly generate, edit, and customize high-quality images.

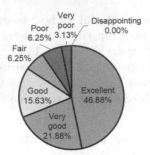

(a) Overall usability and user experience

(b) Integrated image generation and editing

(c) Layer manipulation

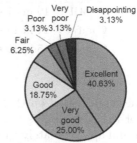

(d) Asset library module and gallery

Fig. 6. User evaluation demonstration

7   Conclusion

CanFuUI design simplifies user workflows and provides basic image editing tools and layer manipulation functionality, together with a convenient asset library. This helps users generate, edit, and create images with ease, leading to an enhanced user experience and satisfaction. Despite the numerous advantages of CanFuUI design, there are still some limitations and drawbacks that require further improvement and optimization as follows:

1. Currently, the range of parameter options in CanFuUI design is relatively limited, thereby restricting user control. Users seeking more fine-grained control over the generation process may wish to specify additional custom parameter choices.
2. CanFuUI design currently lacks real-time adjustments and interactive operations during image generation. Some users may prefer the ability to observe results during the generation process and make immediate adjustments to achieve more satisfactory results.
3. Although we offer certain model selection options, the current range is relatively narrow. A broader selection of diverse models would provide richer styles and effects, catering to a wider variety of user creative demands.

4. For new users who are just beginning to explore AI drawing, additional user tutorials and guidance may be required to help them better understand and master the functionalities and workflows of CanFuUI.

Our future endeavors involve offering customizable tools and editing options, incorporating interactive guidance and visual feedback, introducing intelligent prompts and suggestions, fostering community sharing and collaboration features, expanding output formats and resolutions, and optimizing the user interface's usability and responsiveness. We aim to provide users with a robust AI-based social platform for artistic creation.

Acknowledgements. We would like to thank the Zhejiang Provincial Blended First Class Online and Offline Course "Three-dimensional Character Design" (No. Z202Y22513), the Ministry of Education's Industry School Cooperation Collaborative Education Project "Research on PTA-Based Programming Training and Evaluation Model " (No. 202101151011) as well as the 17th batch Educational Reform Projects of Communication University of Zhejiang: "Cultivation and Practice of Computational Thinking in the Age of AI" for the generous funding support of the work referred to in this paper.

References

1. Stable diffusion ComfyUI. https://github.com/comfyanonymous/ComfyUI. Accessed 07 June 2023
2. Stable diffusion WebUI. https://github.com/db0/stable-diffusion-webui. Accessed 07 June 2023
3. Dhariwal, P., Nichol, A.: Diffusion models beat GANs on image synthesis. In: Ranzato, M., Beygelzimer, A., Dauphin, Y., Liang, P., Vaughan, J.W. (eds.) Advances in Neural Information Processing Systems, vol. 34, pp. 8780–8794. Curran Associates, Inc. (2021)
4. Goodfellow, I., et al.: Generative adversarial nets. In: Ghahramani, Z., Welling, M., Cortes, C., Lawrence, N., Weinberger, K. (eds.) Advances in Neural Information Processing Systems, vol. 27. Curran Associates, Inc. (2014)
5. Ho, J., Jain, A., Abbeel, P.: Denoising diffusion probabilistic models. In: Larochelle, H., Ranzato, M., Hadsell, R., Balcan, M., Lin, H. (eds.) Advances in Neural Information Processing Systems, vol. 33, pp. 6840–6851. Curran Associates, Inc. (2020)
6. Hu, E.J., et al.: LoRA: low-rank adaptation of large language models. In: The Tenth International Conference on Learning Representations, ICLR 2022, Virtual Event, 25–29 April 2022. OpenReview.net (2022). https://openreview.net/forum?id=nZeVKeeFYf9
7. Jo, J., Lee, S., Hwang, S.J.: Score-based generative modeling of graphs via the system of stochastic differential equations. In: Chaudhuri, K., Jegelka, S., Song, L., Szepesvari, C., Niu, G., Sabato, S. (eds.) Proceedings of the 39th International Conference on Machine Learning. Proceedings of Machine Learning Research, vol. 162, pp. 10362–10383. PMLR (2022)
8. Karras, T., Laine, S., Aittala, M., Hellsten, J., Lehtinen, J., Aila, T.: Analyzing and improving the image quality of StyleGAN. In: 2020 IEEE/CVF Conference on Computer Vision and Pattern Recognition (CVPR), pp. 8107–8116 (2020). https://doi.org/10.1109/CVPR42600.2020.00813

9. Rombach, R., Blattmann, A., Lorenz, D., Esser, P., Ommer, B.: High-resolution image synthesis with latent diffusion models. In: Proceedings of the IEEE/CVF Conference on Computer Vision and Pattern Recognition, pp. 10684–10695 (2021)
10. Sohl-Dickstein, J., Weiss, E., Maheswaranathan, N., Ganguli, S.: Deep unsupervised learning using nonequilibrium thermodynamics. In: Bach, F., Blei, D. (eds.) Proceedings of the 32nd International Conference on Machine Learning. Proceedings of Machine Learning Research, Lille, France, vol. 37, pp. 2256–2265. PMLR (2015)
11. Song, Y., Durkan, C., Murray, I., Ermon, S.: Maximum likelihood training of score-based diffusion models. In: Ranzato, M., Beygelzimer, A., Dauphin, Y., Liang, P., Vaughan, J.W. (eds.) Advances in Neural Information Processing Systems, vol. 34, pp. 1415–1428. Curran Associates, Inc. (2021)
12. Zhang, L., Agrawala, M.: Adding conditional control to text-to-image diffusion models (2023)

MFAR-VTON: Multi-scale Fabric Adaptive Registration for Image-Based Virtual Try-On

Shuo Tong(iD) and Han Liu(✉)(iD)

School of Automation and Information Engineering, Xi'an University of Technology,
Xi'an 710048, Shaanxi, China
liuhan@xaut.edu.cn

Abstract. Image-based virtual try-on technology provides a better shopping experience for online consumers. However, existing methods face challenges in effectively capturing high-level semantic information and achieving accurate registration between clothing and the body, particularly in complex body poses or target garments. To address these issues, we propose MFAR-VTON, a novel framework that incorporates a multi-scale enhanced adaptive clothing registration strategy and possesses matching filtering capabilities. Our method enables the generation of highly precise clothing alignment results, leading to seamless integration of try-on images. Additionally, we introduce a deformation energy constraint that effectively preserves intricate garment details. Experimental results demonstrated that MFAR-VTON achieves state-of-the-art performance in terms of accuracy and realism.

Keywords: MFAR-VTON · Virtual try-on · Adaptive clothing registration

1 Introduction

Image-based virtual try-on technology aims to seamlessly replace a person's clothing with in-shop target garment. The virtual try-on pipeline has evolved from initial two-stage [1, 5, 6] to the mainstream three-stage models [2, 14, 15]. The former includes the clothing warping and try-on modules. ACGPN [2] proposes a three-stage model, which introduce a post-try-on segmentation prediction module to provide improved guidance. However, both pipelines strongly depend on the clothing warping stage. Despite advancements in generating realistic try-on results, challenges persist in accurately aligning the clothing and preserving fine clothing details. In traditional garment deformation module, the extracted features are directly fed into the correlation layer to predict Thin-Plate Splines (TPS) parameters, which can only estimate low-complexity parameter transformations, generate rough clothing alignment. Moreover, previous methods compute the features correlation descriptors at a single scale, leading to the loss of some important semantic information.

To address these challenges, we propose the Multi-scale Fabric Adaptive Registration try-on network (MFAR-VITON). In this framework, we have designed a novel geometric matching module called the Multi-scale Neighborhood Consensus Warp Module

F. Zhao and D. Miao (Eds.): AIGC 2023, CCIS 1946, pp. 139–147, 2024.
https://doi.org/10.1007/978-981-99-7587-7_12

(MNCWM). It can extract more comprehensive and rich contextual information, effectively describing the details and shape features of the clothing by performing correlation matching of global semantic patterns across multiple feature scales. Inspired by the NCNet [3], our registration network employs 4D convolutions to refine the correlation feature maps, effectively filtering out geometrically inconsistent correlations, preventing incorrect matches between the clothing and the human body. Furthermore, we design a new end-to-end fabric deformation energy smoothing loss to address the issue of unconvincing distortion in clothing textures. In summary, the main contributions of this work are as follows:

1. We propose a novel image-based virtual try-on network called the Multi-scale Fabric Adaptive Registration Virtual Try-On Network (MFAR-VTON), which generates accurate clothing alignment results and state-of-the-art try-on results.
2. We propose a geometric alignment module, named the Multi-scale Neighborhood Consensus Warp Module (MNCWM), which incorporates multiple scales and employs matching filtering techniques to achieve seamless integration of clothing with body parts.
3. We introduce a fabric warping constraint that enhances the capability to handle complex textures on garments.
4. Experimental results demonstrate that our method outperforms the state-of-the-art approaches in both qualitative and quantitative evaluations.

2   Related Work

2.1   Trainable Image Alignment

Based on deep learning, trainable image alignment methods rely on geometric models, such as affine and thin-plate spline transformations, to estimate geometric transformation parameters through pairwise feature matching. Rocco et al. [3] proposed the Neighborhood Consensus Network (NCNet) by leveraging the idea of semi-local constraint to eliminate ambiguous feature matches. Building upon the NCNet, Li et al. [4] further improved the alignment performance by introducing a self-similarity module.

2.2   Image-Based Virtual Try-On

Image-based virtual try-on methods, which do not rely on 3D scanning devices for data acquisition, have gained significant attention in the academic community. To achieve natural generation effects, it is necessary to warp the clothing based on the reference person's characteristics. Techniques such as VITON [5] and CP-VTON [6] introduced the Thin-Plate Spline (TPS) warping method to deform the clothing. Subsequent models have extended this approach, but traditional clothing warping modules struggle with accurate clothing alignment, leading to issues such as misalignment and ghosting in the generated clothing. Methods based on appearance flow estimation, as in some approaches [7, 8], simulate geometric transformations by estimating dense flows between the source and target clothing regions, enabling flexible clothing warping. However, methods that rely on dense flow often exhibit unstable warping results and may lead to inconsistencies in rendering. In contrast, our proposed method performs dense semantic correspondence at multiple feature scales, capturing global dense semantic correlations, and filters out ambiguous and erroneous alignment results, resulting in more robust generated result.

3 Proposed Method

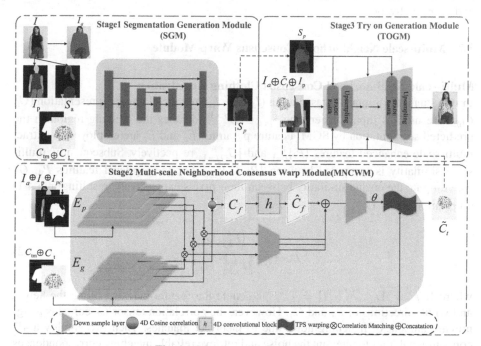

Fig. 1. Overview of MFAR-VTON

Figure 1 illustrates the overall architecture of MFAR-VTON. Given a person image $I \in \mathbb{R}^{H \times W \times 3}$ and a target clothing image $C_t \in \mathbb{R}^{H \times W \times 3}$ as inputs, it aims to seamlessly transfer C_t to the corresponding regions of the reference person I, while preserving the identity details of I unchanged. Our pipeline consists of three stages: (1) Segmentation Generation Module (SGM), which employs a U-net [9] to predict the post-try-on segmentation $S_p \in \mathbb{L}^{H \times W}$; (2) Multi-scale Neighborhood Consensus Warp Module (MNCWM), which performs clothing warping based on multi-scale neighborhood consensus; and (3) Try-on Generation Module (TOGM), which generates the final try-on result $I' \in \mathbb{R}^{H \times W \times 3}$. In the following sections we will describe the specific implementation details of the three-stage pipeline.

3.1 Segmentation Generation Module

As illustrated in Fig. 1, SGM takes the inputs of the cloth-agnostic human segmentation map $S_a \in \mathbb{L}^{H \times W}$, dense pose $I_p \in \mathbb{R}^{H \times W \times 3}$, target garment C_t, and cloth mask $C_{tm} \in$

$\mathbb{L}^{H \times W}$, and employs a U-net architecture to predict the post-try-on segmentation S_p. S_p determines whether the final generated content needs to be preserved or be generated. The total loss l of the SGM is defined as follows:

$$\mathcal{L}_{SSGM} = \lambda_1 \mathcal{L}_{CGAN} + \lambda_2 \mathcal{L}_C \tag{1}$$

where $\mathcal{L}_C$ denotes the pixel-wise cross entropy loss, $\mathcal{L}_{CGAN}$ denotes the conditional adversarial loss, λ_1 and λ_2 are hyperparameters. In this experiment, they are set to 10 and 1, respectively.

3.2 Multi-scale Neighborhood Consensus Warp Module

Multi-scale Neighborhood Consensus Clothing Registration
In this stage, the human representation (I_a, S_{pc}, I_p) and the clothing representation (C_t, C_{tm}) are used as inputs, where the $S_{pc} \in \mathbb{L}^{H \times W}$ denotes the clothing region in the predicted segmentation by SGM. Feature pyramids E_p and E_c are employed to extract multi-scale enhanced features $\{P_l\}_{l=1}^4$ and $\{G_l\}_{l=1}^4$, respectively. Subsequently, multi-scale similarity is computed between $\{P_l, G_l\}_{l=1}^3$. For the top-level features P_4, G_4 we compute the cosine similarity between their pixels exhaustively, resulting in a 4-D correlation tensor $C_{i,j,k,l}^f$:

$$C_{i,j,k,l}^f = \frac{\left\langle P_{i,j}^4, G_{k,l}^4 \right\rangle}{\left\| P_{i,j}^4 \right\|_2 \left\| G_{k,l}^4 \right\|} \tag{2}$$

where $\{i, k\} = 1, ..., h$, $\{j, l\} = 1, ..., w$ denote the feature indexing along the height and width directions of P_4 and G_4. However, the correlation tensor C_f contains a significant amount of matching noise due to incorrect matches. Then, we employ a 4D convolutional filter to filter out the noise and retrieve reliable matching correspondences Subsequently, we apply soft mutual nearest neighbor filtering to impose global constraints on the matches, obtaining $\overline{C}_{ijkl}^f = r_{ijkl}^{P_4} r_{ijkl}^{G_4} \hat{C}_{ijkl}^f$, where $r_{ijkl}^{P_4} = \hat{C}_{ijkl} / \max_{ab} \hat{C}_{abkl}$, $r_{ijkl}^{G_4} = \hat{C}_{ijkl} / \max_{cd} \hat{C}_{ijcd}$. Finally, the multi-scale correlation feature maps are added to $\overline{C}_{ijkl}^f$ and used as the input to the regression layer to predict TPS transformation parameters $\theta \in \mathbb{R}^{2 \times 5 \times 5}$. θ are used to warp the target clothing, generating warped target cloth $\tilde{C}_t \in \mathbb{R}^{H \times W \times 3}$.

Fabric Smoothing Constraint Based on Deformation Energy
The deformation function of TPS interpolation is as follows:

$$f(x, y) = a_0 + a_1 x + a_2 y + \sum_{i=1}^n \omega_i \phi(r_i) \tag{3}$$

where a_0, a_1, a_2 denote the affine transformation parameters, ω_i is the elastic component in elastic deformation, and $\phi(r_i)$ denotes the radial basis function. The cloth deformation energy, denoted as $E(f) = \iint_{R^2} \left[\left(\frac{\partial^2 f}{\partial x^2} \right) + 2 \left(\frac{\partial^2 f}{\partial x \partial y} \right) + \left(\frac{\partial^2 f}{\partial y^2} \right) \right] dx dy$, is defined as a measure of the deformation degree. To ensure the realism and natural smoothness of the distorted clothing, it is necessary to impose constraints on the distortion energy. When the radial basis function is chosen as the TPS $\phi(r) = r^2 \log r$, with a proportionality coefficient of α, $E(f)$ satisfies the following equation [10]:

$$E(f) = \alpha \omega^T K \omega$$

$$K = \begin{bmatrix} \phi(r_{11}) & \cdots & \phi(r_{1n}) \\ \vdots & \ddots & \vdots \\ \phi(r_{n1}) & \cdots & \phi(r_{nn}) \end{bmatrix}, w = [w_1, \cdots, w_n]^T \quad (4)$$

we convert it into an end-to-end trainable smooth distortion loss, thereby transforming the problem into $\arg \min_\omega (\alpha \omega^T K \omega)$. The total loss in the MNCWM is defined as follows, where β_1, β_2, and α are regularization parameters, in this experiment, they are set to 1, 0.1, and 0.005, respectively.

$$\mathcal{L}_{MNCWM} = \beta_1 \left\| \tilde{C}_t - I_c \right\|_{1,1} + \beta_2 \mathcal{L}_{vgg} \left(\tilde{C}_t, I_c \right) + \alpha \omega^T K \omega \quad (5)$$

3.3 Try-On Generator Module

In this stage, we use the predicted segmentation map S_p as guidance to fuse the warped clothing $\tilde{C}_t$, dense pose I_p, and clothing-agnostic representation I_a, generate the final try-on result I'. The try-on module consists of several SPADE residual blocks [11] and upsampling layers. Each scale's SPADE normalization layer predicts modulation parameters by S_p. Additionally, the multi-scale inputs $(\tilde{C}_t, I_p, I_a)$ are concatenated with the activation before each residual block for optimization. We utilize the same losses as SPADE and pix2pixHD [12], and the total loss is defined as:

$$\mathcal{L}_{TOM} = \gamma_1 \mathcal{L}_{cGAN} + \gamma_2 \mathcal{L}_{vgg} + \gamma_3 \mathcal{L}_{FM} \quad (6)$$

where $\mathcal{L}_{cGAN}$ represents the conditional adversarial loss, $\mathcal{L}_{vgg}$ denotes the perceptual loss, and $\mathcal{L}_{FM}$ represents the feature matching loss. γ_1, γ_2 and γ_3 are the regularization parameters, which are set to 1, 10, and 10, respectively, in this experiment.

4 Experiments

4.1 Dataset

VITON-HD: Our experiments are conducted using the VITON-HD [13] datasets under a resolution of 192×256. The VITON-HD dataset consists of front-view images of women and corresponding front-view clothing images. The dataset includes 11,647 pairs for the training set and 2,033 pairs for the testing set.

4.2 Implementation Details

We trained the SGM stages for 200,000 steps and MNCWM stages for 100,000 steps with a batch size of 8 with Adam optimizer with $\beta_1 = 0.5$ and $\beta_2 = 0.999$. For the TOGM stage, we trained it for 100,000 steps with a batch size of 4 with Adam optimizer with $\beta_1 = 0.5$ and $\beta_2 = 0.9$. The SGM and the MNCWM used the Adam optimizer with a same learning rate of 0.0002, while the discriminator had a learning rate of 0.0002 and 0.0001, respectively. The TOGM used the Adam optimizer with a learning rate of 0.0001, while the discriminator had a learning rate of 0.0004.

4.3 Qualitative Results

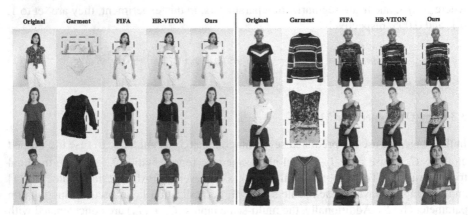

Fig. 2. Qualitative comparison on VITON-HD dataset.

Figure 2 shows a qualitative comparison between MNCWM and the latest baselines, HR-VTON [14], and FIFA [15]. On the left side of Fig. 2, it is evident that MFAR-VTON exhibits exceptional performance in the MNCWM module. In the first row, where the target clothing is similar in color to the background (Please note that we adjusted the contrast of the clothing image in the first row to distinguish it from the background, but the clothing image will not be changed during actual training.), the HR-VITON model mistakenly recognizes the clothing's inner contour as the outer edge. FIFA can only generate blurry result. In contrast, our method can capture features at multiple scales, effectively filter out erroneous registration patterns, improve the model's sensitivity to complex local features and shape variations, and thus improve the precision of clothing registration. In the second row, where the target clothing is not frontally placed, the compared models incorrectly identify the clothing alignment as short sleeves. MFAR-VTON can accurately aligns the target clothing and generates reasonable try-on results. On the right side of Fig. 2, it can be observed that the deformation energy loss has a more natural constraint on clothing texture deformation, which better preserves fine details of the clothing.

4.4 Quantitative Results

We compared the generated quality of the final try-on results of several state-of-the-art virtual try-on models using the Fréchet Inception Distance (FID) and Structural Similarity (SSIM) metrics. The comparison results on the VITON-HD dataset are presented in Table 1. From the results, it can be observed that our method achieves the best performance in terms of SSIM and FID compared to the advanced methods listed in the table.

Table 1. Quantitative comparisons on VITON-HD dataset

Model	SSIM ↑	FID ↓
CP-VTON + [1]	0.750	21.08
ACGPN [2]	0.845	16.64
DCTON [16]	0.830	14.82
HR-VITON [14]	0.864	9.38
MFAR-VTON (ours)	**0.874**	**7.11**

4.5 Ablation Study

We conducted clothing deformation on paired garments and images, and assessed the disparity between the deformation mask results and the ground truth mask using the Intersection over Union (IOU) metric. As shown in Table 2. It can be observed that MNCWM effectively improves the accuracy of clothing alignment.

Table 2. Ablation study of the MNCWM

Method	IOU ↑
w/o MNCWM	0.808
MFAR-VTON	**0.822**

To demonstrate the impact of the deformation energy constraint on clothing deformation, we conducted experiments by removing the deformation energy constraint from the network architecture and comparing the results with the complete MFAR-VTON model. The results are shown in the Fig. 3 With the deformation energy constraint, local clothing deformations are effectively constrained and smoothed, preserving the details of the target clothing. Without this constraint, exaggerated deformations lead to distorted clothing texture results.

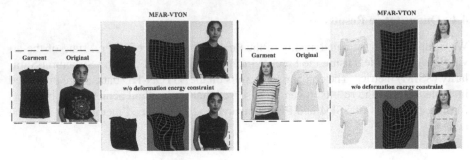

Fig. 3. Ablation study of the deformation energy constraint

5 Conclusion

In this paper, we proposed a novel cloth-adaptive registration try-on architecture, MFAR-VTON, which effectively handles complex poses, accurately aligns clothing deformations, and naturally constrains complex textures of garments to generate photo-realistic virtual try-on results. The experimental results demonstrated that our method outperforms the state-of-the-art approaches both in qualitative and quantitative.

Acknowledgments. This work was supported in part by National Natural Science Foundation of China under Grants 92270117, 61973248.

References

1. Minar, M.R., Tuan, T.T., Ahn, H., Rosin, P.: Cp-vton+: clothing shape and texture preserving image-based virtual try-on. In: Proc. of the IEEE international conference on computer vision workshop (ICCVW). pp. 10–14 (2020)
2. Yang, H., Zhang, R., Guo, X., Liu, W., Zuo, W.: Towards photo-realistic virtual try-on by adaptively generating-preserving image content. In: Proc. of the IEEE conference on computer vision and pattern recognition (CVPR), pp. 7850–7859 (2020)
3. Rocco, I., Cimpoi, M., Arandjelović, R., Torii, A., Pajdla, T., Sivic, J.: Neighbourhood consensus networks. In: Advances in neural information processing systems, pp. 1658–1669 (2018)
4. Li, S., Han, K., Costain, T. W., Howard-Jenkins, H., Prisacariu, V.: Correspondence networks with adaptive neighbourhood consensus. In: Proc. of the IEEE conference on computer vision and pattern recognition (CVPR), pp. 10196–10205 (2020)
5. Han, X., Wu, Z., Wu, Z., Yu, R., Davis, L. S.: Viton: An image-based virtual try-on network. In: Proc. of the IEEE conference on computer vision and pattern recognition (CVPR), pp. 7543–7552 (2018)
6. Wang, B., Zheng, H., Liang, X., Chen, Y., Lin, L., Yang, M.: Toward characteristic-preserving image-based virtual try-on network. In: Proc. of the European conference on computer vision (ECCV), pp. 589–604 (2018)
7. Chopra, A., Jain, R., Hemani, M., Krishnamurthy, B.: Zflow: gated appearance flow-based virtual try-on with 3d priors. In: Proc. of the IEEE conference on computer vision and pattern recognition (CVPR), pp. 5433–5442 (2021)

8. Han, X., Hu, X., Huang, W., Scott, M.R.: Clothflow: a flow-based model for clothed person generation. In: Proc. of the IEEE international conference on computer vision (ICCV), pp. 10471–10480 (2019)
9. Ronneberger, O., Fischer, P., Brox, T.: U-net: Convolutional networks for bio-medical image segmentation. In: Medical Image Computing and Computer-Assisted Intervention (MICCAI), pp. 234–241 (2015)
10. Wahba, G.: Spline models for observational data. Society for industrial and applied mathematics (1990)
11. Park, T., Liu, M.Y., Wang, T.C., Zhu, J.Y.: Semantic image synthesis with spatially-adaptive normalization. In: Proc. of the IEEE conference on computer vision and pattern recognition (CVPR), pp. 2337–2346 (2019)
12. Wang, T.C., Liu, M.Y., Zhu, J.Y., Tao, A., Kautz, J., Catanzaro, B.: High-resolution image synthesis and semantic manipulation with conditional gans. In: Proc. of the IEEE conference on computer vision and pattern recognition (CVPR), pp. 8798–8807 (2018)
13. Choi, S., Park, S., Lee, M., Choo, J.: Viton-hd: High-resolution virtual try-on via misalignment-aware normalization. In: Proc. of the IEEE conference on computer vision and pattern recognition (CVPR), pp. 14131–14140 (2021)
14. Lee, S., Gu, G., Park, S., Choi, S., Choo, J.: High-Resolution Virtual Try-On with Misalignment and Occlusion-Handled Conditions. In: Proc. of the European conference on computer vision (ECCV), pp. 204–219 (2022)
15. Zunair, H., Gobeil, Y., Mercier, S., Hamza, A.: Fill in Fabrics: Body-Aware Self-Supervised Inpainting for Image-Based Virtual Try-On. In: BMVC (2022)
16. Ge, C., Song, Y., Ge, Y., Yang, H., Liu, W., Luo, P.: Disentangled cycle consistency for highly-realistic virtual try-on. In: Proc. of the IEEE/CVF conference on computer vision and pattern recognition (CVPR), pp. 16928–16937 (2021)

An Assessment of ChatGPT on Log Data

Priyanka Mudgal(✉) and Rita Wouhaybi

Intel Corporation, Hillsboro, OR 97124, USA
{priyanka.mudgal,rita.h.wouhaybi}@intel.com

Abstract. Recent development of large language models (LLMs), such as ChatGPT has been widely applied to a wide range of software engineering tasks. Many papers have reported their analysis on the potential advantages and limitations of ChatGPT for writing code, summarization, text generation, etc. However, the analysis of the current state of ChatGPT for log processing has received little attention. Logs generated by large-scale software systems are complex and hard to understand. Despite their complexity, they provide crucial information for subject matter experts to understand the system status and diagnose problems of the systems. In this paper, we investigate the current capabilities of ChatGPT to perform several interesting tasks on log data, while also trying to identify its main shortcomings. Our findings show that the performance of the current version of ChatGPT for log processing is limited, with a lack of consistency in responses and scalability issues. We also outline our views on how we perceive the role of LLMs in the log processing discipline and possible next steps to improve the current capabilities of ChatGPT and the future LLMs in this area. We believe our work can contribute to future academic research to address the identified issues.

Keywords: log data · log analysis · log processing · ChatGPT · log analysis using LLM · large language model · deep learning · machine learning

1 Introduction

In recent years, the emergence of generative AI and large language models (LLMs) such as OpenAI's ChatGPT have led to significant advancements in NLP. Many of these models provide the ability to be fine-tuned on custom datasets [1–3] and achieve the state-of-the-art (SOTA) performance across various tasks. A few of the LLMs such as GPT-3 [4] have demonstrated in-context-learning capability without requiring any fine-tuning on task-specific data. The impressive performance of ChatGPT and other LLMs [5–9,79] in zero-shot and few-shot learning scenarios is a major finding as this helps LLMs to be more efficient [74–78]. With such learning methodologies, the LLMs can be used as a service [10] to empower a set of new real-world applications.

Despite the impressive capability of ChatGPT in performing a wide range of challenging tasks, there remain some major concerns about it in solving real-world problems like log analysis [93]. Log analysis is a vast area, and much

F. Zhao and D. Miao (Eds.): AIGC 2023, CCIS 1946, pp. 148–169, 2024.
https://doi.org/10.1007/978-981-99-7587-7_13

research has been done. It mainly comprises three major categories, namely, log parsing, log analytics, and log summarization. Log parsing is an important initial step of system diagnostic tasks. Through log parsing, the raw log messages are converted into a structured format while extracting the template [11–14]. Log analytics can be used to identify the system events and dynamic runtime information, which can help the subject matter experts to understand system behavior and perform system diagnostic tasks, such as anomaly detection [15–18], log classification [19], error prediction [20,21], and root cause analysis [22,23]. Log analytics can further be used to perform advanced operations e.g., identify user activities, and security analysis e.g., detect logged-in users, API/service calls, malicious URLs, etc. As logs are huge in volume, log summarization enables the operators to provide a gist of the overall activities in logs and empowers the subject matter experts to read and/or understand logs faster. Recent studies leverage pre-trained language models [17,24,25] for representing log data. However, these methods still require either training the models from scratch [26] or tuning a pre-trained language model with labeled data [17,24], which could be impractical due to the lack of computing resources and labeled data.

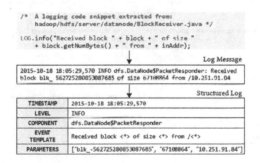

Fig. 1. An example of log code, log message, and structured log from [34]

More recently, LLMs such as ChatGPT [93] have been applied to a variety of software engineering tasks and achieved satisfactory performance [27,28]. With a lack of studies to analyze ChatGPT's capabilities on log processing, it is unclear whether it can be performed well on the logs. Although many papers have performed the evaluation of ChatGPT on software engineering tasks [29,30,33], specific research is required to investigate its capabilities in system log area. We are aware that the LLMs are fast evolving, with new models, versions, and tools being released frequently, and each one is improved over the previous ones. However, our goal is to assess the current situation and to provide a set of experiments that can enable the researchers to identify possible shortcomings of the current version for analyzing logs and provide a variety of specific tasks to measure the improvement of future versions. Hence, in this paper, we conduct an initial level of evaluation of ChatGPT on log data. Specifically, we divide the log processing [32] into three subsections: log parsing, log analytics, and log summarization. We design appropriate prompts for each of these tasks and analyze ChatGPT's capabilities in these areas. Our analysis shows that ChatGPT achieves promising results in some areas, but limited outcomes in others and contains several real-world challenges in terms of scalability. In summary, the major contributions of our work are as follows:

- To the best of our knowledge, we are the first to study and analyze ChatGPT's ability to analyze the log data in multiple detailed aspects.
- We design the prompts for multiple scenarios in log processing and record ChatGPT's response.
- Based on the findings, we outline several challenges and prospects for ChatGPT-based log processing.

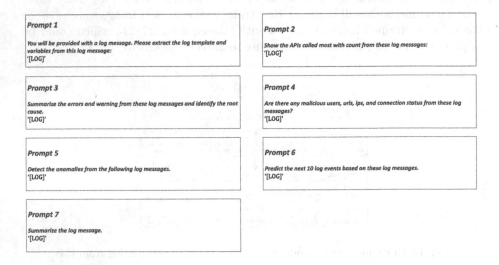

Fig. 2. Various prompt designs to address the research questions.

2 Related Work

2.1 Log Data

With the increasing scale of software systems, it is complex to manage and maintain them. To tackle this challenge, engineers enhance the system observability [31,99] with logs.

Logs capture multiple system run-time information such as events, transactions, and messages. A typical piece of log message is a time-stamped record that captures the activity that happened over time (e.g., software update events or received messages). Logs are usually generated when a system executes the corresponding logging code snippets. An example of the code snippet and generated code is shown in Fig. 1. A system with mature logs essentially facilitates the system behavior understanding, health monitoring, failure diagnosis, etc. Generally, there are three standard log formats, i.e., structured, semi-structured, and unstructured logs [72]. These formats share the same components: a timestamp and a payload content.

Structured logs usually keep a consistent format within the log data and are easy to manage. Specifically, the well-structured format allows easy storing, indexing, searching, and aggregation in a relational database. The unstructured log data achieves its high flexibility at the expense of the ease of machine processing. The characteristic of free-form text becomes a major obstacle for efficient query and analysis on unstructured or semi-structured logs. For instance, to count how often an API version appears in unstructured logs, engineers need to design a complex query with ad-hoc regular expressions to extract the desired information. The manual process takes lots of time and effort and is not scalable.

2.2 Log Processing

Logs have been widely adopted in software system development and maintenance. In industry, it is a common practice to record detailed software runtime information into logs, allowing developers and support engineers to track system behaviors and perform postmortem analysis. On a high level, log processing can be categorized in three types as discussed below.

Log Parsing. Log parsing is generally the first step toward automated log analytics. It aims at parsing each log message into a specific log event/template and extracting the corresponding parameters. Although there are many traditional regular expression-based log parsers, but, they require a predefined knowledge about the log template. To achieve better performance in comparison to traditional log parsers, many data-driven [12,37–42] and deep learning based approaches [24,26] have been proposed to automatically distinguish template and parameter parts.

Log Analytics. Modern software development and operations rely on log monitoring to understand how systems behave in production. There is an increasing trend to adopt artificial intelligence to automate operations. Gartner [97] refers to this movement as AIOps. The research community, including practitioners, has been actively working to address the challenges related to extracting insights from log data also being referred to as "Log Analysis" [96]. Various insights that can be gained are in terms of log mining [85], error detection and root cause analysis, security and privacy, anomaly detection, and event prediction.

Log Mining. Log mining seeks to support understanding and analysis utilizing abstraction and extracting useful insights. However, building such models is a challenging and expensive task. In our study, we confine ourselves to posing specific questions in terms of most API/service calls that can be extracted out of raw log messages. This area is well studied from a deep learning aspect and most of those approaches [49–56] require to first parse the logs and then process them to extract the detailed level of knowledge.

Error Detection and Root Cause Analysis. Automatic error detection from logs is an important part of monitoring solutions. Maintainers need to investigate what caused that unexpected behavior. Several studies [22,43,45–48] attempt to provide their useful contribution to root cause analysis, accurate error identification, and impact analysis.

Security and Privacy. Logs can be leveraged for security purposes, such as malicious behaviour and attack detection, URLs, and IP detection, logged-in user detection, etc. Several researchers have worked towards detecting early-stage malware and advanced persistence threat infections to identify malicious activities based on log data [57–61].

Anomaly Detection. Anomaly detection techniques addresses to identify the anomalous or undesired patterns in logs. The manual analysis of logs is time-consuming, error-prone, and unfeasible in many cases. Researchers have been trying several different techniques for automated anomaly detection, such as deep learning [62–65] and data mining, statistical learning methods, and machine learning [23,66–71].

Event Prediction. The knowledge about the correlation of multiple events, when combined to predict the critical or interesting event is useful in preventive maintenance or predictive analytics that can reduce the unexpected system downtime and result in cost saving [80–82]. Thus, the event prediction method is highly valuable in real-time applications. In recent years, many rule-based and deep learning based approaches [83,88–92] have evolved and performing significantly.

Log Summarization. Log statements are inserted in the source code to capture normal and abnormal behaviors. However, with the growing volume of logs, it becomes a time-consuming task to summarize the logs. There are multiple deep learning-based approaches [19,44,96,98] that perform the summarization, but they require time and compute resources for training the models.

2.3 ChatGPT

ChatGPT is a large language model which is developed by OpenAI [93,94]. ChatGPT is trained on a huge dataset containing massive amount of internet text. It offers the capability to generate text responses in natural language that are based on a wide range of topics. The fundamental of ChatGPT is generative pre-training transformer (GPT) architecture. GPT architecture is highly effective for natural language processing tasks such as translation in multiple languages, summarization, and question answering (Q & A). It offers the capability to be fine-tuned on specific tasks with a smaller dataset with specific examples. ChatGPT can be adopted in a variety of use cases including chatbots, language translation, and language understanding. It is a powerful tool and possesses the potential to be used across wide range of industries and applications.

2.4 ChatGPT Evaluation

Several recent works on ChatGPT evaluation have been done, but most of the papers target the evaluations on general tasks [33,73], code generation [27], deep learning-based program repair [28], benchmark datasets from various domains [29], software modeling tasks [30], information extraction [87], sentiment analysis of social media and research papers [84] or even assessment of evaluation methods [86]. The closest to our work is [35], but they focus only on log parsing.

We believe that the log processing area is huge and a large-level evaluation of ChatGPT on log data would be useful for the research community. Hence, in our work, we focus on evaluating ChatGPT by conducting an in-depth and wider analysis of log data in terms of log parsing, log analytics, and log summarization.

3 Context

In this paper, our primary focus is to assess the capability of ChatGPT on log data. In line with this, we aim to answer several research questions through experimental evaluation.

3.1 Research Questions

Log Parsing RQ1. How does ChatGPT perform on log parsing?

Log Analytics RQ2. Can ChatGPT extract the errors and identify the root cause from raw log messages?

RQ3. How does ChatGPT perform on advanced analytics tasks e.g., most called APIs/services?

RQ4. Can ChatGPT be used to extract security information from log messages?

RQ5. Is ChatGPT able to detect anomalies from log data?

RQ6. Can ChatGPT predict the next events based on previous log messages?

Log Summarization RQ7. Can ChatGPT summarize a single raw log messages?

RQ8. Can ChatGPT summarize multiple log messages?

General RQ9. Can ChatGPT process bulk log messages?

RQ10. What length of log messages can ChatGPT process at once?

To examine the effectiveness of ChatGPT in answering the research questions, we design specific prompts as shown in Fig. 2. We append the log messages in each of the prompts (in place of the slot '[LOG]').

3.2   Dataset

To perform our experiments, we use the datasets provided from the Loghub benchmark [13,34]. This benchmark covers log data from various systems, including, windows and linux operating systems, distributed systems, mobile systems, server applications, and standalone software. Each system dataset contains 2,000 manually labeled and raw log messages.

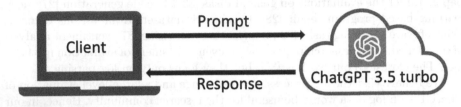

Fig. 3. Flow Diagram.

3.3   Experimental Setup

For our experiments, we are using the ChatGPT API based on the gpt-3.5-turbo model to generate the responses for different prompts [93]. As shown in Fig. 3, we send the prompts appended with log messages to ChatGPT from our system with Intel® Xeon® E3-1200 v5 processor and Intel® Xeon® E3-1500 v5 processor and receive the response. To avoid bias from model updates, we use a snapshot of gpt3.5-turbo from March 2023 [95].

3.4   Evaluation Metrics

As our study demands a detailed evaluation and in some cases, there was no state-of-the-art tool, we evaluated the output by our manual evaluation.

4   Experiments and Results

Each of the subsections below describes the individual evaluation of ChatGPT in different areas of log processing.

4.1   Log Parsing

In this experiment, we assess the capability of ChatGPT in parsing a raw log message and a preprocessed log message and find the answer to **RQ1**. For the first experiment, we provide a single raw log message from each of the sixteen publicly available datasets [34] and ask ChatGPT to extract the log template. We refer to it as first-level log parsing. ChatGPT performs well in extracting the

---

Prompt: You will be provided with a log message. Please extract the log template and variables from this log message:

03-17 16:13:38.819 1702 8671 D PowerManagerService: acquire lock=233570404, flags=0x1, tag="View Lock", name=com.android.systemui, ws=null, uid=10037, pid=2227

ChatGPT Response:

Log Template: {timestamp} {process_id} {thread_id} {log_level} {tag}: {message}
Variables:
- timestamp: 03-17 16:13:38.819
- process_id: 1702
- thread_id: 8671
- log_level: D
- tag: PowerManagerService
- message: acquire lock=233570404, flags=0x1, tag="View Lock", name=com.android.systemui, ws=null, uid=10037, pid=2227

Fig. 4. Log parsing of raw log message.

specific parts of log messages for all sixteen log messages. One of the examples of ChatGPT's response for first-level log parsing is shown in Fig. 4. Next, we preprocess the log message, extract the content, and ask chatGPT to further extract the template from the log message. ChatGPT can extract the template and variables from the log message successfully on all sixteen log messages with a simple prompt. One of the examples of ChatGPT's response is shown in Fig. 5.

4.2 Log Analytics

To evaluate ChatGPT's capability in log analytics, we perform several experiments in each of the categories described in Sect. 2.2.

Log Mining. In this experiment, we are seeking the answer of **RQ2** by investigating if ChatGPT can skim out the knowledge from raw logs without building an explicit parsing pipeline. We perform our experiments in several parts. We provide a subset of log messages containing 5, 10, 20, and 50 log messages from Loghub benchmark [34] and ask ChatGPT to identify the APIs. Figure 6 shows an example of ChatGPT response when a smaller set of log messages were passed. We notice that ChatGPT consistently missed identifying some APIs from the log messages irrespective of the count of log messages, but still shows 75% or more accuracy in all cases. Results are reported in Table 1.

Error Detection and Root Cause Analysis. In this experiment, we explicitly ask ChatGPT [95] to identify the errors, warnings, and possible root causes of

> **Prompt: You will be provided with a log message. Please extract the log template and variables from this log message:**
>
> PowerManagerService: acquire lock=233570404, flags=0x1, tag="View Lock",
>
> **ChatGPT Response:**
>
> Log Template: PowerManagerService: acquire lock={lock}, flags={flags}, tag="{tag}", name={name}, ws={ws}, uid={uid}, pid={pid}
> Variables:
> - lock: 233570404
> - flags: 0x1
> - tag: "View Lock"
> - name: com.android.systemui
> - ws: null
> - uid: 10037
> - pid: 2227

Fig. 5. Log parsing of preprocessed log message.

Table 1. ChatGPT's performance to identify the APIs, errors and root cause from Loghub dataset [34].

Log Message Count	API Count	API captured	API Accuracy (%)	API Response Time (s)	Error Count	Error captured	Error Accuracy (%)	Error Response time (s)
5	5	4	80	2.48	2	2	100	18.49
10	10	8	80	3.96	3	3	100	27.61
20	20	15	75	6.44	5	3	60	36.38
50	50	46	92	5.66	13	5	38.46	46.46

those in the provided log messages and address **RQ3**. Aligning towards our study structure, we first provide five log messages from the Loghub dataset [34] and later increase the size of log messages to ten, twenty, and fifty. Figure 7 shows the identified errors from five log messages and a detailed report for all the combinations with their response time is being reported in Table 1. It is evident from Table 1 that ChatGPT successfully identifies the errors and warnings on a smaller set of log messages than a larger set.

Security and Privacy. In this experiment, we focus on addressing **RQ4** and investigate if ChatGPT can identify the URLs, IPs, and logged users from the logs and extract knowledge about malicious activities. We use the open source dataset from Loghub [100] and follow the same approach of sending the set of five, ten, twenty, and fifty log messages to chatGPT to detect the URLs, IPs, and users from them. We use the 'Prompt 4' from Fig. 2 to ask if there are any malicious activities present in the logs. As shown in Table 2, ChatGPT extracts out the IPs and logged-in users with high accuracy irrespective of the length

Prompt: Show the APIs called most with count from these log messages:

03-17 16:13:46.671 1702 17633 I NotificationManager:
onNotificationVisibilityChanged called
03-17 16:13:46.672 2227 2318 I PhoneStatusBar:
logNotificationVisibilityChanges runInThread over
03-17 16:13:46.764 2227 2794 E KeyguardUpdateMonitor: isSimPinSecure
mSimDatas is null or empty
03-17 16:13:46.765 2227 2794 W KeyguardUpdateMonitor: registerCallback

ChatGPT Response:

API Name	Count
NotificationManager	1
PhoneStatusBar	1
KeyguardUpdateMonitor	2

Fig. 6. ChatGPT response to extract the APIs from log messages.

of log messages. An example of ChatGPT's response is shown in Fig. 8. The detailed report is shown in Table 2.

Table 2. ChatGPT performance to extract urls, IPs, and users from the log messages from Loghub dataset [34].

Log Message Count	URLs Count	URLs captured	URL Accuracy (%)	User Count	User captured	User Accuracy (%)	Response time (s)
5	4	4	100	2	2	100	13.77
10	9	9	100	7	7	100	46.41
20	13	13	100	14	14	100	112.14
50	24	20	83.33	16	14	87.5	163.76

Anomaly Detection
To evaluate ChatGPT's capability to detect anomalies in logs and to address **RQ5**, we use 'Prompt 5' from Fig. 2. As detecting anomalies through log messages would require context, we append 200 log message entries and ask ChatGPT to detect anomalies from it. Without showing any examples to ChatGPT of how an anomaly might look like, it still tries to identify the possible anomalies and provide its analysis in the end. One of the examples is shown in Fig. 9.

Event Prediction
It is interesting to evaluate ChatGPT's performance in predicting future events

Prompt: Summarize the errors and warning from these log messages and identify the root cause.

[Sun Dec 04 04:52:49 2005] [notice] workerEnv.init() ok
/etc/httpd/conf/workers2.properties
[Sun Dec 04 04:52:49 2005] [notice] workerEnv.init() ok
/etc/httpd/conf/workers2.properties
[Sun Dec 04 04:52:52 2005] [error] mod_jk child workerEnv in error state 7

ChatGPT Response:

Date	Log Message	Tag	Root Cause
Sun Dec 04 04:52:49 2005	workerEnv.init() ok /etc/httpd/conf/workers2.properties	Notice	N/A
Sun Dec 04 04:52:52 2005	workerEnv.init() ok /etc/httpd/conf/workers2.properties	Notice	N/A
Sun Dec 04 04:52:52 2005	mod_jk child workerEnv in error state 7	Error	Issue with child workerEnv
Sun Dec 04 04:52:52 2005	mod_jk child workerEnv in error state 6	Error	Issue with child workerEnv
Sun Dec 04 04:53:05 2005	jk2_init() Found child 6750 in scoreboard slot 7	Notice	N/A

Root Cause: There is an issue with the child workerEnv which is causing it to be in an error state (state 7 or state 6).

Fig. 7. ChatGPT response to identify the errors and root cause from set of 5 log messages from Loghub dataset [34].

in log messages. Typically, for future event prediction, a context of past event is required, hence, we append 200 log messages to 'Prompt 6' from Fig. 2 and ask ChatGPT to predict the next 10 messages for simplicity. This experiment addresses the **RQ6**. While ChatGPT predicts the next 10 events in log format, it fails to predict even a single log message correctly when compared with the ground truth. ChatGPT's response is shown in Fig. 10.

4.3   Log Summarization

This experiment is designed to understand if ChatGPT could succinctly summarize logs. We perform this study in two steps. First, To address the **RQ7**, we provide a single log message from each of the sixteen datasets of opensource benchmark [34] to ChatGPT to understand its mechanics. This is useful to understand the log message in natural language. Figure 11 shows one of the log messages from the Android subset of the Loghub dataset [34] and ChatGPT response. It is evident from the response that ChatGPT provides a detailed explanation of the log message. Next, to address the **RQ8**, we provide a set of ten log messages from each of the sixteen subsets of the Loghub dataset [34]

Prompt: Are there any malicious users, urls, ips, and connection status from these log messages.

Jun 14 15:16:01 combo sshd(pam_unix)[19939]: authentication failure;
logname= uid=0 euid=0 tty=NODEVssh ruser= rhost=218.188.2.4
Jun 14 15:16:02 combo sshd(pam_unix)[19937]: check pass; user unknown
Jun 14 15:16:02 combo sshd(pam_unix)[19937]: authentication failure;

ChatGPT Response:

Date	Time	User	IP	Connection Status
Jun 14	15:16:01	Unknown	218.188.2.4	Authentication Failure
Jun 14	15:16:02	Unknown	218.188.2.4	Authentication Failure
Jun 15	02:04:59	root	220.135.151.1	Authentication Failure
Jun 15	02:04:59	root	220.135.151.1	Authentication Failure

Based on the log messages provided, there are no indicators of malicious activities. However, the repeated authentication failures from different IP addresses may suggest potential unauthorized access attempts. It is recommended to monitor and investigate any further unusual activities.

Fig. 8. ChatGPT response to extract urls, IPs, and users from set of 5 log messages from Loghub dataset [100].

to ChatGPT and ask to summarize the logs. ChatGPT generates a concrete summary collectively from the provided log messages as shown in Fig. 12. In Fig. 12, we only show a few log messages for visual clarity. ChatGPT generates an understandable summary for all the sixteen subsets.

5   Discussion

Based on our study, we highlight a few challenges and prospects for ChatGPT on log data analysis.

5.1   Handling Unstructured Log Data

For our experiments, we send the unstructured raw log messages to ChatGPT to analyze its capabilities on various log-specific tasks. Our study indicates that ChatGPT shows promising performance in processing the raw log messages. It is excellent in log parsing and identifying security and privacy information, but encounters difficulty in case of API detection, event prediction, and summarizing. It misses out on several APIs and events from raw log messages.

Prompt: Detect the anomalies from the following log messages.

Jun 14 15:16:01 combo sshd(pam_unix)[19939]: authentication failure;
logname= uid=0 euid=0 tty=NODEVssh ruser= rhost=218.188.2.4
Jun 14 15:16:01 combo sshd(pam_unix)[19939]: authentication failure;
logname= uid=0 euid=0 tty=NODEVssh ruser= rhost=218.188.2.4
Jun 14 15:16:02 combo sshd(pam_unix)[19937]: check pass; user unknown
Jun 14 15:16:02 combo sshd(pam_unix)[19937]: authentication failure;
logname= uid=0 euid=0 tty=NODEVssh ruser= rhost=218.188.2.4
Jun 15 02:04:59 combo sshd(pam_unix)[20882]: authentication failure;

ChatGPT Response:

Anomalies detected in the log messages are as follows:

Jun 15 04:06:20 combo logrotate: ALERT exited abnormally with [1]
This indicates an abnormal exit of logrotate process.

Jun 15 12:12:34 combo sshd(pam_unix)[23397]: authentication failure;
logname= uid=0 euid=0 tty=NODEVssh ruser= rhost=218.188.2.4
Jun 15 12:12:34 combo sshd(pam_unix)[23395]: authentication failure;
logname= uid=0 euid=0 tty=NODEVssh ruser= rhost=218.188.2.4

These log messages indicate multiple authentication failures from the same IP address (rhost). This could signify a security threat or unauthorized access attempts.

Fig. 9. ChatGPT response for anomaly detection for a sample from Loghub dataset [34].

5.2 Performance with Zero-Shot Learning

We perform our experiments with zero-shot learning. Our experimental results show that ChatGPT exhibits good performance in the areas of log parsing, security, and privacy, and average performance in the case of API detection, incident detection, and root cause identification. As ChatGPT supports few-shot learning, it remains an important future work to select important guidelines to set effective examples and evaluate ChatGPT's performance with them.

5.3 Scalability - Message Cap for GPT

Most of the intelligent knowledge extraction from logs depends on processing a large amount of the logs in a short period. As ChatGPT 3.5 can only process limited tokens at once, it poses a major limitation in feeding the bigger chunk of log data. For our experiments, we could only send 190 to 200 log messages appended (addressing **RQ9 and RQ10**) with the appropriate prompt at once. As most of

Prompt: Predict the next 10 log events based on these log messages.

Jun 14 15:16:01 combo sshd(pam_unix)[19939]: authentication failure; logname= uid=0 euid=0 tty=NODEVssh ruser= rhost=218.188.2.4
Jun 14 15:16:02 combo sshd(pam_unix)[19937]: check pass; user unknown
Jun 14 15:16:02 combo sshd(pam_unix)[19937]: authentication failure; logname= uid=0 euid=0 tty=NODEVssh ruser= rhost=218.188.2.4
Jun 15 02:04:59 combo sshd(pam_unix)[20882]: authentication failure; logname= uid=0 euid=0 tty=NODEVssh ruser= rhost=220-135-151-1.hinet-ip.hinet.net user=root

ChatGPT Response:

Based on the given log messages, the next 10 log events could be:
1. Jun 17 04:08:47 combo sshd(pam_unix)[28324]: check pass; user unknown
2. Jun 17 04:08:47 combo sshd(pam_unix)[28324]: authentication failure; logname= uid=0 euid=0 tty=NODEVssh ruser= rhost=218.188.2.4
3. Jun 17 04:08:47 combo sshd(pam_unix)[28322]: check pass; user unknown
4. Jun 17 04:08:47 combo sshd(pam_unix)[28322]: authentication failure; logname= uid=0 euid=0 tty=NODEVssh ruser= rhost=218.188.2.4
5. Jun 17 04:08:47 combo sshd(pam_unix)[28329]: check pass; user unknown
6. Jun 17 04:08:47 combo sshd(pam_unix)[28329]: authentication failure; logname= uid=0 euid=0 tty=NODEVssh ruser= rhost=218.188.2.4
7. Jun 17 04:08:47 combo sshd(pam_unix)[28325]: check pass; user unknown
8. Jun 17 04:08:47 combo sshd(pam_unix)[28325]: authentication failure; logname= uid=0 euid=0 tty=NODEVssh ruser= rhost=218.188.2.4
9. Jun 17 04:08:47 combo sshd(pam_unix)[28328]: check pass; user unknown
10. Jun 17 04:08:47 combo sshd(pam_unix)[28328]: authentication failure; logname= uid=0 euid=0 tty=NODEVssh ruser= rhost=218.188.2.4

Fig. 10. ChatGPT response for event prediction from Loghub dataset [34].

Prompt: Summarize the log message.

PowerManagerService: acquire lock=233570404, flags=0x1, tag="View Lock", name=com.android.systemui, ws=null, uid=10037, pid=2227

ChatGPT Response:

On March 17th at 16:13:38.819, the PowerManagerService acquired a lock with certain flags and attributes. The tag for this lock was "View Lock" and it belonged to the package "com.android.systemui". The lock was associated with a specific process ID (PID) and user ID (UID).

Fig. 11. Summary generated by ChatGPT for single log message from Loghub dataset [34].

Prompt: Summarize the log messages.

03-17 16:13:38.811 1702 2395 D WindowManager: printFreezingDisplayLogsopening app wtoken = AppWindowToken{9f4ef63 token=Token{a64f992 ActivityRecord{de9231d u0 com.tencent.qt.qtl/.activity.info.NewsDetailXmlActivity t761}}}, allDrawn= false, startingDisplayed = false, startingMoved = false, isRelaunching = false
03-17 16:13:38.819 1702 8671 D PowerManagerService: acquire lock=233570404, flags=0x1, tag="View Lock", name=com.android.systemui, ws=null, uid=10037, pid=2227
03-17 16:13:38.820 1702 8671 D PowerManagerService: ready=true,policy=3,wakefulness=1,wksummary=0x23,uasummary=0x1,bootcompleted= true,boostinprogress=false,waitmodeenable=false,mode=false,manual=38,auto=-

ChatGPT Response:

The log message shows various activities and operations happening in the system. These include the opening of an app window token, acquisition of a lock by the PowerManagerService, and manipulation of TextView visibility.

Fig. 12. Collective summary generated by ChatGPT for ten log messages from Loghub dataset [34].

the real-time applications would require to continuously send larger chunks of log messages to a system for processing, this limitation of ChatGPT 3.5 may pose a major hindrance in terms of scalability making them less suitable for tasks that require up-to-date knowledge or rapid adaptation to changing contexts. With the newer versions of ChatGPT, the number of tokens may be increased which would make it more suitable for its application in the log processing area.

5.4   Latency

The response time of ChatGPT ranges from a few seconds to minutes when the number of log messages is increased in the prompt. The details about response time are shown in Table 1 and 2. Most of the intelligent knowledge extraction from logs depends on the processing time of the large amount of the logs. With the current state of response time, ChatGPT would face a major challenge in real-time applications, where a response is required in a shorter period. As currently, we have to call openAI API to get ChatGPT's response, with the newer versions of ChatGPT, it may be possible to deploy these models close to applications and reduce the latency significantly.

5.5   Privacy

Log data often contains sensitive information that requires protection. It is crucial to ensure that log data is stored and processed securely to safeguard sensitive information. It is also important to consider appropriate measures to mitigate any potential risks.

6 Conclusion

This paper presents the first evaluation to give a comprehensive overview of ChatGPT's capability on log data from three major areas: log parsing, log analytics and log summarization. We have designed specific prompts for ChatGPT to reveal its capabilities in the area of log processing. Our evaluations reveal that the current state of ChatGPT exhibits excellent performance in the areas of log parsing, but poses certain limitations in other areas i.e., API detection, anomaly detection, log summarization, etc. We identify several grand challenges and opportunities that future research should address to improve the current capabilities of ChatGPT.

7 Disclaimer

The goal of this paper is mainly to summarize and discuss existing evaluation efforts on ChatGPT along with some limitations. The only intention is to foster a better understanding of the existing framework. Additionally, due to the swift evolution of LLMs especially ChatGPT, they would likely become more robust, and some of their limitations described in this paper are remediated. We encourage interested readers to take this survey as a reference for future research and conduct real experiments in current systems when performing evaluations. Finally, with continuous evaluation of LLMs, we may miss some new papers or benchmarks. We welcome all constructive feedback and suggestions to help make this evaluation better.

References

1. Wang, A., Singh, A., Michael, J., Hill, F., Levy, O., Bowman, S.: GLUE: a multi-task benchmark and analysis platform for natural language understanding. In: Proceedings of the 2018 EMNLP Workshop BlackboxNLP: Analyzing and Interpreting Neural Networks for NLP, pp. 353–355 (2018). https://aclanthology.org/W18-5446
2. Wang, B.: Mesh-Transformer-JAX: model-parallel implementation of transformer language model with JAX (2021)
3. Wang, B., Komatsuzaki, A.: GPT-J-6B: a 6 billion parameter autoregressive language model (2021)
4. Brown, T., et al.: Language models are few-shot learners (2020)
5. Tay, Y., et al.: UL2: unifying language learning paradigms (2023)
6. Thoppilan, R., et al.: LaMDA: language models for dialog applications (2022)
7. Fedus, W., Zoph, B., Shazeer, N.: Switch transformers: scaling to trillion parameter models with simple and efficient sparsity. J. Mach. Learn. Res. **23**(1), 5232–5270 (2022)
8. Hoffmann, J., et al.: Training compute-optimal large language models (2022)
9. Zeng, A., et al.: GLM-130B: an open bilingual pre-trained model (2022)
10. Sun, T., Shao, Y., Qian, H., Huang, X., Qiu, X.: Black-box tuning for language-model-as-a-service (2022)

11. Du, M., Li, F.: Spell: streaming parsing of system event logs. In: 2016 IEEE 16th International Conference on Data Mining (ICDM), pp. 859–864 (2016). https://api.semanticscholar.org/CorpusID:206784678

12. He, P., Zhu, J., Zheng, Z., Lyu, M.: Drain: an online log parsing approach with fixed depth tree. In: 2017 IEEE International Conference on Web Services (ICWS), pp. 33–40 (2017)

13. Zhu, J., et al.: Tools and benchmarks for automated log parsing (2019)

14. Khan, Z., Shin, D., Bianculli, D., Briand, L.: Guidelines for assessing the accuracy of log message template identification techniques. In: Proceedings of the 44th International Conference on Software Engineering, pp. 1095–1106 (2022). https://doi.org/10.1145/3510003.3510101

15. Du, M., Li, F., Zheng, G., Srikumar, V.: DeepLog: anomaly detection and diagnosis from system logs through deep learning. In: Proceedings of the 2017 ACM SIGSAC Conference on Computer and Communications Security, pp. 1285–1298 (2017). https://doi.org/10.1145/3133956.3134015

16. Zhang, X., et al.: Robust log-based anomaly detection on unstable log data. In: Proceedings of the 2019 27th ACM Joint Meeting on European Software Engineering Conference and Symposium on the Foundations of Software Engineering, pp. 807–817 (2019). https://doi.org/10.1145/3338906.3338931

17. Le, V., Zhang, H.: Log-based anomaly detection without log parsing. In: Proceedings of the 36th IEEE/ACM International Conference on Automated Software Engineering, pp. 492–504 (2022). https://doi.org/10.1109/ASE51524.2021.9678773

18. Zhang, B., Zhang, H., Le, V., Moscato, P., Zhang, A.: Semi-supervised and unsupervised anomaly detection by mining numerical workflow relations from system logs. Autom. Softw. Eng. **30** (2022). https://doi.org/10.1007/s10515-022-00370-w

19. Ramachandran, S., Agrahari, R., Mudgal, P., Bhilwaria, H., Long, G., Kumar, A.: Automated log classification using deep learning. Procedia Comput. Sci. **218**, 1722–1732 (2023). International Conference on Machine Learning and Data Engineering. https://www.sciencedirect.com/science/article/pii/S1877050923001503

20. Das, A., Mueller, F., Siegel, C., Vishnu, A.: Desh: deep learning for system health prediction of lead times to failure in HPC. In: Proceedings of the 27th International Symposium on High-Performance Parallel and Distributed Computing, pp. 40–51 (2018). https://doi.org/10.1145/3208040.3208051

21. Russo, B., Succi, G., Pedrycz, W.: Mining system logs to learn error predictors: a case study of a telemetry system. Empirical Softw. Eng. **20**, 879–927 (2015). https://api.semanticscholar.org/CorpusID:19032755

22. Gurumdimma, N., Jhumka, A., Liakata, M., Chuah, E., Browne, J.: CRUDE: combining resource usage data and error logs for accurate error detection in large-scale distributed systems. In: 2016 IEEE 35th Symposium on Reliable Distributed Systems (SRDS), pp. 51–60 (2016)

23. Lu, S., Rao, B., Wei, X., Tak, B., Wang, L., Wang, L.: Log-based abnormal task detection and root cause analysis for spark. In: 2017 IEEE International Conference on Web Services (ICWS), pp. 389–396 (2017)

24. Le, V., Zhang, H.: Log parsing with prompt-based few-shot learning (2023)

25. Tao, S., et al.: LogStamp: automatic online log parsing based on sequence labelling (2022)

26. Liu, Y., et al.: UniParser: a unified log parser for heterogeneous log data. In: Proceedings of the ACM Web Conference 2022 (2022). https://doi.org/10.1145/3485447.3511993

27. Feng, Y., Vanam, S., Cherukupally, M., Zheng, W., Qiu, M., Chen, H.: Investigating code generation performance of ChatGPT with crowdsourcing social data. In: 2023 IEEE 47th Annual Computers, Software, and Applications Conference (COMPSAC), pp. 876–885 (2023)
28. Cao, J., Li, M., Wen, M., Cheung, S.: A study on prompt design, advantages and limitations of ChatGPT for deep learning program repair (2023)
29. Laskar, M., Bari, M., Rahman, M., Bhuiyan, M., Joty, S., Huang, J.: A systematic study and comprehensive evaluation of ChatGPT on benchmark datasets (2023)
30. Cámara, J., Troya, J., Burgueño, L., Vallecillo, A.: On the assessment of generative AI in modeling tasks: an experience report with ChatGPT and UML. Softw. Syst. Model. **22**, 781–793 (2023). https://doi.org/10.1007/s10270-023-01105-5
31. Sridharan, C.: Distributed Systems Observability: A Guide to Building Robust Systems. O'Reilly Media Inc. (2018)
32. He, S., et al.: An empirical study of log analysis at Microsoft. In: Proceedings of the 30th ACM Joint European Software Engineering Conference and Symposium on the Foundations of Software Engineering, pp. 1465–1476 (2022)
33. Sridhara, G., Ranjani, H.G., Mazumdar, S.: ChatGPT: a study on its utility for ubiquitous software engineering tasks (2023)
34. He, S., Zhu, J., He, P., Lyu, M.: Loghub: a large collection of system log datasets towards automated log analytics (2020)
35. Le, V., Zhang, H.: An evaluation of log parsing with ChatGPT (2023)
36. Fu, Q., Lou, J., Wang, Y., Li, J.: Execution anomaly detection in distributed systems through unstructured log analysis. In: 2009 Ninth IEEE International Conference on Data Mining, pp. 149–158 (2009)
37. Tang, L., Li, T., Perng, C.: LogSig: generating system events from raw textual logs. In: Proceedings of the 20th ACM International Conference on Information and Knowledge Management, pp. 785–794 (2011). https://doi.org/10.1145/2063576.2063690
38. Shima, K.: Length matters: clustering system log messages using length of words. arXiv:1611.03213 (2016). https://api.semanticscholar.org/CorpusID:16326353
39. Dai, H., Li, H., Shang, W., Chen, T., Chen, C.: Logram: efficient log parsing using n-gram dictionaries (2020)
40. Nagappan, M., Vouk, M.: Abstracting log lines to log event types for mining software system logs. In: 2010 7th IEEE Working Conference on Mining Software Repositories (MSR 2010), pp. 114–117 (2010)
41. Vaarandi, R.: A data clustering algorithm for mining patterns from event logs. In: Proceedings of the 3rd IEEE Workshop on IP Operations & Management (IPOM 2003) (IEEE Cat. No. 03EX764), pp. 119–126 (2003)
42. Wang, X., et al.: SPINE: a scalable log parser with feedback guidance. In: Proceedings of the 30th ACM Joint European Software Engineering Conference and Symposium on the Foundations of Software Engineering, pp. 1198–1208 (2022). https://doi.org/10.1145/3540250.3549176
43. Zheng, Z., et al.: Co-analysis of RAS log and job log on Blue Gene/P. In: 2011 IEEE International Parallel & Distributed Processing Symposium, pp. 840–851 (2011)
44. Meng, W., et al.: Summarizing unstructured logs in online services (2020)
45. Chuah, E., Jhumka, A., Narasimhamurthy, S., Hammond, J., Browne, J., Barth, B.: Linking resource usage anomalies with system failures from cluster log data. In: 2013 IEEE 32nd International Symposium on Reliable Distributed Systems, pp. 111–120 (2013)

46. Pi, A., Chen, W., Zhou, X., Ji, M.: Profiling distributed systems in lightweight vir-tualized environments with logs and resource metrics. In: Proceedings of the 27th International Symposium on High-Performance Parallel and Distributed Computing, pp. 168–179 (2018)
47. Ren, Z., Liu, C., Xiao, X., Jiang, H., Xie, T.: Root cause localization for unreproducible builds via causality analysis over system call tracing. In: 2019 34th IEEE/ACM International Conference on Automated Software Engineering (ASE), pp. 527–538 (2019)
48. Kimura, T., et al.: Spatio-temporal factorization of log data for understanding network events. In: IEEE INFOCOM 2014-IEEE Conference on Computer Communications, pp. 610–618 (2014)
49. Steinle, M., Aberer, K., Girdzijauskas, S., Lovis, C.: Mapping moving landscapes by mining mountains of logs: novel techniques for dependency model generation. VLDB **6**, 1093–1102 (2006)
50. Lou, J., Fu, Q., Yang, S., Xu, Y., Li, J.: Mining invariants from console logs for system problem detection. In: 2010 USENIX Annual Technical Conference (USENIX ATC 2010) (2010)
51. Kc, K., Gu, X.: ELT: efficient log-based troubleshooting system for cloud computing infrastructures. In: 2011 IEEE 30th International Symposium on Reliable Distributed Systems, pp. 11–20 (2011)
52. Awad, M., Menascé, D.: Performance model derivation of operational systems through log analysis. In: 2016 IEEE 24th International Symposium on Modeling, Analysis and Simulation of Computer and Telecommunication Systems (MAS-COTS), pp. 159–168 (2016)
53. Tan, J., Kavulya, S., Gandhi, R., Narasimhan, P.: Visual, log-based causal tracing for performance debugging of MapReduce systems. In: 2010 IEEE 30th International Conference on Distributed Computing Systems, pp. 795–806 (2010)
54. Beschastnikh, I., Brun, Y., Ernst, M., Krishnamurthy, A.: Inferring models of concurrent systems from logs of their behavior with CSight. In: Proceedings of the 36th International Conference on Software Engineering, pp. 468–479 (2014)
55. Mariani, L., Pastore, F.: Automated identification of failure causes in system logs. In: 2008 19th International Symposium on Software Reliability Engineering (ISSRE), pp. 117–126 (2008)
56. Ulrich, A., Hallal, H., Petrenko, A., Boroday, S.: Verifying trustworthiness requirements in distributed systems with formal log-file analysis. In: 2003 Proceedings of the 36th Annual Hawaii International Conference on System Sciences, p. 10 (2003)
57. Oprea, A., Li, Z., Yen, T., Chin, S., Alrwais, S.: Detection of early-stage enterprise infection by mining large-scale log data. In: 2015 45th Annual IEEE/IFIP International Conference on Dependable Systems and Networks, pp. 45–56 (2015)
58. Balzarotti, D., Stolfo, S., Cova, M.: Research in Attacks, Intrusions and Defenses. Springer, Heidelberg (2012). https://doi.org/10.1007/978-3-642-33338-5
59. Yoon, E., Squicciarini, A.: Toward detecting compromised MapReduce workers through log analysis. In: 2014 14th IEEE/ACM International Symposium on Cluster, Cloud and Grid Computing, pp. 41–50 (2014)
60. Yen, T., et al.: Beehive: large-scale log analysis for detecting suspicious activity in enterprise networks. In: Proceedings of the 29th Annual Computer Security Applications Conference, pp. 199–208 (2013)
61. Gonçalves, D., Bota, J., Correia, M.: Big data analytics for detecting host misbehavior in large logs. In: 2015 IEEE Trustcom/BigDataSE/ISPA, vol. 1, pp. 238–245 (2015)

62. Du, M., Li, F., Zheng, G., Srikumar, V.: DeepLog: anomaly detection and diagnosis from system logs through deep learning. In: Proceedings of the 2017 ACM SIGSAC Conference on Computer and Communications Security, pp. 1285–1298 (2017)
63. Bertero, C., Roy, M., Sauvanaud, C., Trédan, G.: Experience report: log mining using natural language processing and application to anomaly detection. In: 2017 IEEE 28th International Symposium on Software Reliability Engineering (ISSRE), pp. 351–360 (2017)
64. Meng, W., et al.: LogAnomaly: unsupervised detection of sequential and quantitative anomalies in unstructured logs. In: IJCAI 2019, pp. 4739–4745 (2019)
65. Zhang, X., et al.: Robust log-based anomaly detection on unstable log data. In: Proceedings of the 2019 27th ACM Joint Meeting on European Software Engineering Conference and Symposium on the Foundations of Software Engineering, pp. 807–817 (2019)
66. He, S., Zhu, J., He, P., Lyu, M.: Experience report: system log analysis for anomaly detection. In: 2016 IEEE 27th International Symposium on Software Reliability Engineering (ISSRE), pp. 207–218 (2016)
67. Cândido, J., Aniche, M., Deursen, A.: Log-based software monitoring. PeerJ Comput. Sci. **7**, e489 (2021)
68. Tang, D., Iyer, R.: Analysis of the VAX/VMS error logs in multicomputer environments-a case study of software dependability. In: Proceedings Third International Symposium on Software Reliability Engineering, pp. 216–217 (1992)
69. Lim, C., Singh, N., Yajnik, S.: A log mining approach to failure analysis of enterprise telephony systems. In: 2008 IEEE International Conference on Dependable Systems and Networks with FTCS and DCC (DSN), pp. 398–403 (2008)
70. Xu, W., Huang, L., Fox, A., Patterson, D., Jordan, M.: Detecting large-scale system problems by mining console logs. In: Proceedings of the ACM SIGOPS 22nd Symposium on Operating Systems Principles, pp. 117–132 (2009)
71. Xu, W., Huang, L., Fox, A., Patterson, D., Jordan, M.: Online system problem detection by mining patterns of console logs. In: 2009 Ninth IEEE International Conference on Data Mining, pp. 588–597 (2009)
72. Gandomi, A., Haider, M.: Beyond the hype: big data concepts, methods, and analytics. Int. J. Inf. Manag. **35**, 137–144 (2015). https://www.sciencedirect.com/science/article/pii/S0268401214001066
73. Chang, Y., et al.: A survey on evaluation of large language models (2023)
74. Chang, M., Ratinov, L., Roth, D., Srikumar, V.: Importance of semantic representation: dataless classification. In: Proceedings of the 23rd National Conference on Artificial Intelligence, vol. 2, pp. 830–835 (2008)
75. Larochelle, H., Erhan, D., Bengio, Y.: Zero-data learning of new tasks. In: Proceedings of the 23rd National Conference on Artificial Intelligence, vol. 2, pp. 646–651 (2008)
76. Palatucci, M., Pomerleau, D., Hinton, G., Mitchell, T.: Zero-shot learning with semantic output codes. In: Proceedings of the 22nd International Conference on Neural Information Processing Systems, pp. 1410–1418 (2009)
77. Lampert, C., Nickisch, H., Harmeling, S.: Learning to detect unseen object classes by between-class attribute transfer. In: 2009 IEEE Conference on Computer Vision and Pattern Recognition, pp. 951–958 (2009). https://api.semanticscholar.org/CorpusID:10301835
78. Miller, E., Matsakis, N., Viola, P.: Learning from one example through shared densities on transforms. In: Proceedings of the IEEE Conference on Computer

Vision and Pattern Recognition, CVPR 2000 (Cat. No. PR00662), vol. 1, pp. 464–471 (2000)

79. Scao, T., et al.: BLOOM: a 176B-parameter open-access multilingual language model (2023)
80. Fu, X., Ren, R., Zhan, J., Zhou, W., Jia, Z., Lu, G.: LogMaster: mining event correlations in logs of large-scale cluster systems. In: 2012 IEEE 31st Symposium on Reliable Distributed Systems, pp. 71–80 (2012)
81. Tama, B., Comuzzi, M.: An empirical comparison of classification techniques for next event prediction using business process event logs. Expert Syst. Appl. **129**, 233–245 (2019). https://www.sciencedirect.com/science/article/pii/S0957417419302465
82. Shmueli, G., Koppius, O.: Predictive analytics in information systems research. MIS Q. **35**, 553–572 (2011). http://www.jstor.org/stable/23042796
83. van der Aalst, W.M.P., Pesic, M., Song, M.: Beyond process mining: from the past to present and future. In: Pernici, B. (ed.) CAiSE 2010. LNCS, vol. 6051, pp. 38–52. Springer, Heidelberg (2010). https://doi.org/10.1007/978-3-642-13094-6_5
84. Leiter, C., et al.: ChatGPT: a meta-analysis after 2.5 months (2023)
85. Pettinato, M., Gil, J., Galeas, P., Russo, B.: Log mining to re-construct system behavior: an exploratory study on a large telescope system. Inf. Softw. Technol. **114**, 121–136 (2019). https://www.sciencedirect.com/science/article/pii/S0950584919301429
86. Mao, R., Chen, G., Zhang, X., Guerin, F., Cambria, E.: GPTEval: a survey on assessments of ChatGPT and GPT-4 (2023)
87. Li, B., et al.: Evaluating ChatGPT's information extraction capabilities: an assessment of performance, explainability, calibration, and faithfulness (2023)
88. Maggi, F.M., Di Francescomarino, C., Dumas, M., Ghidini, C.: Predictive monitoring of business processes. In: Matthias, J., et al. (eds.) CAiSE 2014. LNCS, vol. 8484, pp. 457–472. Springer, Cham (2014). https://doi.org/10.1007/978-3-319-07881-6_31
89. Francescomarino, C., Ghidini, C., Maggi, F., Petrucci, G., Yeshchenko, A.: An eye into the future: leveraging a-priori knowledge in predictive business process monitoring. In: International Conference on Business Process Management (2017). https://api.semanticscholar.org/CorpusID:206703657
90. Verenich, I., Dumas, M., Rosa, M., Maggi, F., Teinemaa, I.: Survey and cross-benchmark comparison of remaining time prediction methods in business process monitoring. ACM Trans. Intell. Syst. Technol. **10**(4), 1–34 (2019)
91. Evermann, J., Rehse, J., Fettke, P.: Predicting process behaviour using deep learning. Decis. Support Syst. **100**, 129–140 (2017)
92. Tax, N., Verenich, I., La Rosa, M., Dumas, M.: Predictive business process monitoring with LSTM neural networks. In: Dubois, E., Pohl, K. (eds.) CAiSE 2017. LNCS, vol. 10253, pp. 477–492. Springer, Cham (2017). https://doi.org/10.1007/978-3-319-59536-8_30
93. ChatGPT: Optimizing language models for dialogue. https://openai.com/blog/chatgpt
94. ChatGPT: UI. https://chat.openai.com/
95. ChatGPT: ChatGPT 3.5. https://platform.openai.com/docs/models/gpt-3-5
96. LogAI: A Library for Log Analytics and Intelligence. https://blog.salesforceairesearch.com/logai/
97. LogAI: A Library for Log Analytics and Intelligence. https://www.gartner.com/smarterwithgartner/how-to-get-started-with-aiops

98. How to Get Started with AIOps. https://www.zebrium.com/blog/part-1-machine-learning-for-logs
99. Observability. https://en.wikipedia.org/wiki/Observability
100. Zhu, J., He, S., He, P., Liu, J., Lyu, M.: Loghub: a large collection of system log datasets for ai-driven log analytics. In: IEEE International Symposium on Software Reliability Engineering (ISSRE) (2023)

DSQA-LLM: Domain-Specific Intelligent Question Answering Based on Large Language Model

Dengrong Huang[1], Zizhong Wei[1], Aizhen Yue[1], Xuan Zhao[3], Zhaoliang Chen[2],
Rui Li[1], Kai Jiang[1], Bingxin Chang[1], Qilai Zhang[1], Sijia Zhang[1],
and Zheng Zhang[1(✉)]

[1] Inspur Academy of Science and Technology, Jinan, Shandong, China
jiangkai@inspur.com
[2] Inspur Software Co., Ltd., Jinan, Shandong, China
[3] China International Information Center, Beijing, China

Abstract. Question Answering (QA) is crucial for humans to access vast knowledge bases, but there is a lack of attention towards representing raw, unstructured questions and answers in specific fields. Additionally, the efficiency of finding candidate questions based on the trigger question and the generation of reasonable answers have been neglected. In this paper, we introduce Domain Specific Question Answering Language Model (DSQA-LLM), a framework that delivers informative answers within a specific domain. We utilize techniques like question classification, information retrieval, and answer generation. We enhance efficiency and accuracy through the integration of XLNET for question classification and a novel similarity searching method using Sentence-T5. Furthermore, the powerful GPT-3.5-turbo is employed for generating coherent answers. We implemented DSQA-LLM and curated a dataset of 127,840 question-answer pairs. Empirical experiments conducted on real-world questions confirm the effectiveness of our QA system.

Keywords: Question Answering · LLM · XLNET · Sententce-T5 ·
deep learning · natural language processing

1 Introduction

With the expanding volume of domain-specific knowledge bases (KBs), there is a growing interest in accessing these valuable resources effectively. Domain-specific knowledge base-based question answering (DSKB-QA) has gained prominence as a user-friendly solution, utilizing natural language as the query language. The objective of DSKB-QA is to automatically retrieve accurate results and aggregate them based on relevance to user queries. This paper focuses on DSKB-QA in the digital government domain, where each data sample consists of a 4-tuple

---

D. Huang, Z. Wei and A. Yue—Equal contribution.

F. Zhao and D. Miao (Eds.): AIGC 2023, CCIS 1946, pp. 170–180, 2024.
https://doi.org/10.1007/978-981-99-7587-7_14

(type, question, document, and answer). Specifically, our paper addresses the domain-specific extractive QA task, extracting answers from contextual information based on given questions as input.

Question classification is essential for QA systems in NLP, as it assigns labels to questions and narrows down the search range in large datasets. This helps accurately locate and verify answers. Transformer-based models like XLNet have gained attention for their ability to learn global semantic representation and handle large-scale datasets without relying on sequential information. They have significantly improved NLP tasks, including text classification. In this paper, we fine-tune XLNet using our question-type dataset to enhance question classification accuracy.

Similarity query processing is essential in domains like databases and machine learning. Deep learning techniques, including embedding and pre-trained models, have significantly improved similarity query processing for high-dimensional data. Question embedding plays a crucial role in retrieving similar questions, and a recent approach involves fine-tuning large language models like Sentence-T5 for a candidate question retriever. In our embedding module, we also leverage Sentence-T5 to enhance the precision and effectiveness of question search in our government-related dataset. To achieve efficient similarity search of dense vectors, we utilize cosine distance specifically designed for similarity question search. By calculating similarity, we retrieve the most similar queries and obtain candidate document-level answers accordingly.

Document summarization is essential for condensing text while preserving important information. With the abundance of public text data, automatic summarization techniques are becoming increasingly important. Large language models like GPT-3 possess strong natural language understanding and generation capabilities. Comparisons with traditional fine-tuning methods show that GPT-3 exhibits excellent memory and semantic understanding. Furthermore, analysis confirms that these large language models generate answer summaries that are comparable to those produced by human experts. To improve the conciseness, readability, and logical consistency of answers derived from original documents, we have integrated these models into our question answering system.

In summary, this paper presents the following contributions:

1) Introduction of DSQA-LLM, a domain-specific question answering system designed to provide relevant and concise answers to trigger questions within a specific domain.
2) Proposal of a novel technique that combines text classification, sentence embedding, and answer generation, utilizing both traditional fine-tuning models and large language models (LLMs) to enhance accuracy and reasonableness.
3) Extensive experiments on domain-specific question answering datasets to demonstrate the effectiveness of our approach.

The subsequent chapters are structured as follows: Sect. 2 provides an overview of related work on DSQA-LLM. Section 3 presents our proposed approach and implementation details. Section 4 outlines the experiments conducted and presents the results. Finally, Subsect. 5 concludes our work.

2   Related Work

2.1   Question Classification

Question classification techniques can be categorized into Statistics-based, NN-based, Attention-based, and Transformer-based methods. Statistics-based techniques, such as Naive BayesSupport Vector Machine [1], offer accuracy and stability. Recent methods like XGBoost [2] show promise in this area. NN-based techniques, such as TextCNN [3], employ neural networks for text classification. Attention-based techniques like HAN [5] have achieved success in text classification by leveraging informative components and addressing imbalances in few-shot scenarios Transformer-based models, like ALBERT [6], and BART [7], excel at handling large-scale datasets and capturing bidirectional context, demonstrating excellent performance in text classification tasks.

2.2   Question Embedding

Deep learning techniques, such as XLNet [15], RoBERTa [17], SimCSE [8], and Sentence-T5 [9], have been effective in modeling sentence similarity. These Transformer-based models have achieved impressive performance in tasks like question answering. The Transformer model, introduced by Vaswani et al. [11], is successful in sequence-to-sequence tasks. Cer et al. [12] and Radford et al. [13] employed Transformer encoder and decoder for transfer learning and language modeling. BERT [14], with contextualized representations, is a notable advancement. XLNet improves upon BERT through the Transformer-XL architecture [16]. SimCSE and Sentence-T5 stand out by introducing the contrastive loss. Among these models, Sentence-T5 demonstrates innovative design choices and training strategies, outperforming SimCSE with superior performance.

2.3   Answer Extraction and Generation

Advancements in deep neural networks have accelerated extractive summarization. Sequential neural models like recurrent neural networks [24] and pre-trained language models are widely used [18]. However, limited exploration has been done with large language models, such as ChatGPT. Studies have explored the application of large language models in text summarization. Goyal et al. [19] compared GPT-3-generated summaries with traditional methods. Zhang et al. [20] also examined ChatGPT's performance in extractive summarization and proposed an extract-then-generate pipeline. LLMs have also been used for summarization evaluation, outperforming previous methodologies.

3   Approach

3.1   Overview

Our Question Answer system, DSQA-LLM, consists of three phases: "Question Processing", "Similarity Searching", and "Answer Processing". When given a

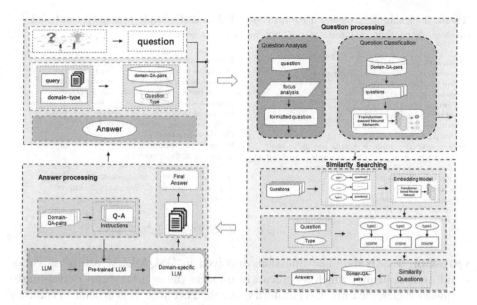

Fig. 1. Overview of our approach

question, DSQA-LLM follows these steps. Firstly, the question analyzer generates the formatted question, Qf, based on specific rules. The Question Classification Module identifies the question type T using a classification model trained on labeled questions from the government-related domain. Qf and T are then passed to the second phase. In the second phase, candidate answers are obtained using an embedding module. Similarity questions (QQP) and question-answer pairs (QAP) from the government-related domain are transformed into formatted representations using a question analyzer. The Sentence-T5 model is fine-tuned on the QQP dataset to obtain embeddings for targeted questions and questions in QAP. By matching the candidate embeddings corresponding to question type T, similarity questions are obtained, and candidate answers from the knowledge dataset QAP are acquired. A list of documents serving as candidate answers is then passed to the third phase. In the third phase, the final answer is derived using an LLM model. DSQA-LLM fine-tunes the LLM model with domain-specific prompts to address miscellaneous and redundant answers obtained in the previous phase. With the candidate answers and fine-tuned LLM model, a reasonable answer is generated, which is validated for accuracy before being presented to the user. Keep in mind that this is an overview, and certain details are omitted due to page limitations (Fig. 1).

3.2 Question Processing Phase

We introduce the popular classification model XLNet into our question process phase, and design the question process as follows.

Table 1. Example few-shot Prompts for answer extraction and generation

Prompt1 (few-shot):
Please extract Summary information for the following Content:
Content: {content1}
Summary: {summary1}
Content: {content}
Summary: {?}
Prompt2 (few-shot):
Please extract summary information from the following content, using the examples provided as guidance, note that not to make up irrelevant information yourself:
Content: {content1}
Summary: {summary1}
Content: {content}
Summary: {?}
......

Table 2. Example prompt for optimal answer selection

Please evaluate the relevance of Summary 1, Summary 2, and Summary 3 in relation to the corresponding Text. A fully relevant summary should include information that is important to the content and should not include other irrelevant information. Afterward, select one of the following options (A, B, C):
Content: {content1}
Summary 1: {summary1}
Summary 2: {summary2}
Summary 3: {summary3}
A: Summary 1 is more relevant. B: Summary 2 is more relevant. C: Summary 3 is more relevant.

Your choice (enter A, B, or C): ?

Question Analysis. The questions involved in DSQA-LLM are often colloquial and confusing, which significantly impacts the accuracy and performance of question classification and similarity search. Therefore, it is crucial to conduct specific pre-processing to generate formatted questions for subsequent use. We employ pattern matching rules to identify the main focus of the questions and remove any unnecessary information.

Question Type Classification. To understand the domain-specific information sought by the question and establish constraints on relevant data, DSQA-LLM uses the XLNet model for question type classification. The model incorporates the segment recurrence mechanism and relative encoding scheme of Transformer-XL, providing improved performance for longer text sequences. The final hidden state of [CLS] in XLNet is used as the representation for the entire sequence, and a softmax classifier predicts the probability of the label [15]. The parameters of XLNet are fine-tuned by maximizing the log-probability of the correct label. The loss function for the classification task is defined as follows,

$$\mathcal{L}_\theta = \max_\theta \mathbb{E}_{s \sim S_t} \left[\sum_{t=0}^{T} \log p_\theta(X_{s_t} | X_{s<t}) \right] \quad (1)$$

where X_t and $X_{s<t}$ represent the t_{th} element and the first $t-1$ elements of a permutation X.

3.3 Similarity Searching Phase

As the reformulated question is submitted to the similarity searching phase, which retrieves a ranked list of relevant candidate answers for the third phase. Our Similarity Searching process consists of two modules: question embedding module and similarity searching module.

Question Embedding Module. To improve the uniformity of sentence embeddings for similarity searching, DSQA-LLM utilizes contrastive learning. This approach, known for its effectiveness in tasks like Semantic Textual Similarity (STS), involves fine-tuning Sentence-T5 representations using a contrastive loss function [8]. During training, positive examples (related sentences) are encouraged to be closer to the input sentence, while all other examples in the batch are treated as negatives. The contrastive loss is computed using in-batch sampled softmax and the similarity score calculated by the function f [26]. It utilizes paired examples (s_i, s_i^+) where s_i represents the input sentence and s_i^+ is a related sentence, along with additional negative examples in the form of s_j^-. The loss function can be described as follows:

$$\mathcal{L} = -log\frac{exp(f(s_i, s_i^+)/\tau)}{\sum_{j \in D} exp(f(s_i, s_j^+)/\tau) + exp(f(s_i, s_j^-)/\tau)}, \qquad (2)$$

Searching Module. Efficiently searching for similar questions in our knowledge base is crucial to minimize search costs in our system. To achieve this, we utilize cosine distance for type-specific similarity search, enabling the construction of a type-dominated search module. When a question is inputted, our system identifies its type using the "Question Type Classification" module. We then retrieve the corresponding embeddings using the embedding module and obtain candidate questions based on the targeted question type. This approach allows for the efficient retrieval of related questions. Furthermore, leveraging the labeled question-answer pairs in our DSQA-LLM dataset, we employ a simple matching technique to obtain candidate answers.

Answer Extraction and Generation. Abstractive summarization involves generating a summary, referred to as Y, by considering the input source document, represented as X. The source document is composed of individual sentences, creating a representation as $X = \{X_1, X_2, ..., X_T\}$. To generate the summary $Y = \{Y_1, Y_2, ..., Y_T\}$, a generative language model utilizes the following probability [26]:

$$p(Y|X, \theta) = \prod_t^m p(Y_t|Y_{<t}, X, \theta). \qquad (3)$$

Large language models have demonstrated impressive task performance, even with limited training data, thanks to in-context learning (ICL). In the standard ICL approach [26], a language model M is trained on a set of example input-output pairs, represented as $\{(x_1, y_1), (x_2, y_2), ..., (x_m, y_m)\}$, where x is the input text and y is the expected output. The objective is to predict the answer text $\hat{y}$ given a query text using this training. This prediction is achieved by calculating the likelihood of each candidate answer y_j using a scoring function f that incorporates the entire input sequence and the language model M.

$$\hat{y} = \underset{y_j \in Y}{argmax} \sum_j f_M(y_j, C, x). \qquad (4)$$

Table 3. Overview Of Metrics.Qes Clas denotes classification evaluation metrics; Sim Sea denotes similarity searching evaluation metrics; Ans Gene denotes answer generation evaluation metrics

#Qes Clas		#Sim Sea		#Ans Gene
$Precision = \frac{v_j}{\sum_{i=1}^{T} v_i}$	(5)	$MRR = \frac{1}{\|Q\|} \sum_{i=1}^{\|Q\|} \frac{1}{rank_i}$	(8)	ROUGE-1
$Recall = \frac{v_i}{\sum_{j=1}^{T} v_j}$	(6)	$MAP = \frac{1}{\|Q\|} \sum_{i=1}^{\|Q\|} AveP(C_i, A_i)$	(9)	ROUGE-2
$F1 - Score = \frac{2 \times Precision \times Recall}{Precision + Recall}$	(7)	$AveP(C_i, A_i) = \frac{\sum_{k=1}^{n}(P(k) \cdot rel(k))}{min(m,n)}$	(10)	ROUGE-L

The set $C = \{I, s(x_1, y_1)...s(x_m, y_m)\}$ represents the collection of explanations and input-output pairs used as prompts in this formulation. Additionally, this research explores how in-context learning affects extractive summarization. In our DSQA-LLM system, we generate summaries using prompt-based approaches, including few-shot prompts following OpenAI API guidelines. In Table 1, we employ various prompts with the LLM to generate summaries. These prompts, such as "prompt1", "prompt2", and others listed, were accompanied by content and summary examples. By inputting the desired content into the LLM, we obtained three summaries: "summary1", "summary2", and "summary3". To evaluate their quality and determine the optimal summary, we utilized an evaluation LLM and collected feedback using the prompt specified in Table 2.

4 Experiment

4.1 Datasets

We gather a training dataset consisting of 67,840 categorized questions from various government fields for training the question classification model. An additional 14,650 questions (NT1) are utilized for testing purposes. To assess the accuracy of our approach in identifying similar questions, we compile a set of 83,790 labeled true similarity pairs and 32,174 false pairs (NT2). Additionally, we curate a collection of 127,840 question-answer pairs (NT3) to evaluate answer extraction capabilities. Further information regarding the distribution of related pairs in the training and test datasets can be found in Table 4.

4.2 Metrics and Baseline

Metrics. We evaluate our Question Type Classification with precision, recall, and F1-Score. Similarity Searching is assessed using MRR and MAP. Answer processing is evaluated using ROUGE [25] metrics, specifically ROUGE-1, ROUGE-2 and ROUGE-L. These metrics can be found in Table 3 (Table 5).

Table 4. Overview Of dataset

Dataset	#NT1	#NT2	#NT3
Training	67840	83790	127840
Test	14650	32174	/

Table 5. Result of question classification.

Metrics	Presicion (%)	Recall (%)	F1-score (%)
BLSTM-2DCNN	86.09	86.39	86.24
HAN	88.89	86.71	87.79
ALBERT	97.24	98.81	98.20
BART	97.98	98.60	98.29
RoBERTa	98.67	98.81	98.74
XLNet	**99.24**	99.06	**99.15**

Baseline. To demonstrate the effectiveness of the proposed QA system, VDQA-LLM, we compare our methods with the corresponding state-of-art efforts.

- **Question Classification.** We compare our approach, which utilizes the XLNet model, to other 5 typical text classification models, i.e. BLSTM-2DCNN [4], HAN [5], ALBERT [6], BART [7] and RoBERTa [17].
- **Similarity Searching.** We compare our Sentence-T5 based approach to other 5 typical sentence embedding models, i.e. BERT [14], SBERT [10], SRoBERTa [10], SimCSE-BERT [8], and SimCSE-RoBERTa [8].
- **Answer Extraction and Generation.** We have conducted a thorough investigation of several state-of-the-art summary generation models. These models include Seq2seq [23], Seq2seq + Att [24], BERT [14], RoBERTa [17], LLAMA [22], GLM [21], and GPT-3.5-turbo. To explore various learning strategies for Language Learning Models (LLMs), we employ few-shot learning approaches.

4.3 Results

Question Classification. Our XLNet-based approach achieves outstanding performance in question classification, boasting an impressive F1-Score of 99.15%. This remarkable accomplishment can be attributed to the XLNet-based approach's ability to overcome the limitations of previous autoregressive models by incorporating bidirectionality. Transformer-based methods consistently outperform traditional neural network-based methods due to their multi-head attention layer and self-attention module. Notably, ALBERT, BART, RoBERTa, and XLNet outshine other methods, achieving an F1-Score of over 98%. This highlights the effectiveness of BERT in text classification tasks.

Similarity Searching. The sentence embedding model's evaluation metrics include Precision, Recall, and F1-Score. Among BERT-based methods, Sentence-T5, SimCSE-BERT, and SimCSE-RoBERTa achieve impressive F1-Scores of 99.43%, 98.57%, and 97.98% respectively. The utilization of contrastive loss by these models enhances feature extraction and contributes to their superior performance. Notably, Sentence-T5 outperforms both SimCSE-BERT and

SimCSE-RoBERTa, highlighting T5's advantages in extracting sentence features. When it comes to similarity search, our Sentence-T5-based model surpasses other baselines in Mean Average Precision (MAP) and Mean Reciprocal Rank (MRR). Compared to BERT, SBERT, SRoBERTa, SimCSE-BERT, and SimCSE-RoBERTa, our model achieves significant MAP improvements of 0.097, 0.070, 0.036, 0.042, and 0.015 respectively. Similarly, the MRR improvements are 0.097, 0.086, 0.060, 0.031, and 0.022 respectively (Table 6).

Table 6. Result of question searching.

Metrics	MAP	MRR
BERT	0.870	0.892
SBERT	0.897	0.903
SRoBERTa	0.904	0.929
SimCSE-BERT	0.925	0.958
SimCSE-RoBERTa	0.952	0.971
Sentence-T5	**0.967**	**0.989**

Table 7. Result of answer generation

Metrics	ROUGE-1	ROUGE-2	ROUGE-L
Seq2seq	17.9	7.8	13.7
Seq2seq + Att	11.2	13.2	20.5
BERT	31.0	16.5	22.8
RoBERTa	38.6	17.4	25.4
Vicuna-7b	39.2	19.4	27.5
GLM-6b	40.3	20.1	28.9
GPT-3.5-turbo	**43.2**	**21.6**	**33.5**

Answer Extraction and Generation. As presented in Table 7, LLMs demonstrate superior extractive capabilities in summarization compared to traditional methods, with BERT-based models outperforming seq2seq models in terms of ROUGE scores. This is due to their extensive training on large amounts of textual data and the advantages of transformer architectures. GPT-3.5-turbo outperforms Vicuna-7b and GLM-6b in few-shot learning scenarios, indicating its strong performance in extractive summarization. Despite being designed as a generation model, GPT-3.5-turbo exhibits deep understanding of problem formulation and semantic meaning. Its decoder-only structure sets it apart from encoder-decoder models like BERT. Fine-tuning also improves the performance of GLM-6b and Vicuna-7b in few-shot learning scenarios.

5   Conclusion

In this paper, we propose DSQA-LLM, a novel technique for QA in the vertical government-related domain. DSQA-LLM combines LLM and FAISS to improve precision in searching and obtaining accurate answers. Our framework integrates popular NLP techniques such as XLNet, Sentence-T5, and GPT-3.5-turbo to further enhance the precision and optimize searching time. Through the implementation of our prototype framework, DSQA-LLM, and extensive empirical experiments, we validate the effectiveness of our approach. The results demonstrate precise and efficient Question Processing, Similarity Searching, and Answer Generation.

References

1. Joachims, T.: Text categorization with support vector machines: learning with many relevant features. In: Nédellec, C., Rouveirol, C. (eds.) ECML 1998. LNCS, vol. 1398, pp. 137–142. Springer, Heidelberg (1998). https://doi.org/10.1007/BFb0026683
2. Chen, T., Guestrin, C.: XGBoost: a scalable tree boosting system. In: Proceedings of the 22nd ACM SIGKDD International Conference on Knowledge Discovery and Data Mining, pp. 785–794 (2016)
3. Kim, Y.: Convolutional neural networks for sentence classification. arXiv preprint arXiv:1408.5882 (2014)
4. Zhou, P., Qi, Z., Zheng, S., et al.: Text classification improved by integrating bidirectional LSTM with two-dimensional max pooling. arXiv preprint arXiv:1611.06639 (2016)
5. Yang, Z., et al.: Hierarchical attention networks for document classification. In: Proceedings of the 2016 Conference of the North American Chapter of the Association for Computational Linguistics: Human Language Technologies (2016)
6. Lan, Z., Chen, M., Goodman, S., et al.: ALBERT: a lite BERT for self-supervised learning of language representations. arXiv preprint arXiv:1909.11942 (2019)
7. Lewis, M., Liu, Y., Goyal, N., et al.: BART: denoising sequence-to-sequence pre-training for natural language generation, translation, and comprehension. arXiv preprint arXiv:1910.13461 (2019)
8. Gao, T., Yao, X., Chen, D.: SimCSE: simple contrastive learning of sentence embeddings. arXiv preprint arXiv:2104.08821 (2021)
9. Ni, J., Ábrego, G.H., Constant, N., et al.: Sentence-T5: scalable sentence encoders from pre-trained text-to-text models. arXiv preprint arXiv:2108.08877 (2021)
10. Reimers, N., Gurevych, I.: Sentence-BERT: sentence embeddings using Siamese BERT-networks. arXiv preprint arXiv:1908.10084 (2019)
11. Vaswani, A., Shazeer, N., Parmar, N., et al.: Attention is all you need. In: Advances in Neural Information Processing Systems, vol. 30 (2017)
12. Cer, D., Yang, Y., Kong, S., et al.: Universal sentence encoder. arXiv preprint arXiv:1803.11175 (2018)
13. Radford, A., Narasimhan, K., Salimans, T., et al.: Improving language understanding by generative pre-training (2018)
14. Devlin, J., Chang, M.W., Lee, K., et al.: BERT: pre-training of deep bidirectional transformers for language understanding. arXiv preprint arXiv:1810.04805 (2018)
15. Yang, Z., Dai, Z., et al.: XLNet: generalized autoregressive pretraining for language understanding. In: Advances in Neural Information Processing Systems, vol. 32 (2019)
16. Dai, Z., Yang, Z., Yang, Y., et al.: Transformer-XL: attentive language models beyond a fixed-length context. arXiv preprint arXiv:1901.02860 (2019)
17. Liu, Y., Ott, M., Goyal, N., et al.: RoBERTa: a robustly optimized BERT pre-training approach. arXiv preprint arXiv:1907.11692 (2019)
18. Liu, Y., Lapata, M.: Text summarization with pretrained encoders. arXiv preprint arXiv:1908.08345 (2019)
19. Goyal, T., Li, J.J., Durrett, G.: News summarization and evaluation in the era of GPT-3. arXiv preprint arXiv:2209.12356 (2022)
20. Luo, Z., Xie, Q., Ananiadou, S.: ChatGPT as a factual inconsistency evaluator for abstractive text summarization. arXiv preprint arXiv:2303.15621 (2023)

21. Du, Z., Qian, Y., Liu, X., et al.: GLM: general language model pretraining with autoregressive blank infilling. arXiv preprint arXiv:2103.10360 (2021)
22. Touvron, H., Lavril, T., Izacard, G., et al.: LLaMA: open and efficient foundation language models. arXiv preprint arXiv:2302.13971 (2023)
23. Cho, K., Van Merriënboer, B., Gulcehre, C., et al.: Learning phrase representations using RNN encoder-decoder for statistical machine translation. arXiv preprint arXiv:1406.1078 (2014)
24. Nallapati, R., Zhou, B., Gulcehre, C., et al.: Abstractive text summarization using sequence-to-sequence RNNs and beyond. arXiv preprint arXiv:1602.06023 (2016)
25. Lin, C.Y.: ROUGE: a package for automatic evaluation of summaries. In: Text summarization Branches Out, pp. 74–81 (2004)
26. Zhang, H., Liu, X., Zhang, J.: Extractive summarization via chatGPT for faithful summary generation. arXiv preprint arXiv:2304.04193 (2023)

Research on the Construction of Intelligent Customs Clearance Information System for Cross-Border Road Cargo Between Guangdong and Hong Kong

Chengguo Han[1(⊠)], Bin Wang[2], and Xinyu Lai[2]

[1] Guangdong University of Science and Technology, 99 Xihu Road, Nancheng District, Dongguan City, Guangdong Province, China
1819777573@qq.com

[2] Shenzhen Taizhou Technology Co., Ltd., 4028 Jintian Road, Futian District, Shenzhen City, Guangdong Province, China

Abstract. In order to achieve high-efficiency "single entry, dual declaration" customs clearance between Guangdong (GD) and Hong Kong (HK), this paper conducts a comprehensive investigation and analysis of the declaration process of cross-border road cargo, as well as the correlation between key data such as Harmonized System codes and commodity elements in documents from both regions. By comparing the customs supervision processes and information systems of both regions, we propose a solution for the construction of Intelligent customs clearance information system. This system integrates the customs systems of GD and HK, sharing basic data such as vehicles and goods from the road manifest and declaration merchants in both regions. It achieves automation and intelligence of customs clearance operations. To address problems such as diverse document formats and manual recognition and entry of paper texts, we propose solutions based on text image processing, Optical Character Recognition (OCR), and multi-format Excel recognition. Based on big data analysis of customs declarations and using Natural Language Processing (NLP) and Deep Learning methods, our proposed system achieves data sharing and automatic conversion of customs codes for road manifests, customs declarations, and other data between GD and HK. The Intelligent customs clearance information system is implemented using a SaaS architecture on both PC and WeChat Mini-Program platforms to enable users in GD and HK to complete cargo manifest and customs declaration at any time and place. The research results are widely applied in China (Shenzhen) International Trade Single Window and the Qianhai HK Modern Service Cooperation Zone.

Keywords: Cross-border Road Cargo · Intelligent Customs Clearance · Deep Learning · Information System · Guangdong and Hong Kong

F. Zhao and D. Miao (Eds.): AIGC 2023, CCIS 1946, pp. 181–190, 2024.
https://doi.org/10.1007/978-981-99-7587-7_15

1   Introduction

According to data from the Guangdong (GD) branch of the General Administration of Customs, the trade volume between GD and Hong Kong (HK) has increased significantly from 279.85 billion RMB in 1997 to 11.7 trillion RMB in 2021, with an average annual growth rate of 6.1%. In terms of transportation methods, from 1997 to 2021, cross-border trade between GD and HK by road maintained an average annual growth rate of 7.6%, and its proportion increased from 66.1% to 92.6%. Efficient and relatively low-cost cross-border road transportation has become the preferred channel for import and export logistics between GD and HK. Therefore, smooth, convenient, and efficient road transportation between GD and HK is of great significance to deepening trade cooperation between the two regions and enhancing the vitality and competitiveness of regional international trade. GD and HK are two independent entities with different customs jurisdictions in international economic and trade legal relations. Compared with domestic transportation, the key point that affects the efficiency of cross-border transportation between GD and HK is the clearance time between the two regions. The customs of both regions determine the tariffs and tax rates of specific types of products based on their respective Harmonized System (HS) codes, and implement different regulations, customs duties, and value-added tax collection. The customs declaration systems of both regions are not interconnected. The clearance declaration process and regulatory requirements are different, and the customs inspection results are not recognized by each other. Cross-border trucks between GD and HK face many difficulties, such as repeated entry of clearance documents, multiple declaration processes, lack of data sharing, and the use of multiple declaration systems. Taking the example of cross-border trade goods being transported from GD to overseas via HK, the clearance process is shown in Fig. 1.

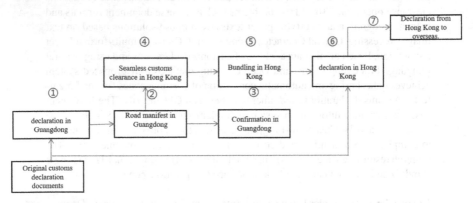

Fig. 1. Clearance procedures for cross-border road cargo from GD via HK to Overseas.

Therefore, the integration of customs declaration systems between GD and HK, automation of cross-border clearance document production, intelligent matching of HS code coordinated by both customs, and data sharing of clearance documents are the main contents of the construction of the GD and HK Intelligent Customs Information System. By constructing this system and achieving "single-entry, dual-declaration," the efficiency of customs clearance can be improved while reducing costs. This research has great economic benefits and social value in promoting the digital transformation of cross-border trade between GD and HK.

Recent research in this field has been increasingly in-depth both domestically and internationally. Wu Shengqin (2021) studied the inspiration of the EU's free flow of goods principle for the Guangdong-Hong Kong-Macao Greater Bay Area and proposed that the mechanism for resolving trade disputes is a core issue [1]. Peng Ling (2019) focused on the research of the cooperation between Guangdong Free Trade Zone and the development of Guangdong-Hong Kong-Macao cooperation, aiming to expand the service trade cooperation [2]. Shi Zhaoying et al. (2019) explored cross-border clearance issues from a logistics perspective, aimed to enhance transparency, cooperation, optimize clearance procedures, reduce costs, and improve efficiency [3]. Gongbei Customs Tariff Department Research Group (2020) studied the facilitation of customs clearance and taxation cooperation among GD, HK, and Macao under the framework of the Greater Bay Area and put forward five suggestions for facilitating customs clearance and taxation for tax objects, commodity classification, differential treatment, and price declaration [4]. In terms of the intelligent and automated production and auditing of clearance documents, Xu Zhongjian et al. (2021) studied the HS code product classification method based on deep learning technology [5]. Jen-Hsiang Cheng (2019) provides an overview of various data mining techniques used in health informatics, particularly with big data [6]. Wang Junjuan (2021) analyzed the experience of countries along the "Belt and Road" and studied the trade facilitation effects of China's cross-border e-commerce development. Based on six dimensions including logistics environment, payment environment, clearance environment, legal environment, policy environment, and regulatory environment, constructed a level indicator system for trade facilitation [7]. Yao L (2019) studied text classification based on convolutional neural networks, providing enhanced algorithms for the automatic splitting and extraction of customs declaration elements [8]. Zavala RM et al. (2018) studied a mixed Bi-LSTM-CRF neural network for knowledge recognition in electronic health documents to extract complex text information [9]. Yi Wang et al. (2022) proposed an improved deep neural network for image recognition based on feature fusion and transfer learning. The method adopted a multi-scale convolutional neural network for feature extraction and combines attention mechanism and residual connection to achieve feature fusion [10].

2   Scheme of Intelligent Customs Clearance Information System between GD and HK

Artificial intelligence (AI) and machine learning (ML) technologies have great value in the customs clearance field. By analyzing large amounts of data, AI and ML algorithms can identify patterns and anomalies that might be missed by human operators. This allows for more accurate and efficient processing of customs declarations, reducing errors and processing times.

Furthermore, AI and ML can be used to automate repetitive tasks such as document classification, invoice matching, and cargo risk assessment. This not only speeds up the customs clearance process but also reduces the workload on human operators, allowing them to focus on more complex tasks that require greater expertise. In addition, AI and ML can help improve risk management by identifying potential risks based on historical data and predicting future trends. This allows for better targeting of inspections and more effective allocation of resources.

The research and development of the Intelligent Customs Information System includes four main components: the data source layer, AI calculation layer, the AI application layer, and the provision of AI application services to the Intelligent Customs Information System. The overall architecture is shown in Fig. 2.

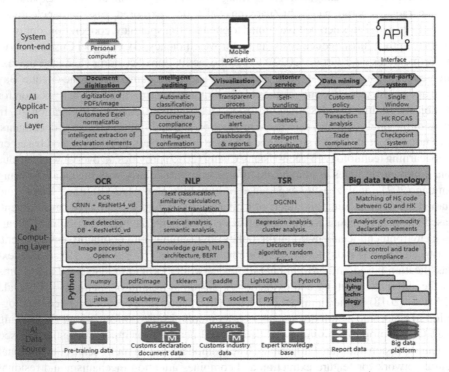

Fig. 2. Architecture diagram of intelligent customs clearance information system.

The scheme mainly includes the following four aspects:

1. Intelligent document production: This component uses artificial intelligence technologies such as Optical Character Recognition (OCR) and Natural Language Processing (NLP) to improve the efficiency of generating documents such as road manifests and customs declaration forms. It can automatically recognize and extract data from various formats such as PDFs, scanned images, and Excel files, and combine multiple documents simultaneously.

2. "Single Entry, Dual Declaration ": This component involves sharing and converting data between road manifests and customs declaration forms in GD and HK. Complete GD customs declaration forms can be automatically converted into seamless and efficient HK customs declaration forms. This process includes automatic product

classification, information extraction, merging of similar product items, and automatic currency conversion.
3. Intelligent review: This component includes intelligent review and error correction functions for documents such as road manifests and customs declaration forms, as well as their related materials (packing lists, invoices, contracts, etc.) and application forms.
4. System integration: This component involves integrating the cross-border clearance process with enterprise internal information systems, GD's "single window" system [11], HK's Road Cargo System (ROCARS), and other systems using interface methods to achieve full coverage of the entire clearance process between GD and HK.

Overall, the Intelligent Customs Clearance Information System improves the efficiency and accuracy of cross-border trade, reduces costs, and promotes digital transformation in the clearance process between GD and HK.

3   The Implementation of AI Machine Learning

The implementation of AI machine learning consists of three main stages: AI import, AI learning library collection, and AI learning. The functions and technical principles of each stage are briefly described as follows:

The function of the AI import stage is to parse customer customs declaration files. After receiving the import files from customers, various technologies such as PDF reading, image OCR, Excel conversion, Table Structure Recognition (TSR), Named Entity Recognition (NER), etc., are used to convert non-standard and unstructured files into structured customs declaration information.

The function of the AI learning library collection stage is to collect information that is inconsistent with the original file after AI import and some manually labeled template information. These pieces of information are then aggregated into the AI learning library to enhance the accuracy of AI import.

The AI learning stage uses the differences between AI imported data and the original files as learning data, fine-tunes relevant models, and generates a structured knowledge base. This knowledge base can be used to enhance the AI import function when customers import data for the second time, thereby improving the overall effectiveness.

An AI customs declaration import system that only includes the AI import stage cannot learn in real time. This study creatively adds the AI learning library collection and AI learning stages to create a complete closed-loop AI learning platform. The result is a three-dimensional AI system platform that covers all aspects of the process.

The AI review and error correction function is achieved through technologies such as file import, intelligent sorting of documentation, information comparison, logic rule verification, and historical information comparison.

Overall, this allows for an efficient and accurate customs declaration process through the use of advanced AI technologies, which greatly improves the productivity and effectiveness of the Intelligent Customs Clearance Information System between GD and HK.

4   The Key Applications of AI

4.1   Intelligent Classification of Hong Kong Commodities Based on Machine Translation and Text Similarity

Different from the traditional manual classification of goods with HS codes, the intelligent commodity classification system in this study consists of two parts: The first part uses machine translation to translate English product names into Chinese. The second part calculates the best commodity classification by conducting text similarity calculation between the Chinese product name and the historical product database.

For the first part, machine translation is used to translate English product names into Chinese. Due to the shorter length of English product names, this project adopts a simpler and faster encoder-decoder framework. Processing historical declaration data to obtain pre-training data, and training a customs declaration translation model to translate English product names in HK.

For the second part, the best commodity classification is determined by conducting text similarity calculation between the Chinese product name and the historical product database. This project collected the commodity classification results from the past few years as well as the machine-translated results, preprocessed the data, and organized it into training data for text similarity. The matched results are shown in Table 1 below.

Table 1. Example of the intelligent classification process for customs goods in Hong Kong.

The English product names in customs-related documents	Machine translation	Text similarity matching	Obtain the HK HS code inked to the Englsh product names
PET UNDERWEAR	宠物内衣	PET内衣	39269090
RESIN CRAFT	树脂工艺	树脂工艺	39264000
REPLACEMENT EARPADS	更换耳塞	耳套	63079090
MAC BOOK PRO PI967	MAC BOOK PRO PI967	苹果笔记本电	84713090
SMART WATCH	智能手表	智能手表	85176900
PHONE	电话	手机	85171310
SHIFT KNOB	换档旋钮	换档旋钮	39269090
MINI PC PI967	微型电脑PI967	迷你电脑	84714900

(*continued*)

Table 1. (*continued*)

The English product names in customs-related documents	Machine translation	Text similarity matching	Obtain the HK HS code inked to the Englsh product names
FIBER FUSION SPLICER PI967	光纤焊接机PI967	拼接机	84059400
SMART WATCHES	智能手表	智能手表	85176900
KEY BAND COCOA	基带可可	智能手环	85249100
BARCODE SCANNER	条码扫措仪	条码扫措器	84719000
SMART CARD	智能卡	智能卡	85235200
BARCODE SCANNER	条码扫措仪	条码扫措器	84719000
SMART WATCH	智能手表	智能手表	85176900
SMART WATCH	智能手表	智能手表	85176900
COSMETIC BAG	化妆袋	化妆品袋	42023220
AIR CONDITIONING PARTS	空调部件	空气调节部件	84159090
MANNEQQINS	人体模型	人体模型	96180005
AIR CONDITIONING PARTS	空调部件	空气调节部件	84159090
AIR CONDITIONING PARTS	空调部件	空气调节部件	84159090
DOCUMENT CAMERA	文档照相机	照相机	85258920
SMART BRACELET	智能手镯	智能手环	85249100

4.2 Intelligent Recognition of Customs Documents and Generation of Customs Declaration Draft in GD and HK

Common customs documents come in various file formats such as PDF, Excel, images, Word, and so on. Each document may involve multiple types of documentation information such as packing lists, invoices, contracts, preliminary customs declarations, etc. Additionally, the same format under the same type of documentation can also have different forms, text formatting, and descriptive content provided by different clients. The system needs to read and extract these non-standard and unstructured files to obtain standardized structured data.

The main requirement of this system is to convert PDF and image files into tables and text, perform structural analysis on complex tables, and parse the contents of complex texts. The difficulty lies in the fact that different clients use different styles for their tables and texts, which are often complex and lack a unified or standard template. Therefore, the program needs to intelligently determine how to extract information and structure it. To solve this core problem, this study uses artificial intelligence algorithms to learn from client samples and manual data entry, allowing for the importation of documentation from different clients. During this process, the system automatically collects client data-entry behavior, compares file contents, obtains labeled information for supervised learning samples, and uses the Dynamic Graph CNN (DGCNN) combined with text feature extraction to extract structured information from complex tables. Bi-LSTM + CRF neural networks are used to extract information from complex texts, and finally, the customs declaration information is summarized.

4.3 Intelligent Extraction of Commodity Declaration Element Information

The Random Forest algorithm, which is a type of ensemble learning algorithm, is used for locating and extracting table data items. Random Forest belongs to the Bagging (Bootstrap AGgregation) method in ensemble learning. It consists of many decision trees with no association between them.

When classifying new input samples, each decision tree in the forest makes its own judgment and classification. Each decision tree obtains its own classification result, and the final result of the Random Forest is determined by selecting the classification that has the highest frequency among all decision trees.

The extraction of commodity declaration element information from customs declarations mainly involves the following five steps:

1. Locating the text content of the declaration elements.
2. Locating the declaration elements in the table content based on domain expert knowledge.
3. Structured extraction of declaration element information.
4. Extraction of text information using the Bi-LSTM + CRF network based on Named Entity Recognition (NER).
5. NER is used to recognize and classify named entities such as product names, purpose, brand, voltage, model, etc. from the given unstructured text. Based on historical declaration information, mass text data is automatically generated, and the Bi-LSTM + CRF neural network is trained for each category to perform classification prediction and extract the necessary elements. As shown in Fig. 3.

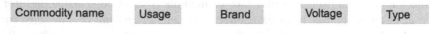

Connector , Industrial communication , Overseas brand : TE, 25V, 40-42096-00831T

Fig. 3. Example of Named Entity Recognition (NER) task.

5 Conclusion

The Intelligent Customs Clearance System for cross-border trucks is implemented on a SaaS architecture, providing two operating modes: PC and WeChat mini-program on mobile phones. As shown in Fig. 4 and 5. This allows for convenient customs declaration and query operations anytime and anywhere in GD and HK. The system covers over 10,000 cross-border trucks in GD and HK, with more than 10 million customs declarations and cargo manifests submitted through the system in 2022. The adoption of this system has reduced customs declaration fees by 40% and improved customs clearance efficiency by 30%.

The research results in this study have been widely applied in China's (Shenzhen) international trade single window [12] and the Qianhai Cooperation Zone for Modern Service Industries in GD and HK, promoting the intelligent and digital upgrade of cross-border trade supply chain services between GD and HK. These improvements have helped to enhance the business environment at the ports.

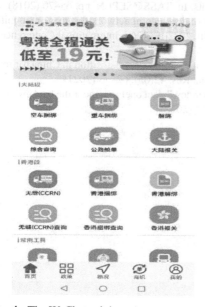

Fig. 4. The WeChat mini-program user interface.

Fig. 5. The operation interface for manifest declaration.

Funding. The work is supported by Research on the Cross-border Intelligent Customs Clearance Operation Mechanism for Goods in Guangdong, Hong Kong, and Macao (Project Number: GD23XGL050).

References

1. Wu, S.: The enlightenment of the EU's free movement of goods principle on the Guangdong-Hong Kong-Macao Greater Bay Area. Overseas Abs. **01**, 31–32 (2021)

2. Peng, L.: Research on the integration of Guangdong free trade zone and Hong Kong-Macau cooperation development. Coop. Econ. Sci. Technol. **05**, 104–106 (2019)
3. Shi, Z., Pan, C.: Discussion on the land customs clearance issues in the Guangdong-Hong Kong region from a logistics perspective. Market Modernization **16**, 73–75 (2019)
4. Gongbei Customs Tariff Department Project Team: Research on the customs taxation facilitation cooperation among Guangdong, Hong Kong and Macao under the framework of the Greater Bay Area. Customs Econ. Res. **41**(05), 12–19 (2020)
5. Xu, Z., Li, X.: Research on product classification method based on HS code using deep learning. Modern Comput. Prof. Ed. **637**(01), 13–21 (2021)
6. Cheng, J.-H., Kuo-Jui, Lu, H.-K.: A review of data mining using big data in health informatics. J. Ambient Intell. Humanized Comput. **10**(1), 1–15 (2019)
7. Wang, J.: The trade facilitation effect of the development of cross-border e-commerce in China: based on the experience analysis of countries along the "Belt and Road." Comm. Econ. Res. **02**, 70–73 (2021)
8. Yao, L., Mao, C., Luo, Y.: Graph convolutional networks for text classification. Proc. AAAI Conf. Artif. Intell. **33**, 7370–7377 (2019)
9. Zavala, R.M.R., Martinez, P., Segura-Bedmar, I.: A hybrid Bi-LSTM-CRF model for knowledge recognition from eHealth documents. In: TASS@SEPLN, pp. 65–70 (2018)
10. Wang, Y., Li, X., Wu, J., Zhang, Y.: An improved deep neural network for image recognition based on feature fusion and transfer learning. In: 2022 IEEE 6th International Conference on Signal Processing (ICSP) (2022)
11. Wang, X.: Research on the development mode of customs informatization in China for forty-five years: from H761 to H2018. Customs Econ. Res. **42**(05), 62–73 (2021)
12. Wu, Q., Xu, F.: Development path of "single window" data opening under the background of digitization. Inform. Technol. Constr. **07**, 62–64 (2022)

Analyzing Multilingual Automatic Speech Recognition Systems Performance

Yetunde E. Adegbegha[1], Aarav Minocha[2], and Renu Balyan[1(✉)] (iD)

[1] State University of New York, Old Westbury, NY 11568, USA
balyanr@oldwestbury.edu
[2] Great Neck South High School, Great Neck, NY 11020, USA

Abstract. Understanding spoken language, or transcribing the spoken words into text, was one of the earliest goals of computer language processing and falls under the realm of speech processing. Speech processing in itself predates the computer by many decades. Speech being the most important and most common means of communication for most people, is always in need of necessary technology advances. Therefore, in the recent decades there has been great interest in techniques including automatic speech recognition (ASR), text to speech etc. This research is focused around English and the scope needs to be expanded to other languages as well. In this study we explore several open-source ASR systems that offer multilingual (English and Spanish) models. We discuss various models these ASR systems offer, evaluate their performance. Based on our manual observations and using automatic evaluation metrics (the word error rate) we find that Whisper models perform the best for both English and Spanish. In addition, it supports a multilingual model that has the ability to process audio that consists of words from both English and Spanish.

Keywords: Automatic Speech Recognition · Whisper · Vosk · Word Error Rate

1 Introduction

Speech Processing is the study of speech signals and the computer processing methods of these signals in a digital representation [1]. Speech processing is essential in today's technological driven world and creates a more natural human-machine interaction. Speech processing is being used in numerous industries to enhance user experiences and simplify communication. New technologies in the field pave the way for voice activated systems that can enrich our interactions with digital platforms, increase accessibility on the internet for those with disabilities, and help people who speak different languages interact [2]. From language translations to understanding audio through voice biometrics, speech processing is crucial to improving communication in a digital age. The increasing need for new technologies such as smart assistants and real-time translations has made the integration of speech processing in our daily life a necessity in order to foster a new age of modern communication.

Speech recognition technology allows computers to take spoken audio as input, interpret it and generate text (referred to as transcription for the rest of the paper) as an output.

F. Zhao and D. Miao (Eds.): AIGC 2023, CCIS 1946, pp. 191–204, 2024.
https://doi.org/10.1007/978-981-99-7587-7_16

Automatic Speech Recognition (ASR) systems aim at converting a speech signal into a sequence of words either for text-based communication purposes or for device controlling [3]. Research in ASR and speech synthesis has gained a lot of importance and attracted a great deal of attention over the past few decades [4]. Technological curiosity about the mechanisms for mechanical realization of human speech capabilities, and the desire to automate simple tasks inherently requiring human-machine interactions have generated interest in studying the ASR systems [4]. Some of the major growing applications in this field include speech enhancement, speaker recognition and verification, spoken dialog systems, emotion and attitude recognition, speech segmentation and labeling, and audio-visual signal processing. With the number of applications using these voice-based systems, special care needs to be taken while building these systems as failures of ASR systems may result in serious risks to users. For example, in the health domain an ASR system error can pose risk for the patient if the patient is not understood correctly by the ASR [5]. Therefore, further research and a closer investigation is needed to understand the importance of being correctly understood or the consequences of being misunderstood by speech recognition systems [6]. Research has shown that ASR systems exhibit racial bias, and there has been concern over these systems not working equally for everyone [7–9]. Therefore, even though the focus of this study is not identifying the bias in ASR systems towards a particular population, we try to identify if ASR systems perform at the same level for languages other than English, particularly Spanish in this study. There are several well-known ASR systems that have been studied and tested for English, but we found only a few studies that have explored and analyzed the transcripts generated for Spanish using these ASR systems or evaluated the performance of these systems for Spanish [10, 11].

The Goal of the current study was to generate English and Spanish transcriptions from an existing set of recorded videos in the health domain. This study forms a part of a bigger NSF-funded project that is developing a culturally sensitive health intelligent tutoring system (ITS) for the Hispanic population. In order to achieve the said goal, some of the research questions (RQ) that were answered in this study are:

RQ1: What open-source ASR systems exist that can transcribe English as well as Spanish videos?
RQ2: What models within these systems can be used to generate transcriptions for recorded videos based on the performance of the models/systems for the two languages?
RQ3: What evaluation measures can be used to automatically evaluate the system/model's performance?

2   Open-Source ASR Systems

2.1   Whisper

Whisper is a general-purpose, multitasking speech recognition model, trained on 680,000 hours of labeled audio and the corresponding transcripts collected from the internet. This training data constitutes 438,000 hours of English audio and the matching English transcripts; 125,000 hours represents X → English translation data, and the remaining 117,000 hours represent non-English audio and the corresponding transcript, covering 99 other languages. The model was trained using an encoder-decoder transformer [12] as it scales well [13].

2.2 Vosk

Vosk is an open-source and free Python toolkit for offline/online speech recognition. Vosk supports two models - big and small; small models are ideal for limited tasks such as mobile applications. Big models are for high-accuracy transcription and apply advanced AI algorithms. Vosk models provide continuous large vocabulary transcription, zero-latency response with streaming API, reconfigurable vocabulary, and speaker identification. The system can result in poor accuracy due to numerous reasons including bad audio quality, vocabulary mismatch, accent, coding and software bugs [14].

2.3 Kaldi

Kaldi is an open-source toolkit for speech recognition that is written in C + + and is licensed under the Apache License v2.0. More details about Kaldi are available on their website (http://kaldi-asr.org). Kaldi is intended for use by speech recognition researchers and professionals; it is a research speech recognition toolkit that implements many state-of-the-art algorithms. Kaldi has speech activity detection (SAD), speaker identification (SID), language Model (LM), diarization (DIAR) and ASR models with 3 of them being English ASR models [15].

2.4 Julius

Julius is an open-source, high-performance speech recognition decoder for academic research and industrial applications. It supports processing of both audio files and a live audio stream. Julius supports standard language models such as the statistical N-gram model, rule-based grammars, and Hidden Markov Model (HMM) as an acoustic model. Julius in itself is developed as a language-independent decoding program and a recognizer of a language can be developed given an appropriate language and acoustic model for the target language. Julius currently has Japanese and English language/acoustic models [16].

2.5 Mozilla DeepSpeech

DeepSpeech is an open-source Speech-To-Text engine using a model trained by machine learning techniques such as recurrent neural network (RNN) [17]. It uses Google's TensorFlow to make the implementation easier, open and universal [18, 19] A pre-trained English model is available for use [19].

3 Methods

3.1 Data

Eleven short videos of varying length, recorded by a doctor in both English and Spanish were used as the base for the current study. These videos were transcribed to obtain text that was processed further for different purposes using various natural language processing (NLP) techniques. The topic of each video varied but the theme of every video is about cancer survivorship.

3.2 Data Preprocessing

There was no data preprocessing needed for generating the automatic or the human expert transcriptions and the video files were fed into the ASR system or given to the human expert as is for the transcriptions. However, for performing the evaluation of the automatically generated transcription, the data had to be aligned sentence by sentence for both the expert/reference transcriptions and the system-generated/ hypothesis transcriptions as per the requirements of one of the packages (ASR-evaluation) used for the transcription evaluation. This package output was used for an in-depth analysis of the errors produced at the sentence level. However, another package (JiWER) used for evaluating the transcriptions was more flexible and did not require any form of preprocessing on the transcriptions before evaluation or for computing the statistics. More details for the packages are discussed later in the 'Automatic Evaluation' section (see Sect. 5).

3.3 Human Transcription

Human transcriptions were created by a fluent English and Spanish speaker and the transcriptions were later also validated and verified by another speaker fluent in both the languages. The descriptives (number of words and number of sentences) for the transcriptions were generated using SpaCy, an open-source NLP python library are shown in Table 1.

Table 1. Data Descriptives for the English and Spanish Expert/Reference Transcriptions

Transcript	Transcript Description	# of sentences (English/Spanish)	# of words (English/Spanish)
1	Visual Symptoms	5/7	97/79
2	Tamoxifen Side Effects	9/5	217/167
3	Survivorship Care	5/8	243/203
4	SE After Surgery	8/8	198/133
5	PT Side Effects	3/13	162/337
6	PT Breast Cancer Basics	3/9	112/167
7	PT Intro-Mi Guia	2/5	38/68
8	Peripheral Neuropathy	6/8	126/171
9	Osteoporosis	5/6	117/193
10	Depression	9/5	111/131
11	Cardiac symptoms	6/5	120/70

This is to be noted that there are differences between the data descriptives for English and Spanish transcriptions, even though the videos were on the same topics. Some of these differences result due to varying length of videos for the two languages, which leads to different numbers of sentences. In addition, the other differences are caused due to linguistic differences between the two languages. For example, a word in English may not have a single word equivalent in Spanish but is represented by multiple words or vice versa, which results in a difference between the number of words in the two language transcripts.

3.4 Automatic Transcription

We explored several ASR systems discussed in Sect. 2 for generating automatic transcriptions but only two open-source ASR (Vosk and Whisper) fulfilled the requirements of this study and as a result were used for all the experiments in this study to automatically transcribe our data (videos). Vosk developed by Alpha Cephi supports 27 languages and dialects and Whisper by OpenAI supports 99 languages. We used these ASR systems to transcribe English and Spanish videos. We experimented with the different models that were provided by the two systems. The models finally used for this study were determined based on varying levels of accuracy and speed.

3.5 Experiments

We conducted two experiments in this study, the first experiment was to determine the best open-source ASR for the requirements of our project and the second experiment was to determine the error rates for the ASR-generated transcription to determine the transcription quality. In the first experiment, we implemented several different models provided by the ASR systems to choose the best model. The same dataset was used to test each model's accuracy and speed. In the second experiment, we evaluated and measured the accuracy of the ASR-generated transcriptions using the available error rate evaluation metrics. We manually transcribed the data, (i.e., the videos) to obtain the reference or the expert transcriptions as discussed previously in Subsect. 3.3. We used different ASR evaluation metrics to compute the accuracy of transcripts (i.e., human transcription vs. the automatic transcripts generated by the ASR systems).

Experiment 1: Transcription Models. Several models from Vosk and Whisper were explored to transcribe the data for English and Spanish. For English transcription, Vosk has multiple English models, however, we used 'vosk-model-en-us-0.22' model as this model fulfilled the requirements of our study and was close to what we needed. Whisper supports four English-only models (tiny-en, base-en, small-en, and medium-en). We used the 'model-medium-en' due to its performance and lower error rates.

For Spanish transcription, Vosk supports two models, the 'vosk-model-small-es-0.42' and the 'vosk-model-es-0.42'. We used the 'vosk-model-es-0.42' instead of the 'vosk-model-small-es-0.42', which is a Lightweight wideband model for Android and RPi. It is important to note that the small model is ideal for some limited tasks on mobile applications, while the big models are for the high-accuracy transcription on the server and apply advanced AI algorithms. Since we were not working on mobile applications, we preferred the larger model for this study. Whisper has four models (tiny, base, small, and medium). We decided to use the 'medium' model because it has better punctuations and spellings, and accurately detected the video lengths compared to the other models. While using this model, one has to explicitly state what language the transcriptions are expected for, because, unlike English, it does not have models trained specifically for Spanish. However, if the language is not explicitly stated, the system detects the language being spoken in the audio or video file and considers the model to be used accordingly.

Experiment 2: ASR Evaluation. The accuracy of each transcription generated by the ASR systems was evaluated through different metrics obtained from the JiWER Python package. These metrics were the Word Error Rate (WER), Match Error Rate (MER), Word Information Loss (WIL), Word Information Preserved (WIP), and Character Error Rate (CER). Another Python package (ASR-evaluation) was also tested for evaluation. This package returned the sentence error rate (SER) and word error rate (WER) but required more data preprocessing and computed information on fewer features. In addition, this package had higher WER as compared to the JiWER package and therefore was not used for initial evaluation. It was also observed that the ASR-evaluation package may be more helpful for deeper analysis for sentence-level transcription evaluation rather than the full transcription.

4   Results

4.1   Transcription Models Performance

For the Spanish transcription, Whisper is the best option to fulfill our project requirements. Vosk and Whisper have similar levels of accuracy with their Spanish models. Whisper also has punctuations to indicate the end of the sentences; whereas, Vosk does not provide punctuations in the transcript and does not have a Spanish model that one can use to include punctuations in the transcripts.

Whisper Spanish (Small vs. Medium model). The Whisper small model is the default model. It has a similar level of accuracy as the Whisper medium model. The manual analysis of the two model outputs indicate that the medium model is a little better for Spanish transcription as compared to the small model. The medium model performed well for all the transcriptions except for Transcript 10 (related to Depression) shown in Fig. 1. The small model for Spanish could perfectly transcribe the name spoken in the video; however, the medium model for Spanish could not transcribe it correctly. It performed poorly than the small model because of differences in paragraphs and space-related issues.

Vosk has multiple English models, yet, only the 'Vosk-model-en-us-0.22' was able to generate superior transcriptions as compared to the other models. This model was however not suitable for our needs as it performed well with a generic US-English accent. Whereas, the speaker in our videos has a non-US accent leading the model to perform poorly. As a result, due to the model's highly inaccurate transcriptions, we decided not to use Vosk for transcribing the English videos. Whisper has four English-only models (tiny-en, base-en, small-en, and medium-en). The default English model is the 'small-en' model but we decided to use the model 'medium-en' after manually analyzing the transcriptions from the two models. The comparison between the Vosk and Whisper English models and the poor performance of Vosk can be seen clearly in Fig. 2. The difference transcriptions are marked in red, where Vosk indicates incorrect

WHISPER –model medium

Buenos días, soy la doctora ▓▓▓▓▓ y ahora vamos a hablar de algunos efectos a largo plazo después del tratamiento del cáncer del seno.
Uno de ellos es función del corazón.
Si uno siente que se ahoga, que no puede respirar bien, que le late abnormal el corazón, es importante ver a sus doctores y posiblemente ver a un cardiólogo.
Esto puede ser consecuencia de los tratamientos de cáncer del seno.

WHISPER –model small

Buenos días, soy la doctora ▓▓▓▓▓ y ahora vamos a hablar de algunos efectos a largo plazo después del tratamiento del cáncer del seno. Uno de ellos es función del corazón. Si uno siente que se ahoga, que no puede respirar bien, que le late al normal el corazón, es importante ver a sus doctores y posiblemente ver a un cardiólogo. Esto puede ser consecuencia de los tratamientos del cáncer del seno.

Fig. 1. The transcriptions generated by the Whisper Spanish (medium and small model). The differences in the two models are highlighted and the name of the doctor (which the medium model id not transcribe correctly) has been redacted for confidentiality purposes.

transcription and the Whisper red color transcription represents correct or what was actually spoken in the video. We use the 'Vosk-model-en-us-0.22', a generic US accent model, and the Whisper 'medium-en' model in this example. The Vosk model cannot transcribe accurately because the speaker in the video has a non-US accent. However, the Whisper model can transcribe regardless of the accent. In the first sentence, the speaker introduces herself, which Whisper transcribes correctly as "this is <<Name of the doctor>>.", whereas Vosk transcribes it as "spark oppressed meyer" which is a far cry from what is said. Vosk has no other model that came this close to transcribing our data (see Fig. 2).

In order to reaffirm that it was indeed the non-US accent because of which the Vosk model performed poorly, we transcribed another video (with a generic US accent) randomly selected from the internet. The Vosk model performs well in this case (see Fig. 3) and is able to transcribe words like 'tidbit' and 'inflation' correctly whereas for a non-generic US accent, the model incorrectly transcribed simple words like 'thank you' as 'think'.

Vosk Transcription

hi spark oppressed meyer today a minute talk about one of the long term side effects of breast cancer treatment it can be manifest is irritability sleeping a long time not carrying out not taken care of oneself and these are manifestations some types of depression your body has undergone a lot of changes are not only physically mentally but also for my family it's important to bring this to be attention of your primary care physician so the different factors that contributed to the state of mind can be teased away and make better it's important to get the appropriate referral
just one one feels depressed after breast cancer treatment think|

Whisper.

Hi, this is ▓▓▓▓▓. Today I'm going to talk about one of the long-term side effects of breast cancer treatment. It can be manifest as irritability, sleeping a long time, not taking care of oneself, and these are manifestations sometimes of depression. Your body has undergone
a lot of changes, not only physically, mentally, but also hormonally. It's important to bring this to the attention of your primary care physician so the different factors that contributed to this state of mind can be teased away and made better. It's important to get the appropriate referrals when one feels depressed after breast cancer treatment. Thank you.

Fig. 2. The transcriptions generated by the Vosk and Whisper English models. The name of the doctor was no where near what it should have been (which has been redacted for confidentiality purposes in Whisper). Red colored words show how poorly the Vosk performs as compared to the Whisper model for English.

Whisper English (Small vs. Medium model). The Whisper English medium and small models both accurately transcribe the data. However, there are minor differences between

198   Y. E. Adegbegha et al.

the funny thing about the big economic news of the day the fed raising interest rates half a
percentage point was that there was only really one tidbit of actual news in the news and the
interest rate increase wasn't it you knew it was coming i knew it was common wall street news
come and businesses knew it was common so
on this fed day on this program something a little bit different jay powell in his own words five of
'em his most used economic words from today's press conference were number one of course
it's the biggie two per cent inflation inflation inflation inflation inflation inflation Ih dealing with
inflation pals big worry the thing keeping him up at night price stability
he is the fed's whole ballgame right now pau basically said as much today we're number two

Fig. 3. The transcriptions generated by the Vosk English models for an audio with generic US accent. Vosk does a far better job than it did with no-generic US accent.

the transcriptions returned by the two models as can be seen in Fig. 4. For example, the first line of the medium model ends with 'University'; however, the small model ends with 'at'. This does not impact manual evaluation of the transcription but this results in poor performance during the automatic evaluation. The 'Medium' models overall performs better than the 'Small' model in most transcriptions except it was observed that for Transcript 5 (related to PT Side Effects) it performed differently than the 'small' model because of inconsistent word representations; for example, the small model transcript has the word 'post-menopausal', whereas the medium model transcribes, it as 'postmenopausal'. Both these transcriptions are correct, but the ground truth or a reference transcription will favor the model with a matching word during the automatic evaluation.

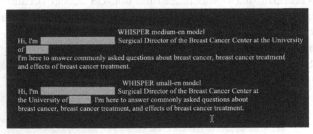

Fig. 4. The transcriptions generated by the Whisper English (medium and small model). There are no differences in the two models transcriptions other than the formatting related, the name of the doctor (the medium model hyphenated the name) and University has been redacted to hold the identity of the person.

Whisper Multilingual Model. Whisper generated all English text with the model 'large' for an audio that contained a mix of both English and Spanish. This was possibly due to the poor audio quality. On the contrary, with the model 'large-v2' in Whisper was able to correctly transcribe an audio file with both English and Spanish (Multilingual). The Whisper multilingual 'large-v2' model performs better than the multilingual 'large' model at transcribing the audio and detecting the languages in the audio. In one of our experiments, the 'large' model transcribed the audio in the language that was dominant rather than the two languages spoken in the audio. It is important to note that when the multilingual audio starts running, the model first detects the language. The 'large-v2' model has given us consistent, accurate results even though it may also detect the dominant language. The multilingual model also supports language-specific models, therefore depending upon what multilingual model is selected, Whisper transcribes audio in the respective language (English or Spanish in our case) correctly. While using the multilingual model, if the English language is explicitly stated in the command, Whisper transcribes the audio to English regardless of the language in the audio. It first translates the audio from the actual language to English and then transcribes it; it behaves similarly if the Spanish language is explicitly stated in the command.

5   Transcription Evaluation

5.1   ASR Evaluation

The purpose of evaluating ASR systems is to simulate human judgment of the performance of the systems in order to measure their usefulness and assess the remaining difficulties especially when comparing systems; the standard metric of ASR evaluation is the Word Error Rate (WER), which is defined as the proportion of word errors to words processed [3]. The WER is based on how much the output (typically a string of words) called the Hypothesis, returned by the ASR system differs from a reference transcription generated by a human expert. The WER is computed using Eq. (1), where I = number of insertions, D = number of deletions, S = number of substitutions, C = number of correct words and N = number of words in the reference.

$$WER = \frac{S+D+I}{S+D+C} = \frac{S+D+I}{N} \tag{1}$$

The Python Jiwer package was used to automatically calculate the WER, Match Error Rate (MER), Word Information Loss (WIL), Word Information Preserved (WIP), and Character Error Rate (CER). The measures are computed with the use of the minimum-edit distance between one or more reference and hypothesis sentences. Although WER is the most popular and commonly used metric to evaluate ASR, it has certain drawbacks [2, 20, 21]. Therefore, many researchers have proposed alternative measures to solve the evident limitations of WER. Andrew et al. [22] introduced Relative Information Lost (RIL) and WIL. WIL value indicates the percentage of words that were incorrectly predicted between a set of ground-truth sentences and a set of hypothesis sentences [23]. WIL is an approximation measure of RIL and is based on HSDI counts. RIL, which is based on Mutual Information (I, or MI), is calculated using the Shannon Entropy H [3]. The CER value indicates the percentage of characters that were incorrectly predicted [23]. The lower the value, the better the performance of the ASR system with a CER of 0 being a perfect score. MER value indicates the percentage of words that were incorrectly predicted and inserted [22, 23]. The lower the value, the better the performance of the ASR system with a MER of 0 being a perfect score. WIP value indicates the percentage of words that were correctly predicted between a set of ground-truth sentences and a set of hypothesis sentences [23, 24]. The higher the value, the better the performance of the ASR system with a WIP of 1 being a perfect score.

5.2 ASR Transcriptions Error Rates

Table 2 shows the evaluation metric results for English Transcription performed by Whisper using the 'medium-en' model. The evaluation metrics in the table includes the metrics returned by the JiWER package namely WER, MER, CER, WIL, and the WIP. The average for each of these metrics for the whole corpus of transcriptions is WER (23.72%), MER (23.49%), CER (4.55%), WIL (35.80%), and WIP (64.20%).

Table 2. The error rates and word information scores for Whisper English model

Transcript	Transcript Description	WER(%)	MER(%)	CER(%)	WIL(%)	WIP(%)
1	Visual Symptoms	21.98	21.50	4.97	34.20	65.80
2	Tamoxifen Side Effects	24.88	24.42	6.14	37.80	62.20
3	Survivorship Care	19.17	19.01	3.97	28.86	71.14
4	SE After Surgery	19.89	19.89	4.12	30.60	69.39
5	PT Side Effects	25.48	25.16	4.87	37.36	62.64
6	PT Breast Cancer Basics	20.75	20.75	3.76	32.76	67.24
7	PT Intro- Mi Guia	40.54	40.54	5.81	59.12	40.88
8	Peripheral Neuropathy	22.58	22.22	3.99	33.23	66.77
9	Osteoporosis	19.30	19.30	3.03	30.61	69.39
10	Depression	23.56	23.15	4.74	35.01	64.99
11	Cardiac symptoms	22.81	22.41	4.66	34.21	65.79

Table 3 shows the same evaluation metrics as in Table 2 but for Spanish transcription for the Whisper using the 'medium' model and Vosk using the 'Vosk-model-es-0.42'. We observed that Whisper outperforms the Vosk model in all the transcriptions accuracy except for Transcripts 7 and 10. Vosk was better due to formatting (paragraphs and spaces) after we compared it with reference transcript. These numbers will change if the formatting in the reference transcript changes. However, in Transcript 10, Vosk in addition to the formatting issues, transcribed the name of the doctor correctly, but Whisper could not.

5.3 WER Related Challenges

The fundamental problem with the WER is that it weighs every word equally. For example, a determiner and an adjective will be treated the same, even though as humans we know that not every word is important and some errors matter more than others. Because the context determines some of these factors, it is difficult to develop a test that can be broadly applied. In addition to ignoring the importance of words, the WER does not give any partial credit. Even if a mis-transcribed word mismatches by just one character, WER treats it as incorrect or a mismatch. The WER does not account for speaker labels and punctuations, which may be important in some cases. Another issue to

Table 3. The error rates and word information scores for Spanish transcriptions (Whisper vs Vosk models)

Transcript	Transcript Description	WER (%) Whisper/ Vosk	MER (%) Whisper/ Vosk	CER (%) Whisper/ Vosk	WIL (%) Whisper/ Vosk	WIP (%) Whisper/ Vosk
1	Visual Symptoms	22.08/31.17	21.52/29.63	5.99/8.87	35.17/46.59	64.83/53.41
2	Tamoxifen Side Effects	29.70/41.21	28.49/38.86	11.30/17.68	45.09/58.20	54.90/41.80
3	Survivorship Care	21.89/27.36	21.57/26.44	5.22/7.65	33.32/41.18	66.68/58.82
4	SE After Surgery	25/28.79	24.26/28.15	10.27/11.55	36.72/46	63.28/54
5	PT Side Effects	21.92/26.73	20.98/25	7.34/9.86	32.81/37.40	67.19/62.60
6	PT Breast Cancer Basics	22.29/23.49	21.26/22.54	7.46/9.98	32.30/36.37	67.70/63.63
7	PT Intro- Mi Guia	**27.27/25.76**	**25.71/24.29**	5.99/7.91	**40.62/39.20**	**59.38/60.80**
8	Peripheral Neuropathy	26.04/32.54	25/30.22	9.25/12.68	38.26/45.46	61.74/54.54
9	Osteoporosis	25.93/33.33	25.52/32.14	17.04/18.40	35.21/45.90	64.79/54.10
10	Depression	**33.85/26.92**	**33.08/26.12**	9.97/7.80	**51.26/42.01**	**48.74/57.99**
11	Cardiac Symptoms	27.94/29.41	26.76/27.40	8.38/9.14	42.37/43.41	57.63/56.59

consider for accuracy is that a verbatim transcript is likely to include many meaningless words such as "umms", "uhs", duplicates and false starts, which do not add anything meaningful to the text [25]. Some of the high error rates seen in Tables 2 and 3 can be attributed to several of WER-related issues as were also observed in our transcripts during the manual analysis of the transcripts.

6 Discussion

In this study, we conducted 2 experiments to answer the three research questions (RQ1–RQ3) discussed in the Introduction (Sect. 1).

6.1 RQ1: What Open-Source ASR Systems Exist that can Transcribe English as well as Spanish Videos?

Several open-source ASR systems including DeepSpeech, Julius, Kaldi, Vosk and Whisper were explored in the study and after some initial research and analysis we selected two open-source ASR systems namely Whisper and Vosk for further experiments in this study. Both these systems can transcribe both English and Spanish audio. We did not continue experiments with DeepSpeech, as it is no longer supported and there have been no new versions or releases since 2020 [19]. Julius has support for English, Thai, Chinese, Korean but no popular version of Spanish has been found. A paper cited only once since 2014 used it to propose the Spanish version. Lack of available Spanish data for our population was another reason for not being able to train our own model for Julius and Kaldi. Whisper outperforms Vosk in both the languages English as well as Spanish. In addition, the Whisper transcriptions are more readable than Vosk because Whisper models return punctuations and capitalization as humans.

6.2   RQ2: What Models within these Systems can be Used to Generate Transcriptions for Recorded Videos based on the Performance of the Models/systems for the Two Languages?

All the English and Spanish models for Whisper and Vosk can generate transcription for recorded videos. However, each model's accuracy level varies; the Vosk English models could only partially transcribe the recorded videos as they are trained with US accents. The requirements for our project need an ASR that considers all accents, as the target population for the project is Hispanic individuals who are less likely to have a generic US accent. Whisper, however, can accurately transcribe videos regardless of the speaker's accent. The Vosk and Whisper Spanish models were able to transcribe the videos accurately. We decided that the Whisper 'medium-en' model for English and the Whisper 'medium' model for Spanish were the best available options for transcribing the videos.

6.3   RQ3: What Evaluation Measures can be Used to Automatically Evaluate the System/model's Performance?

The standard metric for the ASR evaluation is the WER. However, other metrics can be used to evaluate the systems, as WER also has flaws and is not perfect. There are other metrics like the WIL, MER, CER, and WIP. We used the Python Jiwer package to automatically calculate these metrics to compare and determine the best models/systems.

6.4   Limitations

Like any study, this study also has a few limitations that need further research and some future work. 1) So far, Whisper provides the most suitable transcription that best serves our purpose. Although Whisper supports 99 languages, based on our analysis the English models are better trained than the Spanish. In this study we focused mainly on the models that independently support either English or Spanish. However, our future work seeks to find a multilingual ASR because the project's target population (Hispanic) is bilingual (who speak both Spanish and English). Whisper has a large model, which is primarily multilingual, so we will be exploring this model in our future work. 2) Even though the WER is the standard ASR evaluation metric, it has considerable issues, and there have been criticisms against solely relying on it [2, 21]. The accuracy of the ground truth (i.e., manual transcript) also greatly affects the WER. Humans make errors, the manual transcript could be incorrect, but the WER works with the assumption that the ground truth is perfect. Even while working under the assumption that the manual is perfect, the grammar affects the error rate. For example, "Dr." and "Doctor" is correct; however, if the ground truth uses "Doctor" and the hypothesis uses "Dr." the WER calculates that as an error, but they are both right. We will need to explore better and efficient measures to evaluate ASR accuracy in our future studies.

7 Conclusions

A lot of ASR systems exist where the research focus has been mainly English, and minimal research is available for multilingual ASR systems. Therefore, this paper explored the ASR systems with the focus on their capability of handling multiple languages as well as multilingual audio. The ASR systems explored in the study either did not support Spanish (language of our interest for the study) or did not perform well. The ASR system by Whisper was the only system that supported both the languages (English and Spanish) and performed well. In addition, Whisper supports a model that is also capable of handling mixed audio, which is our anticipated data from the target population. The evaluation metrics mostly are based of WER and we discussed several challenges encountered while using WER for evaluation, indicating there needs to be more research for ASR evaluation, better and more efficient techniques are needed in this area.

Acknowledgment. This work was supported by grants from the National Science Foundation (NSF; award# 2131052 and award# 2219587). The opinions and findings expressed in this work do not necessarily reflect the views of the funding institution. Funding agency had no involvement in the conduct of any aspect of the research.

References

1. Garza-Ulloa, J.: Introduction to cognitive science, cognitive computing, and human cognitive relation to help in the solution of artificial intelligence biomedical engineering problems. In: Applied Biomedical Engineering Using Artificial Intelligence and Cognitive Models, pp. 39–111 (2022)
2. Kong, X., Choi, J.Y., Shattuck-Hufnagel, S.: Evaluating automatic speech recognition systems in comparison with human perception results using distinctive feature measures. In: 2017 IEEE International Conference on Acoustics, Speech and Signal Processing (ICASSP) pp. 5810–5814. IEEE (2017)
3. Errattahi, R., El Hannani, A., Ouahmane, H.: Automatic speech recognition errors detection and correction: a review. Procedia Comput. Sci. **128**, 32–37 (2018)
4. Juang, B.H., Rabiner, L.R.: Automatic Speech Recognition–a Brief History of the Technology Development. Georgia Institute of Technology. Atlanta Rutgers University and the University of California. Santa Barbara (2005). https://doi.org/10.1016/B0-08-044854-2/00906-8
5. Topaz, M., Schaffer, A., Lai, K.H., Korach, Z.T., Einbinder, J., Zhou, L.: Medical malpractice trends: errors in automated speech recognition. J. Med. Syst. **42**(8), 153–154 (2018)

6. Mengesha, Z., Heldreth, C., Lahav, M., Sublewski, J., Tuennerman, E.: "I don't think these devices are very culturally sensitive."—impact of automated speech recognition errors on African Americans. Front. Artif. Intell. **4**, 169. (2021)
7. Koenecke, A., Nam, A., Lake, E., Nudell, J., Quartey, M., Mengesha, Z., et al.: Racial disparities in automated speech recognition. Natl. Acad. Sci. **117**(14), 7684–7689 (2020)
8. Harwell, D.: "The Accent Gap". The Washington Post. https://www.washingtonpost.com/gra phics/2018/business/alexa-does-not-understand-your-accent/ (2018). Last accessed 14 Aug 2023
9. Tatman, R.: Gender and dialect bias in YouTube's automatic captions. In: Proceedings of the First ACL Workshop on Ethics in Natural Language Processing, pp. 53–59 (2017)
10. Zea, J.A., Aguiar, J.: "Spanish Políglota": an automatic Speech Recognition system based on HMM. In: 2021 Second International Conference on Information Systems and Software Technologies (ICI2ST), pp. 18–24. IEEE (2021)
11. Hernández-Mena, C.D., Meza-Ruiz, I.V., Herrera-Camacho, J.A.: Automatic speech recognizers for Mexican Spanish and its open resources. J. Appl. Res. Technol. **15**(3), 259–270 (2017)
12. Vaswani, A., et al.: Attention is all you need. Advances in neural information processing systems (2017)
13. Radford, A., Kim, J. W., Xu, T., Brockman, G., McLeavey, C., Sutskever, I.: Robust Speech Recognition via Large-Scale Weak Supervision. ArXiv (2022)
14. Vosk Documentation. https://alphacephei.com/vosk/. Last accessed 14 Aug 2023
15. Povey, D., et al.: The Kaldi speech recognition toolkit. In: IEEE 2011 workshop on automatic speech recognition and understanding (No. CONF). IEEE Signal Processing Society (2011)
16. Lee, A., Kawahara, T.: Recent development of open-source speech recognition engine Julius. In: Asia-Pacific Signal and Information Processing Association Annual Summit and Conference (APSIPA ASC), pp.131–137. Asia-Pacific Signal and Information Processing Association (2009)
17. Hannun, A., et al.: Deep Speech: Scaling up end-to-end speech recognition (2014)
18. DeepSpeech Documentation. https://deepspeech.readthedocs.io. Last accessed 14 Aug 2023
19. DeepSpeech Python Library. https://pypi.org/project/deepspeech. Last accessed 14 Aug 2023
20. Maier, V.: Evaluating ril as basis for evaluating automated speech recognition devices and the consequences of using probabilistic string edit distance as input. 3rd year project. Sheffield University (2002)
21. Szymański, P., et al.: WER we are and WER we think we are. In: Findings of the Association for Computational Linguistics: EMNLP 2020, pp. 3290–3295. Online. Association for Computational Linguistics (2020)
22. Morris, A.C.: Maier, V., Green, P.: From WER and RIL to MER and WIL: improved evaluation measures for connected speech recognition. Interspeech (2004)
23. TouchMetrics Homepage. https://torchmetrics.readthedocs.io. Last accessed 14 Aug 2023
24. Morris, A.C.: An information theoretic measure of sequence recognition performance. IDIAP (2003)
25. Kincaid,J: Challenges in Measuring Automatic Transcription Accuracy. https://medium.com/ descript/challenges-in-measuring-automatic-transcription-accuracy-f322bf5994f (2018)

Exploring User Acceptance of AI Image Generator: Unveiling Influential Factors in Embracing an Artistic AIGC Software

Biao Gao[1,2]([✉]), Huiqin Xie[2], Shuangshuang Yu[2], Yiming Wang[2], Wenxin Zuo[2], and Wenhui Zeng[2]

[1] Jiangxi Publishing and Media Group Co., Ltd, Nanchang 330038, China
biaogao.edu@outlook.com
[2] Jiangxi University of Finance and Economics, Nanchang 330013, China

Abstract. This study aims to explore the user acceptance of AI Image Generators and offers an analysis based on their distinct features. Drawing from the Technology Acceptance Model (TAM), we examined the impact of factors such as creativity, accuracy, image quality, interactivity, and efficiency on users' willingness to adopt. Furthermore, we evaluated the mediating effects of perceived usefulness and perceived ease of use. Data analysis reveals that both perceived usefulness and ease of use play a significant role in users' intentions to adopt the AI Image Generator technology. The features of the AI Image Generator significantly influence users' perceptions of its usefulness and ease of use. Additionally, perceived ease of use has a notable effect on perceived usefulness. The insights from this research offer valuable guidance for the design, development, and promotion of AI Image Generators, facilitating the industry in catering better to users' needs and promoting its adoption in the art world.

Keywords: TAM · AIGC · AI Image Generator · Acceptance

1 Introduction

In recent years, the surge in Artificial Intelligence (AI) has catalyzed revolutionary changes across a myriad of sectors, encompassing healthcare, education, transportation, commerce, politics, finance, security, and even warfare [5]. This influence unmistakably extends to the realm of digital arts and creative content generation.

Delving deeper into the technology, AI Image Generators, such as Midjourney or Stable Diffusion, are underpinned by neural networks or machine learning algorithms. These neural networks, mimicking a rudimentary brain, comprise numerous interlinked elementary units. Predicated on the task at hand, these networks analyze the artworks uploaded to their database, identifying imagery, techniques, and stylistic hallmarks, subsequently leveraging this newfound knowledge in creative endeavors [14].

The booming market for generative AI has witnessed the emergence of a multitude of AI Image Generators. A thorough investigation into AI Image Generators, as AI-driven software applications, is imperative to pinpoint the key elements affecting user

F. Zhao and D. Miao (Eds.): AIGC 2023, CCIS 1946, pp. 205–215, 2024.
https://doi.org/10.1007/978-981-99-7587-7_17

receptivity and intent to use. By illuminating these factors, we can offer valuable insights for enhancing user experience, refining software design, and promoting wider adoption among artists and creative professionals.

The primary objective of this study is to delve into the user acceptance of AI Image Generators and probe the influencing factors that propel their adoption. Specifically, this research aims to:

First, investigate the influence of key factors such as creativity, accuracy, quality of illustration, interactivity, and efficiency on users' intent to accept.

Second, examine the mediating effects of perceived usefulness and perceived ease of use on the relationship between the identified factors and user acceptance.

Third, offer insights and recommendations for the development and enhancement of AI-driven art content generation software, fostering its integration into the arts community.

To accomplish these research objectives, we will employ a mixed-methods approach, marrying quantitative surveys with qualitative interviews. Our subjects will comprise artists, designers, and creative professionals with experience in using AI Image Generators. Data analysis techniques, such as regression analysis and thematic analysis, will be utilized to study the interrelationships between variables and uncover underlying patterns and themes.

2   Literature Review

2.1 AI-based Art Content Generation and Relevant Theoretical Frameworks

The intersection of computer science, specifically AI, and artistry is primarily attributed to the emergence of theoretical frameworks like Generative Adversarial Networks (GANs) and Variational Autoencoders (VAEs) [2].

Key theoretical frameworks employed in AI-based art content generation encompass GANs, VAEs, and Convolutional Neural Networks (CNNs).

Firstly, the Generative Adversarial Network (GAN) is a deep learning architecture devised for the creation of novel images and artworks. Comprising a generator and discriminator, the GAN operates as a tandem of neural networks. While the generator processes the image, the discriminator evaluates it, fostering a continuous symbiotic tug-of-war of learning. Through iterative training, GANs have proven adept at crafting astonishingly lifelike synthetic images, as showcased by creations like DeepArt and DeepDream. This innovation has profoundly accelerated AI's application in the visual art creation process [2].

Secondly, Variational Autoencoders (VAEs) function as probabilistic models tailored for learning and generating data distributions. Within generative artistry, VAEs can craft new images mirroring specific characteristics by learning the latent space representations of input imagery [10].

Lastly, the Convolutional Neural Network (CNN) is a deep learning model intrinsically designed to handle grid-like structured data such as images. Through convolutional layers, CNNs autonomously and adaptively decipher spatial hierarchical features, proficiently recognizing images and videos. Based on this technology, AI-driven art content generation achieves feats like style transfer [7].

While these frameworks differ in their generative techniques and structures, their unified goal is the creation of new artistic content. Typically rooted in deep learning and supported by robust neural networks, they can be synergistically combined to accomplish more intricate tasks.

2.2 AI-based Art Content Generation and Relevant Theoretical Frameworks

User acceptance refers to the user's attitude and acceptance of a technology, product, or service. Its research can assess and predict users' attitudes towards new technologies. Key models proposed in past literature, like the Technology Acceptance Model (TAM), the Unified Theory of Acceptance and Use of Technology (UTAUT), and the Extended Expectation-Confirmation Model (EECM), are employed to examine user intentions towards new technology. Of all the information system usage theories, TAM, introduced by Davis in 1989, stands out as one of the most influential and commonly applied theoretical frameworks for understanding technological acceptance. The TAM, built on the foundation of the Theory of Reasoned Action (TRA) and the Theory of Planned Behavior (TPB), aims to predict user intention through two perceived attributes [4].

Numerous studies reveal that a user's perception of the usefulness of a technology or system plays a significant role in influencing its acceptance and adoption behaviors. For the AI Image Generator, when users deem it as a valuable tool that delivers relevant and real-time information, personalized recommendations, and efficient drawing capabilities, they're more inclined to embrace it. Thus,

H1: Perceived usefulness significantly affects the intention to accept AI drawing tools.

Research indicates that users' perceptions of the ease of use of a technology or system critically affect their acceptance and adoption intentions. When users find the AI Image Generator user-friendly, having an intuitive interface and straightforward operation, it positively impacts their acceptance intentions. This perceived ease of use is vital in shaping user attitudes and behavioral intentions towards new technologies. Therefore,

H2: Perceived ease of use significantly influences the intention to adopt AI drawing tools.

For users of the AI Image Generator, perceived ease of use is paramount as it directly affects their perception of the software's usefulness. When users find the AI Image Generator's interface to be intuitive, easy to learn, and operate, they can better understand and master its features. They'll perceive the AI Image Generator as a practical and valuable tool since they can effortlessly harness its diverse functionalities during usage, thereby enhancing their perception of the software's usefulness. Thus,

H3: Perceived ease of use significantly influences the perceived usefulness of AI drawing tools.

2.3 Characteristics of AI Image Generator

AI Image Generators are distinguished by their adaptability. They can mimic artistic styles, autonomously generate artworks, and assist in creation, thus allowing artists to materialize their ideas more efficiently [11]. Additionally, they possess the traits of versatility and diversity. By referencing a multitude of artworks, they can reproduce the unique techniques and styles of different artists, rendering diverse aesthetic styles [12].

AI-driven artistic content generation has recently garnered significant attention. Research in this domain has explored the capacities and limitations of AI systems, their influence on the creative process, and perceptions towards AI-generated art. However, the user acceptance of AI Image Generators like Midjourney requires further exploration.

Creativity is the key for users to measure the utility of artificial intelligence image generators. This ability includes positive and useful creative ideas, products, and innovative processes. Users' self-efficacy can be improved when they believe in their creative potential [8]. Thus,

H4: Creativity significantly influences the perceived usefulness of AI drawing tools.

Accuracy, given the subjective nature of artistic preferences, is paramount. AI content generators cater to individuals with diverse artistic tastes and preferences. This accuracy in reflecting user preferences enhances user trust and satisfaction, influencing their intention to use the software. Thus,

H5: Accuracy substantially affects the perceived usefulness of AI drawing tools.

The quality of the generated image is a critical factor in assessing the usefulness of AI Image Generator software. This quality is reflected in the clarity, detail, realism, and artistic effect of the images. When users feel that the images generated by the software are of exceptional quality and meet their artistic standards, it reinforces their perception of the software's utility. Therefore,

H6: The quality of the generated artwork significantly affects the perceived usefulness of AI drawing tools.

The quality of the artwork generated is an important factor affecting how users perceive the ease of use of the AI Image Generator. The quality can be measured in terms of image clarity, detail, realism, and artistic effect. When users perceive the artwork generated by the software as high-quality and in line with their artistic standards, they find it easier to master and use. Thus,

H7: The quality of the generated artwork significantly impacts the perceived ease of use of AI drawing tools.

Interactivity is a key feature of the AI Image Generator. Through touch screens, gesture recognition, or drawing pads, users can directly participate in the drawing process, receiving real-time feedback. This interactivity strengthens the bond between users and the software, elevating the user experience and fostering greater engagement and satisfaction. Its interactive capabilities attract users to be actively involved in the creation process. With an intuitive interface and instantaneous feedback, users can explore and experiment more freely, which enhances their connection and engagement with the software. Therefore,

H8: Interactivity significantly affects the perceived ease of use of AI drawing tools.

Efficiency is another pivotal attribute of the AI Image Generator, encompassing both software performance and streamlined workflows. The software excels when processing large images or intricate effects, maintaining high speeds and smooth operations. Additionally, the AI Image Generator offers a user-friendly workflow, enabling users to complete their creative tasks with increased efficiency. This heightened efficiency inclines users to adopt the software, relishing the convenience and outcomes it offers.

The software's efficiency significantly influences how users perceive its ease of use. Therefore,

H9: Efficiency significantly affects the perceived ease of use of AI drawing tools.

The following proposed conceptual framework illustrates the relationships among variables in this study (Fig. 1).

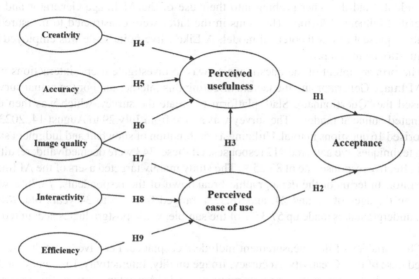

Fig. 1. Conceptual model.

3 Research Methodology

For this study, we predominantly employed a questionnaire survey method to gather sample data. After meticulous selection from existing measurement scales, modifications were made, and new scales were integrated to construct the research questionnaire. Prior to its official distribution, a preliminary survey was conducted on a smaller scale. Feedback and suggestions from various sources led to revisions, culminating in the finalized questionnaire version. Once data collection was complete, we utilized the SPSS statistical software to undertake a descriptive statistical analysis, which provided insights into the demographics of our sample and their experiences with the AI Image Generator. Additionally, we rigorously verified the reliability and validity of our sample data, ensuring the accuracy of our research findings.

We use TAM as a research method because it has a significant advantage. The structure of the TAM model is relatively simple, easy to understand and operate, and is supported by extensive empirical research. It can study people's acceptance or willingness to accept new technologies and information systems. It is not only easy to measure but also can be extended and modified according to the needs of specific research. On the whole, TAM is very suitable for the research method of this study.

4 Data Analysis and Results

4.1 AI-based Art Content Generation and Relevant Theoretical Frameworks

The questionnaire design was rooted in a comprehensive review of pertinent literature. It was bifurcated into two sections: one delving into the basic demographics of the respondents, and the other probing into their use of the AI Image Generator and the associated influential factors. The items in the latter were constructed to measure the variables present in our theoretical model. A Likert five-point scale was employed for the questionnaire setup.

The primary intent of the questionnaire was to investigate users' interactions with the AI Image Generator and the factors affecting this interaction. For the actual survey, we used the "Questionnaire Star" platform to create the survey, which was then disseminated online at random. The survey was active from July 29 to August 14, 2023. It comprised 16 questions in total. Utilizing a combination of snowball and judgment sampling techniques, we gathered 412 responses. Of these, 343 were deemed valid, resulting in an effective response rate of 83.25%. The study mainly targeted users of the AI Image Generator. In terms of the demographic breakdown of the respondents, 73.78% were between the ages of 18 and 24, and 18.73% ranged from 25 to 34 years old. Moreover, undergraduates made up 57.35% of the sample, while postgraduates accounted for 28.82%.

The variables of this measurement include Acceptance, Perceived usefulness, Perceived ease of use, Creativity, Accuracy, Image quality, Interactivity, Efficiency, and so on, all of which refer to the previous maturity scale. Acceptance set up five measurement items [16], such as "Using AI image generator is a great choice for me." Perceived usefulness has set up a total of 5 measurement items [6], including "Using AI image generator made the process of creating art easier and more effortless for me." Perceived ease of use sets four measurement items [9], such as "Learning to use AI image generator has been easy for me." Creativity set four measurement items [1], such as "AI image generator is a creative tool for creating artwork." Accuracy set up four measurement items [15], such as "The generated images in AI image generator are based on user input prompts." Image quality sets five measurement items [3], such as "The imaging styles generated by AI image generator are comprehensively diversified." Interactivity sets five measurement items, such as "The interactive features of AI image generator meet my artistic requirements." Efficiency sets five measurement items, such as "Using an AI image generator is user-friendly and easy for me to operate[13]."

4.2 Skewness and Kurtosis Statistical Analysis

In data analysis, skewness and kurtosis are typically used to determine the normality of the data. The results show that the absolute values of the skewness for each variable factor in this measurement (with the exception of CR1, CR2, AC1, AC3, and PU1) are all less than 1, and the absolute values for kurtosis are all less than 7. This largely indicates conformity to univariate normality. Specifically, CR1 (with an absolute skewness value of 1.238), CR2 (1.017), AC1 (1.110), AC3 (1.043), and PU1 (1.171) all approach a value of 1. This suggests that these factors align with the trend of univariate normality. However, due to constraints in the time and budget allocated for the survey distribution, there was an insufficiency in the collected sample data, which resulted in a non-significant normality. Subsequent experiments could address this issue by increasing the sample size.

Additionally, the Cronbach's Alpha values for each sub-item are all greater than 0.75 (ranging from 0.754 to 0.874), verifying that the internal validity of each measurement variable is sound.

4.3 Variable Correlation Analysis

The data analysis results show that the absolute values of the correlation coefficients for each measured variable are all greater than 0.4 and fall within the 0.491 to 0.809 range, indicating a moderate correlation among the variables.

Furthermore, some variables exhibit absolute correlation coefficients greater than 0.7. For instance, the correlation coefficient between creativity and interactivity has an absolute value of 0.711; between interactivity and usefulness, it's 0.809. This indicates a high degree of correlation between these variables.

4.4 Variable Correlation Analysis

This path analysis consists of four sets of evaluations. In the first set, perceived usefulness and perceived ease of use are the independent variables, with acceptance intention as the dependent variable. The second set employs perceived ease of use as the independent variable, focusing on perceived usefulness as the dependent variable. The third set has creativity, accuracy, and drawing quality as independent variables, targeting perceived usefulness as the dependent variable. For the fourth set, drawing quality, interactivity, and efficiency serve as independent variables, with perceived ease of use being the dependent variable.

The first set of path analysis. Typically, R2 (the square of the multiple correlation coefficient) indicates the extent to which a model fits the data. The data analysis results show that the measured data has an R2 value of 0.653, significantly greater than 0. This demonstrates that the data from our sample measurement has a robust explanatory power.

A significant ANOVA indicates that at least one independent variable has explanatory power. Given that the ANOVA value in this group of data analysis results is 0.000, which is less than 0.05, this implies that at least one independent variable is related to the dependent variable.

If the p-value derived from the T-test is less than 0.05, it indicates that the corresponding independent variable has a significant impact on the dependent variable. And data reveal that both perceived usefulness and perceived ease of use have p-values of 0.000, significantly less than 0.05. This signifies that both perceived usefulness and ease of use have a notable influence on acceptance intention.

A Durbin-Watson test value close to 2 indicates the data's independence. With a test value of 2.073, which is proximate to 2, the data demonstrates good independence. Moreover, the VIF values for both perceived usefulness and ease of use are less than 5, indicating the absence of multicollinearity.

The second set of path analysis. The data analysis results show that with an R2 value of 0.384, exceeding 0, it suggests that this group's data provides a moderate explanatory strength.

The ANOVA value stands at 0.000, which is below the 0.05 threshold, indicating that at least one independent variable is significantly related to the dependent variable. Notably, the p-value for perceived ease of use is a striking 0.000, significantly less than 0.05, which signifies its substantial influence on perceived usefulness. Additionally, the Durbin-Watson test value for this group is 1.801, closely approximating 2, suggesting a commendable independence of the data. The VIF value for perceived ease of use in this group's measurement is significantly below 5, which implies an absence of multicollinearity.

The third set of path analysis. The data results show that the R2 value is 0.604, which is significantly greater than 0, indicating that the regression model is meaningful for this group. The ANOVA value was 0.000, less than 0.05, indicating that at least one independent variable was significantly correlated with the dependent variable. Specifically, the p-values of Creativity, Accuracy, and Image quality were significantly less than 0.05, between 0.000 and 0.007. This shows that creativity, accuracy, and painting quality have an important impact on perceived usefulness. In addition, the Durbin-Watson test value of this group is 1.735, which is approximately 2, indicating that the data has good independence. In this group of measurements, the VIF values of the independent variables (Creativity, Accuracy, and Image quality) are all lower than 5, indicating that there is no multicollinearity problem.

The fourth set of path analysis. With an R2 value of 0.595, notably exceeding 0, this suggests that the regression test for this group possesses a robust explanatory capability. The ANOVA value registers at 0.000, which is below the 0.05 threshold, implying that at least one independent variable is significantly related to the dependent variable. Moreover, the Durbin-Watson test value for this set is 1.973, hovering close to 2, affirming a commendable data independence. In this group's measurement, the VIF values for the independent variables—Image quality, Interactivity, and Efficiency – all stand below 5 (2.263–3.484), indicating no issues of multicollinearity.

Unstandardized coefficients are used to gauge the significance of regression coefficients and are typically utilized in the regression model equation. In contrast, standardized coefficients are employed to ascertain the extent to which independent variables influence dependent variables. A larger value indicates that the variable exerts a more potent impact on the dependent variable. As shown in Table 1, this test will use the non-standardized coefficient B value and P value in the previous four groups of path analysis to be used in the hypothesis test table and the subsequent path analysis chart.

As shown in Table 1, hypotheses H1, H2, H3, H4, H5, H6, H8, and H9 are confirmed.

As depicted in Fig. 2, the model has been refined based on data from the preceding four path analyses and hypothesis tests. The unstandardized coefficients for the paths all consistently exceed 0.1, specifically ranging between 0.183 and 0.753. This indicates that the corresponding predictor variables significantly influence their outcome variables. Most prominently, the effect of perceived usefulness on the intention to accept stands out, boasting an unstandardized coefficient of 0.753.

Table 1. Hypothesis testing.

The hypotheses	Path coefficients	P-values	Results
H1 Perceived usefulness → Acceptance	.753	***	yes
H2 Perceived ease of use → Acceptance	.230	***	yes
H3 Perceived ease of use → Perceived usefulness	.641	***	yes
H4 Creativity → Perceived usefulness	.253	***	yes
H5 Accuracy → Perceived usefulness	.212	***	yes
H6 Image quality → Perceived usefulness	.338	***	yes
H7 Image quality → perceived ease of use	-.092	n.s	no
H8 Interactivity → Perceived ease of use	.235	***	yes
H9 Efficiency → Perceived ease of use	.705	***	yes

Note: ***: p < 0.001, **: p < 0.01,*: p < 0.05

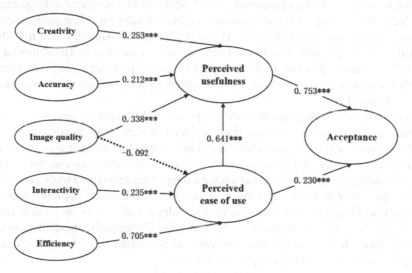

Fig. 2. Path analysis.

5   Conclusions

This study employed TAM to investigate how characteristics of artificial intelligence art generation software (such as creativity, accuracy, paint quality, interactivity, and efficiency) influence perceived usefulness and perceived ease of use. The goal was to predict user acceptance of AI art generation software. Our empirical research yielded several insights.

Hypotheses 1 and 2 of this study confirm positive relationships between perceived usefulness, perceived ease of use, and user intentions towards AI Image Generator. The

TAM framework supports these findings, showing that positive evaluations of technology's benefits and value, as well as smooth visual design and interactions, contribute to user acceptance. Both perceived usefulness and ease of use address user needs and enhance the user experience, leading to greater adoption of AI Image Generators.

Hypothesis 3 confirms that the user's perceived ease of use has a positive impact on perceived usefulness, and this effect is very significant. Users feel relaxed about the use of AI Image Generator, which will increase their sense of efficacy in the use of AI Image Generator.

Hypotheses 4, 5, and 6, we identified creativity, accuracy, and image quality as salient features of the AI Image Generator that significantly impact users' perceived usefulness. Users relish the experience of crafting artwork with a creative tool. Users utilize the AI Image Generator for assisted creation, which in turn provides them with information and inspiration. Thus, user perceptions of feedback usefulness tend to be holistic. In this study, accuracy, and image quality as salient features of the AI Image Generator that significantly impact users' usefulness was confirmed.

Hypotheses 7, 8, and 9 suggest that while image quality positively impacts perceived usefulness, it does not affect perceived ease of use. From this perspective, for perceived usefulness, users using AI Image Generators focus on achieving their goals, focusing on the efficiency and interactivity of AI Image Generators. Among them, hypothesis 8 affirms that interactivity positively affects perceived ease of use. The smoother and easier the interactive function is, the smoother the user experience is, and the enhanced perceived ease of use. Hypothesis 9 confirms efficiency's significant positive impact on perceived ease of use.

This study delves deep into the user acceptance of AI Image Generators, offering a meticulous analysis tailored to the unique attributes of such tools. Beyond the conventional dimensions of perceived usefulness and ease of use, factors such as image quality, interactivity, creativity, and accuracy emerge as pivotal in shaping user adoption. These elements highlight the distinction between AI Image Generators and traditional digital utilities, underscoring the latent value of AI in artistic endeavors. Applying TAM further reaffirms the central roles of perceived usefulness and ease of use. These remain paramount when users evaluate novel technologies. Collectively, these insights offer potent guidance for the design, development, and promotion of AI Image Generators, aiding the industry in better catering to user needs.

Acknowledgments. This work was supported by Jiangxi Provincial University Humanities and Social Sciences Research Project [Grant number GL22223]. This work was also supported by the Science and Technology Research Project of the Jiangxi Provincial Department of Education: [Grant number GJJ2200517].

References

1. Gralewski, J., Jankowska, D.M.: Do parenting styles matter? Perceived dimensions of parenting styles, creative abilities and creative self-beliefs in adolescents. Thinking Skills Creativity **38**, 100709 (2020)
2. Cetinic, E., She, J.: Understanding and creating art with AI: review and outlook. ACM Trans. Multimed. Comput. Commun.and Appl. **18**(2), 1–22 (2022)

3. Chi, T.: Understanding Chinese consumer adoption of apparel mobile commerce: an extended TAM approach. J. Retail. Consum. Serv. **44**, 274–284 (2018)
4. Dillon, A., Morris, M.G.: User acceptance of information technology: theories and models. In: Williams, M. (ed.) Annual Review of Information Science and Technology, vol. 31, pp. 3–32. Information Today, Medford NJ. [Verified 11 Dec 2011; preprint]. http://www.ischool.utexas.edu/~adillon/BookChapters/User%20acceptance.htm (1996)
5. Ford, M.: Rule of the Robots: How Artificial Intelligence will Transform Everything. Basic Books, Hachette (2021)
6. Hendrickson, A.R., Massey, P.D., Cronan, T.P.: On the test-retest reliability of perceived usefulness and perceived ease of use scales. MIS Q. **17**(2), 227 (1993). https://doi.org/10.2307/249803
7. Jing, Y., et al.: Learning graph neural networks for image style transfer. In: Avidan, S., Brostow, G., Cissé, M., Farinella, G.M., Hassner, T. (eds.) Computer Vision – ECCV 2022: 17th European Conference, Tel Aviv, Israel, 23–27 Oct 2022, Proceedings, Part VII, pp. 111–128. Springer Nature Switzerland, Cham (2022). https://doi.org/10.1007/978-3-031-20071-7_7
8. Karwowski, M.: It doesn't hurt to ask … but sometimes it hurts to believe: polish students' creative self-efficacy and its predictors. Psychol. Aesthetics Creativity Arts **5**(2), 154–164 (2011)
9. Kitchakarn, O.: How students perceived social media as a learning tool in enhancing their language learning performance. Turkish Online J. Educ. Technol.-TOJET **15**(4), 53–60 (2016)
10. Larsen, A.B.L., Sønderby, S.K., Larochelle, H., Winther, O.: Autoencoding beyond pixels using a learned similarity metric. In: Proceedings of the 33rd international Conference on Machine Learning, vol. 48, pp. 1558–1566 (2016)
11. Lyu, Y., Wang, X., Lin, R., Wu, J.: Communication in human–AI co-creation: perceptual analysis of paintings generated by text-to-image system. Appl. Sci. **12**(22), 11312 (2022)
12. Mazzone, M., Elgammal, A.: Art, creativity, and the potential of artificial intelligence. Arts **8**(1), 26 (2019)
13. Sternad Zabukovšek, S., Deželak, Z., Parusheva, S., Bobek, S.: Attractiveness of collaborative platforms for sustainable e-learning in business studies. Sustainability **14**(14), 8257 (2022)
14. Trach, Y.: Artificial intelligence as a tool for creating and analysing works of art. Cult. Arts Modern World **22**, 164–173 (2021)
15. Ullah, Z., Ahmad, N., Scholz, M., Ahmed, B., Ahmad, I., Usman, M.: Perceived accuracy of electronic performance appraisal systems: the case of a non-for-profit organization from an emerging economy. Sustainability **13**(4), 2109 (2021)
16. Zhang, X., Liu, Y.: Research on the influencing factors of users' acceptance willingness of yuanyu library from the perspective of configuration. Libr. Theory Pract. (03), 73–85 (2023). (in China)

File Management System Based on Intelligent Technology

Fan Zhang(✉) ⓘ

Shenzhen University, Shenzhen, China
352462119@qq.com

Abstract. With the rapid development of China's economy and the popularization of global Internet technology, China's archives management is facing new opportunities and challenges how to use modern scientific and technological means to establish and perfect the archives management system independently and comprehensively, and prevent the archives risk becomes very important. This paper introduces the whole archives management system based on intelligent technology, and provides reference for archives related departments to improve service and archives risk management level.

1 The Introduction

Archives management as an important part of management work, facing the challenge of advanced intelligent technology comprehensive use of leapfrog development, the traditional archives management mode has been unable to adapt to the current world of computer network information development trend. At present, there are still many deficiencies in domestic archives management system, such as poor confidentiality, low access efficiency, poor maintainability, a large number of data in hibernation, unable to provide users with convenient and inefficient access, etc. all of which are caused by traditional and inefficient manual management methods in the final analysis. With the acceleration of the urbanization process, the population is more and more concentrated, the amount of archival related data and demand is more and more huge, demand frequency is more and more high, whether it is complete, access is efficient, convenient maintenance has become a landmark achievement to measure the information construction of archives management system. In view of this realistic demand, more and more enterprises begin to pay attention to this problem and constantly try to apply advanced programming framework and technology to the design of archives management system. Through the continuous runnin of business process, gradually improve the support strength of relevant software for historical archives and real business, and greatly improve the utilization rate of archives data.

2 Overview of Artificial Intelligence and Archive Management

Artificial intelligence technology is a new subject that studies how to make computer simulate human perception and action. It is developed on the basis of the integration of computer science, economic psychology and other disciplines. The report of the 19th

F. Zhao and D. Miao (Eds.): AIGC 2023, CCIS 1946, pp. 216–223, 2024.
https://doi.org/10.1007/978-981-99-7587-7_18

National Congress proposed to elevate artificial intelligence to a national strategy, and a series of plans such as the Implementation Plan of Internet + Artificial Intelligence Action show that artificial intelligence is a new technology engine driving social and economic development in the new era. The application scenarios of artificial intelligence are integrated with various industries to constantly stimulate new work needs. Artificial intelligence extracts key technologies to achieve deep integration of various industries, such as realizing intelligent application of medical service diagnosis with the help of artificial intelligence technology. Artificial intelligence focuses on deep integration with various industries, enabling the transformation and development of the current real economy. Artificial intelligence related theories and technologies gradually develop and mature, changing the original way of thinking and production structure of the industry. With the in-depth development of artificial intelligence technology, many artificial intelligence application products based on pattern recognition and deep learning have emerged, involving fields such as health care and intelligent education. At present, the market application of ARTIFICIAL intelligence invested in the field of archives remains at a weak level.

2.1 Artificial Intelligence Technology

In the whole artificial intelligence technology system identification technology belongs to the most basic and the most core technology, so far in all the intelligent technology intelligent identification is the most widely used and developed the most mature technology. In all kinds of technical systems adopted at the present stage, intelligent recognition technology is equivalent to the search box on the search engine. With the continuous and rapid development of technology, when retrieving all kinds of archival information and content, not only the text form can be used to expand but also the voice can be used to complete the retrieval. Based on the form of pictures, intelligent analysis and judgment can be carried out. Especially at present, these technologies are widely used in the management of archives in many industries.

2.2 Intelligent Management of Archives Under Artificial Intelligence Technology

The commencement of any industry in any enterprise service archives management will be safety problems as the main concern, if you can't to ensure the safety of the archives information there are all kinds of information and private data can be exposed at any time in front of the crowd especially some of the associated with medical group or a government agency confidential documents, in the management of these files, security is the first step. At this point in the management of the iris recognition and fingerprint recognition technology will be safe hidden trouble to eliminate effectively. For example, when the relevant confidential documents are consulted, the fingerprint and iris of the reader need to be identified through the system. Only the fingerprint and iris can be matched with the previous permission Settings to open these files, and these confidential documents can only be consulted by special personnel, which greatly improves the security.

Intelligent System Identification Technology. In the whole artificial intelligence technology system identification technology belongs to the most basic and the most core technology, so far in all the intelligent technology intelligent identification is the most widely used and developed the most mature technology. In all kinds of technical systems adopted at the present stage, intelligent recognition technology is equivalent to the search box on the search engine. With the continuous and rapid development of technology, when retrieving all kinds of archival information and content, not only the text form can be used to expand but also the voice can be used to complete the retrieval. Based on the form of pictures, intelligent analysis and judgment can be carried out. Especially at present, these technologies are widely used in the archives management of many industries.

Fingerprint Iris Recognition Technology. Any enterprise or institution in any industry will pay attention to the security problem as the most important issue in the management of archives. If the security of archive information cannot be ensured, all kinds of information and private data may be exposed to the masses at any time, especially some confidential documents related to medical groups or government agencies. In the management of these files, security should be first carried out. At this point in the management of the iris recognition and fingerprint recognition technology will be safe hidden trouble to eliminate effectively. For example, when the relevant confidential documents are consulted, the fingerprint and iris of the reader need to be identified through the system. Only the fingerprint and iris can be matched with the previous permission Settings to open these files, and these confidential documents can only be consulted by special personnel, which greatly improves the security.

Network Platform Sharing Technology. In the process of archives management, the work of borrowing, looking up and returning files is relatively complicated. In the traditional way of management, the whole process generally uses paper documents to the higher level for approval. If the content of borrowing involves some confidential documents or confidential data, the approval process will be more complicated. In order to improve the efficiency and quality of all kinds of work, a large number of examination and approval procedures can be effectively simplified if the platform sharing technology is applied scientifically. Various examination and approval work can be completed on the network platform in the form of OA software and computer. In addition, within the group, some files can be made public. These files can be directly displayed on the platform through sharing technology, and internal employees can directly check these contents after logging in with their IDS.

Intelligent Retrieval Technology. Artificial intelligence technology belongs to the new and high technology and high technology content, the original research purpose for such technology will work efficiency and realize the liberation of productive forces, is widely used in archives management can also be a technology allowing these goals to effective implementation, through the technology of auxiliary can lead to significantly higher efficiency of archives management and even to improve the security of the file. At present, the retrieval efficiency can be effectively improved through the use of artificial intelligence technology for all kinds of information retrieval, so that the staff management pressure can be relieved. For example, when different files are type setted, if they can be arranged in a certain order according to the specific value and importance of the files, some important information can be obtained first according to this standard in the later retrieval of these files. And through the intelligent retrieval form not only can be retrieved from a large number of information data in the short term, but also through technology will be the user's retrieval for statistical results and the application of multiple keywords, the user retrieval expected future large data analysis and application requirements, through the system to match the highest file and keywords are recommended.

2.3 The Impact of Artificial Intelligence on Archive Management

Archives intelligent management services require information technology support such as artificial intelligence, and intelligent service infrastructure to diversified intelligence of service methods requires the integration and embedding of artificial intelligence technology. Artificial intelligence brings new opportunities and challenges to personnel file management, which is reflected in the impact of artificial intelligence technology on conventional file positions, and the application of artificial intelligence technology to improve the efficiency of personnel file services. At the end of 2016, physicist Hawking published an article pointing out that the rise of artificial intelligence has caused a wave of unemployment to spread to the middle class. Repetitive exercises in the era of artificial intelligence can be completed by machines. Technological innovation will change the traditional division of labor in the industry, and Facebook only had 13 employees when it bought Instagram. Traditional archival work is mainly the use of collection and collation services, archival collation as a procedural step, artificial intelligence can replace manual work after the introduction, file query and other work in the digital driven by self-service query machine and other auxiliary realization. Artificial intelligence technology has a wide application value in archive informatization, etc., after the emergence of artificial intelligence technology, paperless office changes the traditional file management mode, office automation forms electronic documents into the file management system, through the intelligent retrieval of archives to shorten the retrieval time.

3 Design and Process of Intelligent Models for File Management

The intelligent model of archives management is based on the archiving range, storage term table and classification table to complete automatic classification modeling. Secondly, mature intelligent processing technology (natural language processing technology, OCR character recognition technology) is used to analyze and process the historical archive data that has formed the classification number and storage period, and complete the training of the model. The larger the training data set, the higher the accuracy. At the same time, the model supports repetitive training and self-learning functions, and can be retrained according to the feedback of use. Finally, the trained model can structuralize the archived electronic files, extract the characteristic information of the electronic files, and automatically classify the file types and divide the storage period.

3.1 System Architecture

Intelligent file management is mainly to structure the electronic files of the archive collation library, extract profile file feature information, automatically determine the classification and retention period of archived files.

3.2 Process Flow

Intelligent classification module, used to output automatic classification results for archive files, including the classification results of the file category and the classification result of the file retention period, the classification results are provided in the form of API interfaces. After the profile data is uploaded, the machine automatically selects the time and file type labels. After the electronic file management system requests the intelligent classification and collation interface, the processing process of the intelligent classification and collation algorithm is described below (Fig. 1):

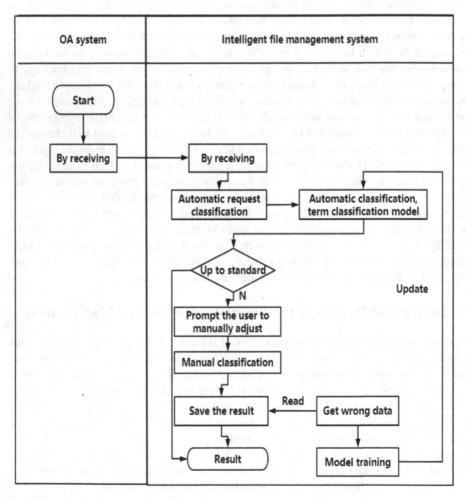

Fig. 1. The processing process of the intelligent classification and collation algorithm

When the OA system initiates the receipt, when the electronic file management system receives the file, call the Automatic Classification Of Collation Data interface to request the classification model of file type and retention period, where the process includes:

(a) Data reading, content extraction. Start data processing
(b) Through the model + Rules predict whether a document is archived
(c) For archived documents, adoption of (10 years, 30 years, permanent) and 0–5 major
    categories (first class) and 41 sub-categories (second class)
(d) Combine data, the interface returns

After the model calculation is completed, the classification, storage period and results are returned to the electronic file management system.

The electronic file management system judges the classification results according to the confidence level, automatically saves the results if the confidence level meets the standard, and highlights the user to manually classify if the confidence level meets the standard.

Intelligent collation data regularly read the manual classification data in the database for model improvement training, after the training is completed to produce a new version of the model, the electronic file management system can select the online model version for classification according to the interface.

4   File Intelligent Management Model Implementation

By exploring the application of artificial intelligence technology in the management of archives, the management of archives has been realized from manual operation to intelligent one-click operation, and the efficiency of file collation has been greatly improved, reducing the overall operating cost of file management. At the same time, it is more conducive to promoting the high quality of archival work to liberate archivists from repetitive, cumbersome and low-value-added work.

4.1   The Efficiency of File Identification and Collation Has Been Significantly Improved

Taking a company as an example, about 6500 management documents are formed every year, and the file management personnel first manually identify, classify and arrange the offline physical documents, organize and catalog, and then synchronize and archive the online electronic documents according to the physical document collation, and strive to achieve one-to-one correspondence, and the whole process takes about 2 people and 3 months to complete. After embedding the intelligent classification model in the electronic file management system, the whole process of the same number of management documents can be easily completed in about 2 people in only 1 month.

4.2   The Results of File Identification and Collation Are More Standardized

Under the same rules for classifying archives and dividing the storage period, there are often cases where different file managers have inconsistent results in the identification of the same type of documents, which causes certain troubles for the search and use of archives. By exploring the intelligent classification model and solidifying the classification standards, the problem of inconsistent classification of the same type of documents due to personnel changes can be effectively avoided, and the uniformity and standardization of file management can be effectively ensured.

4.3 The File Identification and Collation Process is More Optimized

Since the 1980s, the construction of archive information has successively gone through the stage of computer-aided management, the stage of file information management under the dual-set system, the stage of electronic file management system and the stage of network management system, and the file management process has also changed. By embedding the intelligent classification model in the electronic file management system, all the identification and collation work is completed directly in the system at one time, so that the file management process develops from manual operation in batches of collection, identification, classification, arrangement, collation and other steps to one-stop management of electronic documents, which further optimizes the file identification and collation process.

5 Conclusion

Artificial intelligence search technology and its continued maturity provide more power-ful technical support and management concepts for archives management, and contribute to better exchange of information and management of archive resources, especially to improve the quality and efficiency of file management. However, some institutions still face challenges such as backward records management models and the status quo such as single maintenance and outdated technology. There are still major obstacles and dif-ficulties in the knowledge and digital construction of information resources. Therefore, archivists and managers must change their thinking in time, introduce new concepts and innovations, and actively introduce various new technologies that enable them to greatly increase people's awareness of the service and even increase the level of accep-tance and acceptance of new technologies. To meet the needs of users, make better use of resources and improve the value of archives are the important goals of promoting archives management.

In summary, the archival data management business cannot be separated from mod-ern artificial intelligence technology. Artificial intelligence, with its own superior con-venience, interactivity and other outstanding features, has become and will continue to become an important force in the file management business, further optimize the task of file management, and make positive contributions to the sustainable development of the unit. Therefore, the continuous realization of intelligent retrieval and automatic management of archives is another important direction for long-term optimization and development of archives management business in the future.

References

1. Harrington, L.: Primer on artificial intelligence used in electronic health records. AACN Adv. Crit. Care **33**(2), 130–133 (2022)

2. Jaillant, L.: Archives, Access and Artificial Intelligence: Working with Born-Digital and Digitized Archival Collections. Bielefeld University Press (2022)
3. Marasini, A., Shrestha, A., Phuyal, S., Zaidat, O.O., Kalia, J.S.: Role of artificial intelligence in unruptured intracranial aneurysm: an overview. Front. Neurol. **13** (2022)
4. Axle ai Media Search and Editing Platform Teams Up with XenData's New X100 Active Archive. M2 Presswire (2022)
5. Wang, S.Y., Tseng, B., Hernandez-Boussard, T.: Deep learning approaches for predicting glaucoma progression using electronic health records and natural language processing. Ophthalmol. Sci. **2**(2) (2022)

Leveraging Machine Learning for Crime Intent Detection in Social Media Posts

Biodoumoye George Bokolo[1](✉)(iD), Praise Onyehanere[2], Ebikela Ogegbene-Ise[3],
Itunu Olufemi[4], and Josiah Nii Armah Tettey[5]

[1] Department of Computer Science, Sam Houston State University,
Huntsville, TX 77340, USA
bgb023@shsu.edu

[2] Department of Architecture, Technology and Engineering, University of Brighton,
Brighton BN2 4AT, UK
p.onyehanere1@uni.brighton.ac.uk

[3] Department of Computer and Software Engineering, Kennesaw State University,
Kennesaw, GA, USA
eogegben@students.kennesaw.edu

[4] Department of Computer Science and Quantitative Methods, Austin Peay State
University, Clarksville, TN 37044, USA
iolufemi@my.apsu.edu

[5] Department of Computer Science and Engineering, Wright State University,
Dayton, OH 45435, USA
tettey.2@wright.edu

Abstract. Detecting crime intent from user-generated content on social
media platforms has become increasingly important for law enforcement
and crime prevention. This paper presents a comprehensive approach
for crime intent detection from user tweets using machine learning techniques. The study utilizes a dataset of about 400,000 tweets and applies
data preprocessing, feature selection, and model training with logistic
regression, ridge regression classifier, Stochastic Gradient Descent (SGD)
classifier, Random Forests, and support vector machine models. Evaluation metrics such as accuracy, precision, recall, and F1 score are employed
to assess the models' performance. The results reveal that the logistic
regression model achieves the highest accuracy ratio of 0.981 in detecting crime intent from tweets. This research showcases the effectiveness
of machine learning and advanced transformer-based models in leveraging social media data for crime analysis. The findings provide valuable insights into the potential for early detection and monitoring of
crime intent using online platforms, contributing to the field of crime
prevention and law enforcement. The utilization of machine learning
techniques offers new avenues for understanding and analyzing crime-
related sentiments expressed by social media users. By accurately detecting crime intent from user-generated content, law enforcement agencies
can enhance their proactive measures, monitor public sentiment towards
crime, and shape policies and interventions to address public concerns
effectively. The research highlights the significance of leveraging social
media data for crime detection and emphasizes the potential impact of

© The Author(s), under exclusive license to Springer Nature Singapore Pte Ltd. 2024
F. Zhao and D. Miao (Eds.): AIGC 2023, CCIS 1946, pp. 224–236, 2024.
https://doi.org/10.1007/978-981-99-7587-7_19

advanced machine learning models in improving public safety and crime prevention efforts.

Keywords: criminal · crime intent · sentiment analysis · machine learning · digital forensics · social media forensics

1 Introduction

In recent years, the unprecedented growth of social media platforms has resulted in an overwhelming amount of data being generated daily [30]. Twitter, one of the most popular platforms, allows users to express their thoughts, opinions, and emotions in real time, making it a valuable resource for studying public sentiment [20]. Leveraging this vast amount of user-generated content, researchers and organizations have shown a keen interest in utilizing social media data to gain insights into various domains, including crime analysis and sentiment detection [15].

Crime detection and analysis have always been of paramount importance for maintaining law and order within societies. Traditionally, law enforcement agencies relied on conventional methods to identify and prevent criminal activities [5]. According to a report by BBC as seen in Fig. 1, there was a downward trend of crime between 2000 and 2010 however, between 2011 and 2021, the trend was mostly stable which is not a good sign. However, with the advent of social media, a new avenue has emerged for understanding and detecting criminal intent. Twitter, in particular, provides a unique platform for individuals to discuss, share news, and express their emotions regarding various events, including criminal activities [2]. Analyzing the sentiment associated with crime-related tweets can potentially offer valuable insights into public perceptions, concerns, and intentions related to criminal behavior [17].

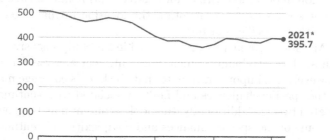

Fig. 1. Violent Crime Statistics in the US Between 2000 and 2021 [10]

The sentiment analysis of Twitter data has gained significant attention due to its potential to capture public sentiment and understand user intentions [1]. Agarwal et al. (2011) discussed the challenges and opportunities in sentiment analysis of Twitter data, highlighting the unique characteristics of the platform such as brevity and informal language. They explored various techniques including lexicon-based approaches, machine learning methods, and rule-based systems. While their research provided valuable insights, it is important to acknowledge some limitations. The study primarily focused on lexicon-based approaches, machine learning methods, and rule-based systems for sentiment analysis. However, it did not extensively explore more advanced techniques such as deep learning models, which have shown promising results in sentiment analysis. Additionally, the study might not have fully accounted for the contextual nuances of crime-related sentiments on Twitter, which can be challenging due to the brevity and informal nature of tweets.

Similarly, Velikovich, Blair-Goldensohn, Hannan, and McDonald (2010) investigated the viability of web-derived polarity lexicons for sentiment analysis, particularly on Twitter data [28]. They generated lexicons using web resources and evaluated their effectiveness in determining sentiment polarity. Although the study highlighted the usefulness of lexicon-based approaches, it is important to consider the limitations of relying solely on pre-built lexicons. Lexicons may not capture all the nuanced expressions of sentiment, especially when it comes to crime-related tweets that may contain slang, abbreviations, or domain-specific language. Consequently, the accuracy and coverage of sentiment analysis in crime-related tweets using lexicons alone could be compromised.

The study by Jurgens, Finethy, and McCorriston (2015) critically analyzed geolocation prediction techniques on Twitter [12]. They reviewed different approaches based on social networks and discussed their strengths and limitations. However, one limitation of geolocation prediction methods is that they heavily rely on user-provided location information, which is often incomplete, inaccurate, or intentionally misleading. This can introduce errors and uncertainty in the geolocation data used for sentiment analysis. Additionally, the study might not have extensively addressed the challenges of accurately associating geolocation information with crime-related sentiments, as tweets discussing crime may not always explicitly mention the location or may refer to broader regions rather than specific coordinates.

Kumar, Morstatter, and Liu (2016) provided a comprehensive overview of Twitter data analytics techniques, including sentiment analysis for crime detection [13]. However, it is important to note that the book's coverage may not delve deeply into the specific challenges and faults associated with sentiment analysis for crime-related tweets. Additionally, the book's content may be generalized and not address some of the specific nuances and complexities of sentiment analysis in the context of crime-intent detection on Twitter.

Furthermore, Zeng, Chen, and Lusch (2010) discussed the concept of social media analytics and its potential in crime analysis [32]. It is important to recognize that sentiment analysis alone may not be sufficient for comprehensive crime analysis. The study's emphasis on sentiment analysis may overlook other critical factors such as network analysis, temporal patterns, or contextual information that can significantly impact crime detection and understanding of public sentiment.

The objective of this paper is to present a novel approach for crime-intent sentiment detection on Twitter data using machine learning techniques. By applying natural language processing (NLP) and sentiment analysis algorithms, we seek to uncover the underlying intent expressed in these tweets, identifying potential indicators of criminal behavior [18].

The proposed research holds significant implications for various stakeholders, including law enforcement agencies, policymakers, and social scientists. The ability to automatically detect crime-related sentiments on Twitter can assist law enforcement agencies in monitoring and predicting criminal activities [29], enabling them to allocate resources more efficiently and effectively. Moreover, policymakers can benefit from a better understanding of public sentiment towards crime [23], which can help shape policies and interventions to address public concerns [22]. Finally, social scientists can leverage this research to gain insights into the collective mindset of individuals regarding criminal behavior [24], fostering a deeper understanding of societal dynamics and behavioral patterns.

2 Methods

This section focuses on sentiment analysis for crime-related discussions using the Sentiment140 dataset, which consists of annotated tweets labeled as positive, negative, or neutral. By leveraging this dataset, we aim to develop robust models to accurately detect crime-intent sentiment expressed on Twitter. We employ a comprehensive approach that includes data collection, preprocessing, tokenization, feature extraction, feature selection, dataset splitting, and model training. Various machine learning algorithms such as Ridge Regression Classifier, SVM, Stochastic Gradient Descent (SGD) classifier, Random Forests, and Logistic Regression are used for training the sentiment analysis model [7]. The methodology ensures reliable evaluation and provides valuable insights into crime-related sentiment and intent expressed on Twitter, with potential applications in law enforcement and public sentiment analysis.

Research studies have demonstrated that positive sentiment expressed in social media posts can be indicative of non-criminal behaviors and conversely, negative sentiment analysis has been effective in predicting criminal intent [19]. However, it is important to acknowledge that relying solely on sentiment analysis may overlook contextual information and other influential factors related to criminal intent.

2.1 About the Dataset

To gather a comprehensive dataset for analysis, we utilized the Sentiment140 dataset [8]. The Sentiment140 dataset, obtained from Go et al. (2009), comprises many annotated tweets labeled as positive, negative, or neutral. This dataset, with its features shown in Table 1 and widely used for sentiment analysis tasks, including crime-intent sentiment detection, was carefully selected to ensure representation and diversity, encompassing tweets on various topics, including crime-related discussions.

Table 1. Dataset Description

Feature	Details
Target	The polarity of the tweet (0 = negative, 2 = neutral, 4 = positive)
ID	The id of the tweet (2087)
Date	The date of the tweet (Sat May 16 23:58:44 UTC 2009)
Query	The query (lyx). If there is no query, then this value is NO_QUERY
UserID	The user that tweeted (robotickilldozr)
Text	The text of the tweet (Lyx is cool)

However, it is crucial to acknowledge potential limitations, such as biases in labeling and the representativeness of real-world Twitter conversations, during the development and evaluation of models [25,31]. Specifically, in this study, positive sentiment was utilized to represent non-criminal intent, while negative sentiment was employed to indicate criminal intent.

Fig. 2. Word Cloud Showing Word Frequency. A: Criminal Intent Sentiment Word Cloud; B: Non-Criminal Intent Sentiment Word Cloud

2.2 Data Preprocessing

The preprocessing steps were applied to the text data to ensure its suitability for analysis. Noise and irrelevant information were eliminated by removing special characters, URLs, and stopwords. Moreover, specific elements unique to Twitter, including hashtags, mentions, and emoticons, were appropriately handled to retain contextual information. This preprocessing step, however, left some odd words like "quot" or "amp" which are ignored because they appear equally in the word cloud frequency for both labels as seen in Fig. 2. In the context of sentiment analysis for crime detection, the labels associated with the sentiment were transformed from positive/negative to criminal/non-criminal, aligning them with the intended classification task. These preprocessing techniques helped to prepare the data for subsequent analysis and facilitated the distinction between criminal and non-criminal intent in the sentiment classification process.

2.3 Feature Extraction

To extract features from the preprocessed and tokenized text data, a variety of techniques were utilized, enabling the representation of each tweet in a numerical format. The text data underwent feature extraction using the sklearn TF-IDF tokenizer instead of the traditional bag-of-words approach. This tokenizer generated vectors that captured the importance of different words within each tweet, considering both their frequency and their rarity in the overall dataset. By incorporating word sequences, the TF-IDF tokenizer provided a more comprehensive representation of the text data, allowing for a deeper understanding of nuanced sentiment and intent [14].

2.4 Tokenization

This was a crucial step after preprocessing the data, involving the splitting of tweets into individual words or tokens [16,27]. The process of tokenization was employed to analyze the sentiment and intent within the text data, particularly when applied to the Sentiment140 dataset. Each token in the dataset was treated as a distinct unit, enabling a granular examination of sentiment and intent. This transformation facilitated quantitative exploration by identifying emotional or subjective words and phrases within the dataset. Tokenization was instrumental in capturing nuanced sentiments and uncovering underlying patterns and trends. Furthermore, it allowed for the analysis of tweet intent by examining tokens related to actions, objects, and descriptive words. The utilization of tokenization on the Sentiment140 dataset enhanced the methodology's effectiveness in sentiment analysis, enabling a comprehensive and detailed examination of the data.

2.5 Dataset Splitting

After preprocessing and feature extraction, the dataset was divided into two separate sets: the training set and the testing set. The training set comprised 80% of the data, while the testing set constituted the remaining 20%. The purpose of this division was to facilitate the development and evaluation of the sentiment analysis model. The training set played a crucial role in training the model. It contained a significant portion of the preprocessed and feature-extracted data,

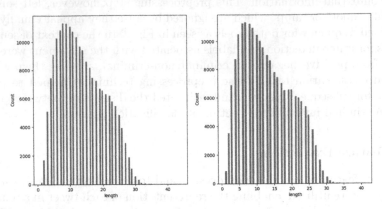

Fig. 3. Word Length Distribution

allowing the model to learn patterns and relationships between the extracted features and the corresponding sentiment labels. As seen in Fig. 3, the distribution of word length for both labels is very similar so any data imbalance will be mitigated when splitting into training and test sets.

On the other hand, the testing set was kept separate from the training set and served as an independent dataset for evaluating the model's performance. By evaluating the model's predictions on the testing set, it was possible to assess its ability to accurately classify sentiments in real-world scenarios. The dataset-splitting process ensured a robust evaluation of the trained model, as it allowed for unbiased testing and provided insights into its performance on unseen data.

2.6   Model Training

In this study, we conducted a comparative analysis of various natural language processing classification models to evaluate their effectiveness in accurately classifying crime intent in Twitter posts. We compared the outcomes generated by these models using the framework shown in Fig. 4 and assessed their classification accuracy.

1. Logistic regression is a classic linear classification algorithm widely used in sentiment analysis [11]. It models the relationship between the input features and the binary sentiment labels using a logistic function. Logistic Regression estimates the probability of a tweet belonging to a particular sentiment class and makes predictions based on a predefined decision boundary [9].
2. Ridge Regression is a regularization technique that extends linear regression by adding a penalty term to the loss function. It helps to mitigate the issue of multicollinearity and improves the model's ability to handle high-dimensional data. The Ridge Regression Classifier was implemented using the sci-kit-learn Python package [21].

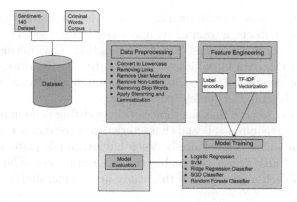

Fig. 4. Proposed Framework

3. Support Vector Machines (SVM) is a versatile algorithm that can handle both linear and non-linear classification tasks [6]. It aims to find an optimal hyperplane that separates the data points belonging to different sentiment classes. SVM maximizes the margin between the classes, promoting better generalization and robustness in sentiment analysis [26].
4. Stochastic Gradient Descent (SGD) Classifier: The SGD classifier was employed as one of the models in this study for depression detection from user tweets. The SGD classifier is a popular linear classifier that utilizes stochastic optimization techniques to update the model parameters iteratively [3].
5. Random Forests: Random Forests, a powerful ensemble learning algorithm, was also employed in this study for depression detection. Random Forests construct an ensemble of decision trees and aggregate their predictions to make accurate and robust classifications [4].

By utilizing these models, we aimed to explore different approaches to detecting crime intent from user tweets. Each model has its strengths and characteristics, allowing us to compare their performance and determine the most effective approach for our specific task.

3 Results

During the model evaluation phase, the sentiment analysis model's performance was assessed using a separate testing dataset. The evaluation involved various metrics such as accuracy, precision, recall, and F1 score. These metrics provided quantitative measures to evaluate the model's effectiveness in predicting sentiment accurately. Additionally, confusion matrices were utilized to analyze the model's classification outcomes in more detail. The evaluation metrics and confusion matrices collectively offered valuable insights into the model's performance, aiding in informed decision-making and facilitating comparisons between different models.

To ensure an unbiased evaluation of the models, the dataset of 404,286 rows was split into two subsets: a training set and an evaluation set. The splitting ratio chosen was 80% for training and 20% for evaluation. This means that approximately 80% of the dataset, which corresponds to 323,428 rows, was allocated for training the models. The remaining 20%, amounting to 80,857 rows, was reserved for evaluating the models' performance.

It is important to note that the choice of the splitting ratio may vary depending on the specific requirements and characteristics of the dataset. The 80-20 split used in this study strikes a balance between having an adequate amount of data for training and ensuring a reasonable size for evaluation. The results of the performance of each model used in this study are enumerated below.

1. Logistic Regression: The Logistic Regression model was evaluated using the testing dataset to assess its performance in sentiment analysis. The model achieved an accuracy of 92.87%, precision of 92.881%, recall of 92.866%, and an F1 score of 92.866%.
   To gain further insights into the model's performance, a confusion matrix was constructed. The confusion matrix, as seen in Table 2 provides a detailed breakdown of the model's predictions by comparing them to the actual class labels.

Table 2. Confusion Matrix for Logistic Regression

	Actual Criminal	Actual Non-Criminal
Predicted Criminal	55815	4883
Predicted Non-Criminal	3769	56819

2. Support Vector Machines (SVM): The SVM model was evaluated using the testing dataset to assess its performance in sentiment analysis. The model achieved an accuracy of 92.564%, precision of 92.560%, recall of 92.560%, and an F1 score of 92.560%. Additionally, the confusion matrix in Table 3 visually demonstrates the model's ability to discriminate between depressive and non-depressive tweets.

Table 3. Confusion Matrix for SVM

	Actual Criminal	Actual Non-Criminal
Predicted Criminal	55582	4797
Predicted Non-Criminal	4226	56681

Table 4. Confusion Matrix for Ridge Regression Classifier

	Actual Criminal	Actual Non-Criminal
Predicted Criminal	55815	4883
Predicted Non-Criminal	3769	56819

3. Ridge Regression Classifier (RRC): The RRC model was evaluated using the testing dataset to assess its performance in sentiment analysis. The model achieved an accuracy of 90.885%, precision of 90.905%, recall of 90.885%, and an F1 score of 90.884%.

This score indicates the overall effectiveness of the model in identifying depressive tweets. Additionally, the confusion matrix in Table 4 visually demonstrates the model's ability to discriminate between depressive and non-depressive tweets.

Table 5. SGD Classifier Confusion Matrix

	Actual Criminal	Actual Non-Criminal
Predicted Criminal	51773	8782
Predicted Non-Criminal	3842	56889

4. SGD Classifier: The SGD classifier yielded promising results for depression detection from user tweets. The model achieved an accuracy of 89.51%, precision of 89.81%, recall of 89.51%, and an F1 score of 89.49%. These metrics were obtained based on the evaluation of the model on the test dataset.

The confusion matrix for the SGD classifier as shown in Table 5 provides a detailed breakdown of the model's predictions.

5. Random Forests: The Random Forests model demonstrated strong performance in detecting depression from user tweets. The model achieved an accuracy of 92.397%, precision of 92.404%, recall of 92.397%, and an F1 score of 92.397% when evaluated on the test dataset.

The confusion matrix for the Random Forests model provides additional insights into the distribution of correct and incorrect predictions made by the model, further confirming its performance as shown in Table 6.

Table 6. Random Forests Confusion Matrix

	Actual Criminal	Actual Non-Criminal
Predicted Criminal	56319	4236
Predicted Non-Criminal	4985	55746

4   Discussion

Table 7. Evaluation Metrics For the Models

Model/Measure	Accuracy	Precision	Recall	F1-Score
Ridge Regression Classifier	90.885%	90.905%	90.885%	90.884%
Logistic Regression	92.870%	92.881%	92.866%	92.866%
Support Vector Machines	92.564%	92.560%	92.560%	92.560%
SGD	89.51%	89.81%	89.51%	89.49%
Random Forests	92.397%	92.404%	92.397%	92.397%

As shown in all previous sections, this paper presents a comprehensive approach for detecting crime intent from user tweets using machine learning techniques. The study utilizes a large dataset of 400, 000 tweets and applies preprocessing steps, including feature selection and data cleaning, to accurately identify tweets indicating criminal intent. Promising results are achieved by employing various models, including logistic regression, ridge regression classifier, SGD classifier, Random forests, and SVM models as seen in Table 7. The evaluation of these models showcases their varying levels of effectiveness in sentiment analysis for crime intent detection, with the logistic regression model demonstrating excellent performance in all metrics with an accuracy of 92.870%.

5   Conclusion

To conclude, these findings have significant implications for sentiment analysis applications in the context of crime detection. The high accuracy, precision, and recall scores of the logistic regression model suggest its potential for accurately identifying sentiment related to criminal intent in textual data. This has important applications in law enforcement, public sentiment analysis, and crime prevention. However, it is important to note that the performance of these models may depend on the dataset and the specific characteristics of the sentiment analysis task related to criminal intent. Continuous refinement and optimization are crucial to enhance their effectiveness and adaptability to different crime detection scenarios. Further research and experimentation on diverse datasets and real-world scenarios are needed to validate their performance and generalizability.

References

1. Agarwal, A., Xie, B., Vovsha, I., Rambow, O., Passonneau, R.J.: Sentiment analysis of twitter data. In: Proceedings of the Workshop on Language in Social Media (LSM 2011), pp. 30–38 (2011)
2. Bendler, J., Brandt, T., Wagner, S., Neumann, D.: Investigating crime-to-twitter relationships in urban environments-facilitating a virtual neighborhood watch (2014)
3. Bottou, L.: Large-scale machine learning with stochastic gradient descent. In: Lechevallier, Y., Saporta, G. (eds.) Proceedings of COMPSTAT 2010, pp. 177–186. Springer, Heidelberg (2010). https://doi.org/10.1007/978-3-7908-2604-3_16
4. Breiman, L.: Random forests. Mach. Learn. **45**, 5–32 (2001)
5. Casey, J.: Implementing community policing in different countries and cultures. Pak. J. Criminol. **2**(4), 55–70 (2010)
6. Cortes, C., Vapnik, V.: Support-vector networks. Mach. Learn. **20**, 273–297 (1995)
7. El Rahman, S.A., AlOtaibi, F.A., AlShehri, W.A.: Sentiment analysis of twitter data. In: 2019 International Conference on Computer and Information Sciences (ICCIS), pp. 1–4. IEEE (2019)
8. Go, A., Bhayani, R., Huang, L.: Twitter sentiment classification using distant supervision. CS224N Proj. Rep. Stanford **1**(12), 2009 (2009)
9. Hasan, M., Rundensteiner, E., Agu, E.: Automatic emotion detection in text streams by analyzing twitter data. Int. J. Data Sci. Anal. **7**, 35–51 (2019)
10. Horton, J.: Us crime: is America seeing a surge in violence? BBC (2022). https://www.bbc.com/news/57581270. Accessed 28 June 2023
11. Hosmer, D.W., Jr., Lemeshow, S., Sturdivant, R.X.: Applied Logistic Regression. Wiley, Hoboken (2013)
12. Jurgens, D., Finethy, T., McCorriston, J., Xu, Y., Ruths, D.: Geolocation prediction in twitter using social networks: a critical analysis and review of current practice. In: Proceedings of the International AAAI Conference on Web and Social Media, vol. 9, pp. 188–197 (2015)
13. Kumar, S., Morstatter, F., Liu, H.: Twitter Data Analytics. Springer, Cham (2014). https://doi.org/10.1007/978-1-4614-9372-3
14. Lavin, M.: Analyzing documents with TF-IDF (2019)
15. Leykin, D., Aharonson-Daniel, L., Lahad, M.: Leveraging social computing for personalized crisis communication using social media. PLoS Curr. **8** (2016)
16. Liu, Y., Zhu, Y., Che, W., Qin, B., Schneider, N., Smith, N.A.: Parsing tweets into universal dependencies. arXiv preprint arXiv:1804.08228 (2018)
17. Mangachena, J.R., Pickering, C.M.: Implications of social media discourse for managing national parks in South Africa. J. Environ. Manage. **285**, 112159 (2021)
18. Medhat, W., Hassan, A., Korashy, H.: Sentiment analysis algorithms and applications: a survey. Ain Shams Eng. J. **5**(4), 1093–1113 (2014)
19. de Mendonça Ricardo, R., Felix de Brito, D., de Franco Rosa, F., dos Reis, J.C., Bonacin, R.: A framework for detecting intentions of criminal acts in social media: a case study on twitter. Information **11**(3), 154 (2020)
20. Neogi, A.S., Garg, K.A., Mishra, R.K., Dwivedi, Y.K.: Sentiment analysis and classification of Indian farmers' protest using twitter data. Int. J. Inf. Manage. Data Insights **1**(2), 100019 (2021)
21. Pedregosa, F., et al.: Scikit-learn: machine learning in python. J. Mach. Learn. Res. **12**, 2825–2830 (2011)

22. Pickett, J.T.: Public opinion and criminal justice policy: theory and research. Annu. Rev. Criminol. **2**, 405–428 (2019)
23. Resko, S., Ellis, J., Early, T.J., Szechy, K.A., Rodriguez, B., Agius, E.: Understanding public attitudes toward cannabis legalization: qualitative findings from a statewide survey. Substance Use Misuse **54**(8), 1247–1259 (2019)
24. Stroh, D.P.: Systems Thinking for Social Change: A Practical Guide to Solving Complex Problems, Avoiding Unintended Consequences, and Achieving Lasting Results. Chelsea Green Publishing (2015)
25. Thieme, A., Belgrave, D., Doherty, G.: Machine learning in mental health: a systematic review of the HCI literature to support the development of effective and implementable ml systems. ACM Trans. Comput.-Hum. Interact. (TOCHI) **27**(5), 1–53 (2020)
26. Tian, Y., Mirzabagheri, M., Bamakan, S.M.H., Wang, H., Qu, Q.: Ramp loss one-class support vector machine; a robust and effective approach to anomaly detection problems. Neurocomputing **310**, 223–235 (2018)
27. Toraman, C., Yilmaz, E.H., Şahinuç, F., Ozcelik, O.: Impact of tokenization on language models: an analysis for Turkish. ACM Trans. Asian Low-Resour. Lang. Inf. Process. **22**(4), 1–21 (2023)
28. Velikovich, L., Blair-Goldensohn, S., Hannan, K., McDonald, R.: The viability of web-derived polarity lexicons (2010)
29. Wang, Y., Yu, W., Liu, S., Young, S.D.: The relationship between social media data and crime rates in the united states. Soc. Media+ Soc. **5**(1), 2056305119834585 (2019)
30. Wong, A., Ho, S., Olusanya, O., Antonini, M.V., Lyness, D.: The use of social media and online communications in times of pandemic COVID-19. J. Intensive Care Soc. **22**(3), 255–260 (2021)
31. Woodbury, S., Gess-Newsome, J.: Overcoming the paradox of change without difference: a model of change in the arena of fundamental school reform. Educ. Policy **16**(5), 763–782 (2002)
32. Zeng, D., Chen, H., Lusch, R., Li, S.H.: Social media analytics and intelligence. IEEE Intell. Syst. **25**(6), 13–16 (2010)

Generation of Visualized Medical Rehabilitation Exercise Prescriptions with Diffusion Models

Juewen Ni[1], Peng Du[2], Qihan Hu[1], Zhenghui Xu[1], Hao Zeng[1(✉)], Hao Xie[1(✉)], Youbing Zhao[1,4] , Gengling Wang[1], Songjin Yang[1], Jian Song[3], and Shengyou Lin[1]

[1] Communication University of Zhejiang, Hangzhou 310018, China
{hao.zeng,xiehao}@cuz.edu.cn
[2] Uber Technologies Inc, 1725 3rd St, San Francisco, CA 94158, USA
[3] Hangzhou Jiuselu Medical Technology Co. Ltd, Hangzhou 310059, China
[4] University of Bedfordshire, Luton LU1 3JU, UK

Abstract. Visualization of medical rehabilitation exercise prescriptions is to provide a more intuitive and understandable way of conveying medical guidance through visual means. Currently, the generation of visualized medical rehabilitation exercise prescriptions is largely based on the manual use of software for hand drawing. However, not only does this production method exhibit the drawbacks of complexity and high labor costs, but it also suffers from low production efficiency. In this study, we present four novel methods that aim to harness the potential of existing Stable Diffusion to generate visualized medical rehabilitation exercise prescription outputs, as well as to exemplify the generation of visualized rehabilitation exercise prescriptions for frozen shoulders. Experimental results demonstrate that our approaches achieve high-quality and more precise visualized rehabilitation exercise prescriptions.

Keywords: Medical Prescription · Visualization · Diffusion Models

1 Introduction

A medical prescription provided by the physician includes instructions on dietary recommendations, medication usage, exercise procedures, and other relevant aspects according to the patient's condition and treatment requirements. Conventional medical prescriptions typically have been presented by the text, often comprising complex professional content that may be difficult for patients to comprehend, particularly in the case of rehabilitation exercises. Failure to properly understand prescriptions can lead to suboptimal recovery outcomes. To enhance recovery effects, it is crucial to express medical prescriptions in a more accessible and visually intuitive format. In comparison to purely textual medical

Many thanks to Mr. Yihui Shen for his generous funding support.

prescriptions, visualized medical prescriptions, or image-based medical prescriptions, offer distinct advantages as it is more readily embraced by patients due to its visual appeal and simplified representation. For instance, patients are not required to comprehend the textual content of rehabilitation exercise prescriptions; rather, they simply need to follow the character movements depicted in the visualized rehabilitation exercise prescriptions. This alleviates the learning difficulties associated with rehabilitation and provides clearer guidance that is universally applicable.

However, the field of visualized medical prescriptions is currently facing certain bottlenecks. For example, most visualized rehabilitation exercise prescriptions are created manually using painting software, which not only entails complexity and high manual costs but also suffers from low production efficiency. Moreover, manual methods are unsuitable for large-scale implementation and struggle to meet the demands of batch production for visualized medical prescriptions. Nowadays, diffusion models [7, 8, 10, 12, 13] have demonstrated remarkable success in generating diverse high-quality images that align with given textual or simple image cues. The application of diffusion models deployed in the field of medical care is of great importance, as it will greatly assist and improve the generation of large batches of visualized rehabilitation exercise prescriptions from textual prescriptions.

In this paper, we propose four methods to use Stable Diffusion [10] and ControlNet [16] to generate visualized rehabilitation exercise prescriptions. These methods offer physicians or other users a rapid means of producing visualized rehabilitation exercise prescriptions from textual prescriptions, thereby saving time in explaining prescriptions to patients. At the same time, most patients will benefit from these methods, as they provide patients with concise, simple, and easily comprehensible rehabilitation exercise guides and help patients actively engage in rehabilitation exercises.

2 Related Work

2.1 Medical Prescription Visualization

Visualization and image-based medical prescriptions have gained significant attention as a research direction. In recent years, numerous research efforts have been dedicated to developing innovative methods and technologies to improve the visualization of prescriptions and improve patients' understanding of medical prescriptions.

In [5], an image-based prescription has been proposed that combines medical knowledge and design principles to improve patient understanding and compliance with medical instructions through visually engaging graphics, symbols and animations. [9] introduces a novel visualization approach to understand and analyze the behavior of medical prescriptions from the perspective of physicians, patients, medicines, and prescriptions, in order to improve the efficiency and effectiveness of treatment. [4] aims to enhance prescription visualization through interactivity and personalization, using interactive tools to create customized image-based prescriptions that take into account individual characteristics, preferences, and needs, ultimately improving information delivery, patient participation, and treatment results.

2.2 Diffusion Models and ControlNet

The latest advancements in diffusion models have significantly enhanced image generation and generative editing capabilities, leveraging the vast text and image databases available on the Internet. The concept of diffusion models was initially introduced in [12] and later improved by [7,8,10,13]. which commonly adopt the U-Net architecture [11] for image generation tasks. For a comprehensive overview of diffusion models, the reader is referred to an extensive survey conducted in [14]. Additionally, a more recent survey focusing on text-to-image related diffusion models can be found in [15].

In order to improve the controllability of the generated images and achieve greater consistency and predictability in the output, ControlNet [16] was introduced. This approach enhances pre-trained large diffusion models by incorporating additional control blocks, enabling the incorporation of various user guide inputs. For instance, ControlNet allows for the generation of images that align with predefined line sketches, edge detection, or character skeletons. Using ControlNet, users have increased flexibility and can guide the image generation process according to their specific requirements. In addition to ControlNet, [6] proposes a novel approach called MoFusion to synthesize long sequences of human motions in 3D from textual and audio input. This approach uses diffusion models and domain-inspired kinematic losses to improve generalizability and realism and can be conditioned on modalities like text and audio. But MoFusion may still have limitations in terms of its ability to generate highly complex rehabilitation exercise motions.

3 Methods

We propose four methods to generate the corresponding visualized rehabilitation exercise prescriptions from textual prescriptions that take advantage of the power of Stable Diffusion and ControlNet. Our four distinct methods are Prompt-Only, Openpose with ControlNet, Canny with ControlNet, and Scribble with ControlNet respectively. In these methods, the visualized rehabilitation exercise prescriptions are generated by Stable Diffusion tools such as WebUI [3] or ComfyUI [2]. Each method is described in detail in the following subsections.

3.1 Prompt-Only

The Prompt-Only method workflow is shown in Fig. 1, we manually extract key movement information from rehabilitation exercise prescriptions and transform it into prompts. These prompts are then entered into the Stable Diffusion, which ultimately generates the corresponding visualized rehabilitation exercise prescriptions as output.

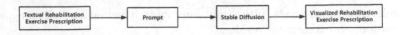

Fig. 1. Prompt-Only method

This method offers the advantage of allowing physicians or other users to use key movement information from textual rehabilitation exercise prescriptions summarized as prompts, then efficiently generate visualized prescriptions. However, there are also notable drawbacks to consider: (1)the generated prompts may not always be highly accurate, resulting in the diffusion model being unable to generate the expected visualized rehabilitation exercise prescriptions; (2)even when the prompts are relatively precise, the Stable Diffusion may struggle to fully understand the intended meaning conveyed by the prompts.

3.2   Openpose with ControlNet

The workflow depicting Openpose with ControlNet method is shown in Fig. 2, we leverage the human skeletal model in combination with prompts to address the limitations of the Prompt-Only method. We adjust the human skeletal model based on the exercise movements described in the textual rehabilitation exercise prescriptions. This human skeletal model is implemented using the Openpose plugin within ControlNet. Additionally, we combine this adjusted skeletal model with prompts generated from textual prescriptions. Then, both the adjusted human skeletal model and the prompts are inputted into the Stable Diffusion and ControlNet. As a result, we obtain the corresponding visualized rehabilitation exercise prescriptions as the final output.

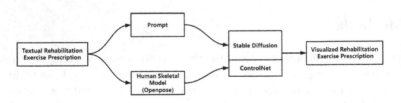

Fig. 2. Openpose with ControlNet workflow

The advantages of this method lie in its ability to generate highly accurate visualized rehabilitation exercise prescriptions by combining the openpose human skeletal model with prompts. However, there are limitations to consider: (1)since the human skeletal model with openpose is based on 2D images, the generated visualizations may not be optimal for certain side-oriented or complex rehabilitation movements; (2)for physicians or other users, there is a time cost associated with learning to adjust the human skeletal model using openpose, which is relatively higher compared to the prompt-only method.

3.3 Canny with ControlNet

The Canny with ControlNet method workflow is illustrated in Fig. 3, we adopt an alternative way distinct from Openpose with ControlNet, utilizing the 3D human body model in conjunction with prompts to overcome the limitations observed in the Prompt Only method. We adjust the 3D human body model following the exercise movements described in the textual rehabilitation exercise prescriptions. Then, we conduct edge detection on the 3D human body model to the outline sketch by utilizing the Canny preprocessor (implemented through the Canny plugin in ControlNet). After that, we input both the outline sketch and the prompt extracted from the textual prescription into the Stable Diffusion and ControlNet. Consequently, this yields the corresponding visualized rehabilitation exercise prescription as the output.

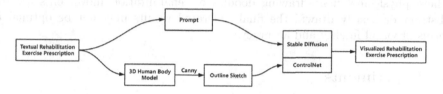

Fig. 3. Canny with ControlNet workflow

Similar to the method used in Openpose with ControlNet, this method leverages the advantages of using a 3D human body modal to depict rehabilitation exercise movements. By transforming the 3D model into an outline sketch and incorporating prompts, this method is capable of generating highly accurate visualized rehabilitation exercise prescriptions, even surpassing the precision achieved by the Openpose with ControlNet method. However, a drawback of this method lies in its relatively complex generation process, which requires physicians or other users to invest a certain amount of time and effort in learning how to use the 3D human body model and the Canny edge detection technique.

3.4 Scribble with ControlNet

The workflow of Scribble with ControlNet method shown in Fig. 4 is remarkably distinctive and intriguing. Following the exercise movements described in the textual rehabilitation exercise prescriptions, we engage in a simple hand-drawing (doodling) process to depict characters and their corresponding actions. Subsequently, this hand-drawing illustration undergoes preprocessing using the scribble preprocessor (this step is accomplished by the scribble plugin in ControlNet), yielding outline sketches in a doodle-like format. Then, we input both the doodle sketches and prompts extracted from the textual prescription into the diffusion model and ControlNet. Finally, producing the corresponding visualized rehabilitation exercise prescriptions as output.

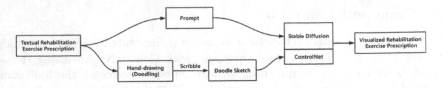

Fig. 4. Scribble with ControlNet workflow

This method offers several advantages. For physicians or other users, the simplicity of hand-drawing doodles to depict rehabilitation exercise movements proves to be highly convenient and straightforward. Then, the integration of prompt enables the rapid generation of visualized rehabilitation exercise prescriptions. However, there are certain limitations to this method. In instances where physicians' hand-drawing doodles of rehabilitation movements are too abstract or hastily drawn, the final generated results may not be optimal in terms of visual fidelity and accuracy.

4   Experiments

A rehabilitation exercise prescription shown in Table 1 has been proven to be effective in relieving symptoms and reducing pain in individuals with frozen shoulders. Then we take the rehabilitation exercise prescription for frozen shoulders as an example and demonstrate the process of generating visualized rehabilitation exercise prescriptions targeting frozen shoulders using our four methods.

We conducted experiments using our proposed methods through the Stable Diffusion WebUI and the ControlNet v1.1.189 plugin. Stable diffusion utilized a pre-trained "anything-v5" checkpoint model. Human skeletal models were generated using the openpose plugin, employing the control_openpose-fp16. The 3D human body models were generated using PoseMy.Art [1]. The canny edge preprocessor (control_canny-fp16) of the ControlNet plugin was employed to generate outline sketch edges. Analogously, the scribble_pidinet preprocessor(control_vp11p_sd15_scribble) of the ControlNet plugin was used to generate doodle sketch edges. The dimensions of intermediate and final images were set to 512×512 pixels and the number of iteration steps was set to 20. These experiments were carried out on a Windows 11 platform with 64GB of memory and a GeForce RTX4090 (24G) graphics card.

Initially, employing the Prompt-Only method, we process the textual rehabilitation exercise prescriptions shown in Table 1. The generated visualized rehabilitation exercise prescriptions are displayed in Fig. 5. For hand swinging exercise, we extracted the content of its exercise procedures as two separate prompts. They are "1boy, standing, side face, full body shot, shirt, arms swinging backward, from the side, white background, sketch, <lora:animeoutlineV4_16:1>[1]" and "1boy, standing, flinging the arm, full body shot, white background, sketch

---

[1] We use a Lora model to generate anime characters.

Table 1. A rehabilitation exercise prescription for a frozen shoulder

No	Exercise Name	Exercise Procedure
1	Hand Swinging Exercise	Stand with your feet apart. Begin by rubbing and massaging the shoulder area with your hands to relax the local muscles. Then, swing your arms back and forth, followed by left and right, gradually increasing the range of motion (from 30 to 90°C relative to the body) and speed (from slow to fast, around 30 to 60 swings per minute). Perform this exercise for 1 to 5 min each time.
2	Circular Motion Exercise	The arms are moved separately, from front to back or from back to front, in a clockwise or counterclockwise direction, drawing circles. The amplitude increases gradually from small to large, reaching the maximum limit, and repeated 50 to 100 times each time.
3	Wall Climbing Exercise	Place the fingers of the affected side against the wall, lift the hand upward to the highest point, and then lower it. Repeat this motion 10–12 times.
4	Cannon Shot Exercise	This exercise can be done while standing or sitting. Clasp your hands into fists and place them on top of your head. Gradually straighten both arms, extending your hands towards the ceiling, until you reach the maximum limit. Repeat this motion 30 to 50 times each time.
5	Extend Arms Exercise	Stand with your feet shoulder-width apart. Extend both arms straight out to the sides, lifting them (abduction) until they form a 90-degree angle with your body. Hold the arms outstretched for 5–10 seconds, and then slowly lower them. Repeat this exercise 30 to 50 times per day.
6	Touch Neck Exercise	You can do this exercise while sitting or standing. Alternately touch the back of your neck with both hands. Repeat this exercise twice a day, with 50–100 repetitions each time.
7	Shrug Shoulders Exercise	You can do this exercise while sitting or standing. Shrug your shoulders, gradually increasing the intensity. Repeat this exercise twice a day, with 50–100 repetitions each time.

<lora:animeoutlineV4_16:1>" respectively. Then, following the Prompt Only method, we obtain two visualized prescriptions of hand swinging exercise as outputs. These generate results exhibit remarkable quality, aligning with the exercise postures and procedures described in hand swinging exercise. However, the visualized prescription generated from the Prompt-Only method may not meet the requirement. For instance, the exercise movement in corresponding generated visualized prescriptions is inconsistent with the exercise movements described in circular motion exercise and shrug shoulder exercise.

Fig. 5. Prompt-Only generations

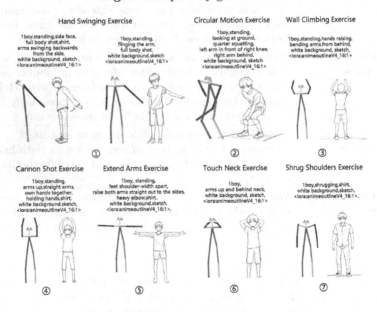

Fig. 6. Openpose with ControlNet generations

Subsequently, we employ the Openpose with ControlNet method, the results are presented in Fig. 6. In particular, incorporating the human skeletal model represented by openpose with prompts led to improved generation results compared to the Prompt-Only method.

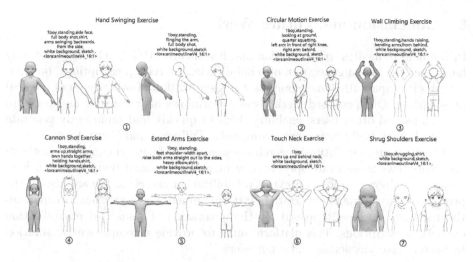

Fig. 7. Canny with ControlNet generations

Then, we utilize the Canny with ControlNet method, the results are illustrated in Fig. 7. It is discernible that the integration of outline sketches, generated by edge detection from 3D human body models, along with prompts, has yielded enhanced generation outcomes as compared to the Prompt Only method as well.

Finally, we use the Scribble with ControlNet method, the results are shown in Fig. 8. Proper hand-drawing doodles, transformed into doodle sketches, in conjunction with prompt generate results that are highly conformed with the original prescriptions.

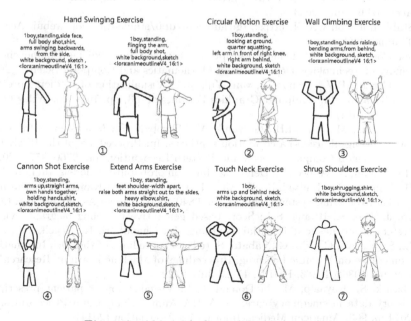

Fig. 8. Scribble with ControlNet generations

5   Conclusion and Future Work

In the field of medical rehabilitation and healthcare, AIGC technology not only facilitates the visual generation of rehabilitation exercise prescriptions, but also extends its capabilities to encompass the visualization of other types of medical prescription. Our proposed methods and workflows offer valuable assistance to physicians and other users, enabling them to quickly and proficiently generate specific human body movements using Sable Diffusion and ControlNet.

Furthermore, our future research endeavors will focus on exploring methods and workflows to generate specific movements of localized body parts, such as hands and elbows, in the context of rehabilitation exercises. We are also in the process of developing an integrated platform named "Scripose" that incorporates the methods we have proposed for the generation of visualized rehabilitation exercise prescriptions. This platform aims to provide a comprehensive solution to better serve physicians and other users.

Acknowledgements. We would like to thank the 12th batch of the Course Teaching Mode Innovation Experimental Zone Project of Communication University of Zhejiang: "Research on Exploring the Integrated Bilingual Teaching Model for Python Programming Courses in the Background of New Engineering Discipline" for the generous funding support of the work referred to in this paper.

References

1. Posemy.art. https://app.posemy.art/. Accessed 07 June 2023
2. Stable diffusion comfyUI. https://github.com/comfyanonymous/ComfyUI. Accessed 07 June 2023
3. Stable diffusion webUI. https://github.com/db0/stable-diffusion-webui. Accessed 07 June 2023
4. der Corput, P.V.: Medicine Prescription Visualization. Medicine Eindhoven University of Technology, Department of Mathematics and Computer Science (2013)
5. der Corput, P.V., Arends, J., van Wijk, J.J.: Visualization of medicine prescription behavior. In: Computer Graphics Forum, vol. 33, pp. 161–170. Wiley Online Library (2014)
6. Dabral, R., Mughal, M.H., Golyanik, V., Theobalt, C.: MoFusion: a framework for denoising-diffusion-based motion synthesis. In: Proceedings of the IEEE/CVF Conference on Computer Vision and Pattern Recognition, pp. 9760–9770 (2023)
7. Ho, J., Jain, A., Abbeel, P.: Denoising diffusion probabilistic models. In: Larochelle, H., Ranzato, M., Hadsell, R., Balcan, M., Lin, H. (eds.) Advances in Neural Information Processing Systems, vol. 33, pp. 6840–6851. Curran Associates, Inc. (2020)
8. Jo, J., Lee, S., Hwang, S.J.: Score-based generative modeling of graphs via the system of stochastic differential equations. In: Chaudhuri, K., Jegelka, S., Song, L., Szepesvari, C., Niu, G., Sabato, S. (eds.) Proceedings of the 39th International Conference on Machine Learning. Proceedings of Machine Learning Research, vol. 162, pp. 10362–10383. PMLR, 17–23 July 2022
9. Ozturk, S., Kayaalp, M., McDonald, C.J.: Visualization of patient prescription history data in emergency care. In: AMIA Annual Symposium Proceedings, vol. 2014, p. 963. American Medical Informatics Association (2014)

10. Rombach, R., Blattmann, A., Lorenz, D., Esser, P., Ommer, B.: High-resolution image synthesis with latent diffusion models. In: Proceedings of the IEEE/CVF Conference on Computer Vision and Pattern Recognition, pp. 10684–10695 (2021)
11. Ronneberger, O., Fischer, P., Brox, T.: U-Net: Convolutional Networks for Biomedical Image Segmentation. In: Navab, N., Hornegger, J., Wells, W.M., Frangi, A.F. (eds.) MICCAI 2015. LNCS, vol. 9351, pp. 234–241. Springer, Cham (2015). https://doi.org/10.1007/978-3-319-24574-4_28
12. Sohl-Dickstein, J., Weiss, E., Maheswaranathan, N., Ganguli, S.: Deep unsupervised learning using nonequilibrium thermodynamics. In: Bach, F., Blei, D. (eds.) Proceedings of the 32nd International Conference on Machine Learning. Proceedings of Machine Learning Research, vol. 37, pp. 2256–2265. PMLR, Lille, France, 07–09 July 2015
13. Song, Y., Durkan, C., Murray, I., Ermon, S.: Maximum likelihood training of score-based diffusion models. In: Ranzato, M., Beygelzimer, A., Dauphin, Y., Liang, P., Vaughan, J.W. (eds.) Advances in Neural Information Processing Systems, vol. 34, pp. 1415–1428. Curran Associates, Inc. (2021)
14. Yang, L., et al.: Diffusion models: a comprehensive survey of methods and applications (2023)
15. Zhang, C., Zhang, C., Zhang, M., Kweon, I.S.: Text-to-image diffusion models in generative AI: a survey (2023)
16. Zhang, L., Agrawala, M.: Adding conditional control to text-to-image diffusion models (2023)

A Literature Analysis of the Application of Artificial Intelligence in Judicial Adjudication

Qi Zhang[1] ⓘ, Sijing Chen[2(✉)] ⓘ, and Tao Song[3] ⓘ

[1] Shanxi University of Finance and Economics, Taiyuan 030012, Shanxi, China
[2] Shanxi Provincial Committee Party School of C.P.C, Taiyuan 030012, Shanxi, China
294795563@qq.com
[3] Editorial Office of China Law Review, Beijing 100161, China

Abstract. The progress of technology has a profound impact on the transformation of social life while judicial activities are no exception. In recent years, artificial intelligence such as big data, blockchains has widely applied to judicial process, which makes lawyers and judges see the advantages of higher efficiency. However, the application of artificial intelligence in legal practice has leads to some disputes such as the fairness of the judgment and the potential of violations of privacy. In this paper, research on the application of artificial intelligence in judicial adjudication was analyzed by the informetric method. About 1224 papers indexed in Thomson Reuter's Web of science has been studied from the perspectives of keywords co-occurrence. The analysis of keywords co-occurrence shows the application of artificial intelligence in both procedural and substantial matters of judicial adjudication. Despite the convenience and high efficiency we have enjoyed during the process of judicial adjudication, we still need to consider ethics problems carefully such as the just, fairness, and transparency of AI technology. We found that these papers can be divided into three categories: The first is the application of AI in procedural matters or the process of judicial adjudication. The second is the application of AI in substantial matters or the decision-making and judgement of judicial adjudication. The third is the controversy of legal values beyond the practice of the application of AI in judicial adjudication.

Keywords: Artificial Intelligence · Judicial Adjudication · Law Suit · Trial

1 Introduction

Human beings have assumed the application of artificial intelligence in judicial adjudication long time ago. In the science fiction The Cyber and Justice Holmes written by Frank Rile in 1955, Professor Neustadt devote his life to compete with cyber justice the

This paper is part work of the project "Research on Implementation Effect Evaluation and Revision Suggestion for Regulations on Science and Technology Progress in Shanxi Province" (Grand Number: 202204031401028); also part work of the project of "The Regional Turn of Global Governance and the Practice of China's Participation in Asian Regional Organizations" (Grand Number: 22JZD040, Project Leader: Pro. ZeWei Yang).

trial for himself. This might be the first literature describes the application of artificial intelligence to court and the confrontation between the value of human beings and the intelligence of robots. In recent years, more and more states pay attention to the construction of intelligent court with abundant policy bases, adequate financial support and sufficient talent resource, while some scientific and technical corporation even invested in research and development of AI lawyer which might reduce the fee of legal service in the future. However, the rationality, width and depth of the application of AI in judicial adjudication remains controversial due to different opinions on the risk of AI. For illustrating these opinions, this paper focus on the literature analysis of the research on the application of AI in judicial adjudication by providing categories and keywords co-occurrence of these research. In addition, we also give a brief review of these research to show representative viewpoint on this topic.

2 Literature Analysis

2.1 Data Collection

We analyzed 1224 papers in Web of Science (Wos) core collection in last five years (2018–2023) through the library of Shanxi University of Finance and Economics by the retrieval model "TS = (AI OR artificial Intelligence) AND TS = (Judgement OR Law Suit OR Court OR Judicial Procedure OR Judicial Adjudication)" on May 10, 2023. By the analyze results of Web of Science (Wos), there are 196 categories related to the application of AI to judicial adjudication and we list top 10 of them by the ranking of quantity of papers. As shown in Fig. 1, these filed are Computer Science Artificial Intelligence, Law, Computer Science Information Systems, Computer Science Information Systems, Engineering Electrical Electronic, Computer Science Interdisciplinary Applications, Computer Science Theory Methods, Engineering Multidisciplinary, Computer Science Software Engineering, Telecommunication, Business. From Fig. 1, we can find that most of these fields are related to AI, Law, and Interdisciplinary Applications which reflects the Multi/Interdisciplinary characters of studies on the AI and law.

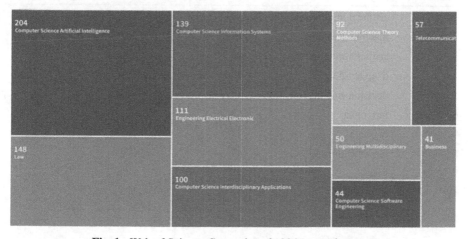

Fig. 1. Web of Science Categories of 1224 research papers

2.2 Keywords Analysis

By the calculation of visualization tool Vosviewer, 1224 paper collected by us contain 5152 key words. We set minimum number of occurrences of a keyword as ten and 74 keywords meet the threshold. We ranked all these 74 keywords by occurrences and total link strength by Vosviewer. As shown in Fig. 1, top ten keywords are artificial intelligence, machine learning, artificial-intelligence, deep learning, decision-making, judgment, big data, classification, bias, information. These high-frequency keywords show that AI is deeply associated with legal practice. Some research papers have presented the form of the application of AI in judicial adjudication such as classification or decision-making of the judgement, while other research papers might be associated with the risk or value conflict such as bias or discrimination due to the design or operating principle of AI.

Besides the analysis of top ten keywords, we also select all keywords to analysis the link between them. We got the knowledge map by Vosviewer which is shown in Fig. 2. Each cluster is set to contain 30 words at most, and a total of 5 cluster were obtained which are purple, red, green, blue and yellow. Each cluster represents different topics or domain clustering. The purple cluster is the main subject related to this study which includes science, law and justice. The red cluster is the main technology means of the application of AI in judicial adjudication such as deep learning, neural networks and validation. The green cluster is the domain of the application of AI in the judicial adjudication which includes judgement, decision-making and the education of students from law schools. The blue cluster is the risk or focus of moral dispute for the application of AI in the judicial adjudication, including fairness, discrimination, privacy, accountability and transparency. The yellow cluster represents the direction of the development or innovation of AI in legal domain in the future such as the increase of degree of automation and getting more trusted by human beings (Table 1).

Table 1. Keywords List Ranked by Occurrences

Rank	Keyword	Occurrences	Total Link Strength
1	artificial intelligence	414	711
2	machine learning	149	357
3	artificial-intelligence	84	228
4	deep learning	79	161
5	decision-making	47	148
6	judgment	63	147
7	big data	47	138
8	classification	47	126
9	bias	39	125
10	information	38	122

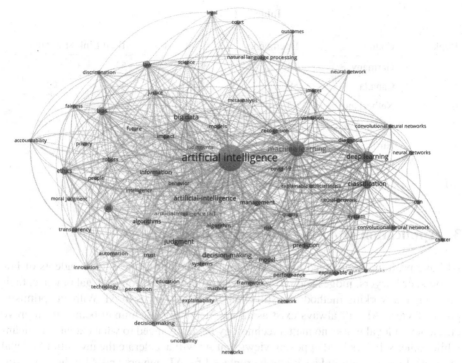

Fig. 2. Keywords co-occurrence map (Color figure online)

2.3 Analysis of Bibliographic Citation

Except the analysis of keywords co-occurrence, we also made analysis of bibliographic citation from the perspective of countries. We set minimum of citation of papers of countries as 5, then we got top 10 countries where highly cited papers come from which is shown in Table 2. According to the rank of total link strength, we could find United States, England and China take the leading position in the research of the application of artificial intelligence in judicial adjudication. This might attribute to a lot of famous organizations and universities located in these countries such as Duke, MIT, Oxford and Zhejiang University which are be proficient at the study of AI. However, on the other hand, all these countries have strong comprehensive national strength, which could support the development of AI technology sustainability.

Table 2. Bibliographic Citation List Ranked by Countries

Rank	Country	Documents	Citations	Total Link Strength
1	USA	303	3507	144
2	England	291	1278	68
3	China	105	1502	66

(*continued*)

Table 2. (*continued*)

Rank	Country	Documents	Citations	Total Link Strength
4	Germany	68	358	53
5	Canada	58	864	47
6	Netherlands	33	772	27
7	South Korea	55	144	27
8	Australia	50	352	25
9	France	21	119	24
10	Lithuania	7	19	16

3 Application of AI in Judicial Adjudication

AI have profoundly changed the legal industry. Whether teachers and students of law schools, or lawyers, judges and prosecutors stand in the forefront of legal practice, their thinking and working methods are still changing by advances in AI. With an optimistic point of view, AI will always exist as an auxiliary tool for human beings to improve efficiency of legal works no matter technology have advanced to what extent [1]. Meanwhile, other scholars hold opposing views that AI will accelerate the involution of legal industry. Lawyers and judges might be replaced by AI lawyers and AI judges and students graduated from law schools might find that getting a job is not easy. Nonetheless, there is no doubt that most states are pulling the application of AI in judicial adjudication with different methods and degrees. In order to illustrate the application scenarios of AI in judicial adjudication and the hidden controversies accurately, we selected some high-cited papers of 1224 papers in WoS to read and review. Then, we divided the application scenarios of AI in judicial adjudication to three categories: The first is the application of AI in procedural matters or the process of judicial adjudication [2]. The second is the application of AI in substantial matters or the decision-making and judgement of judicial adjudication. The third is the controversy of legal values beyond the practice of the application of AI in judicial adjudication.

3.1 The Application of AI in Procedural Matters of Judicial Adjudication

For matters that are not closely related to the final judgment of the case or not directly reference to the judge's final decision-making, working efficiency will increase apparently by the application of AI. Due to worldwide impact of COVID-19, judicial adjudication has been popular for lawyers and judges. In every case, the application of AI is conducive to case registration, case acceptance, payment of litigation fees, the delivery of legal instruments, adducing and cross-examination of evidence, court records by automatic speech recognition [3], which will reduce the cost of litigant parties and the labor intensity of court clerk. In short, the application of AI in the platform of legal disputes is beneficial to all legal workers.

In addition to the settlement of specific cases, advances in AI technology have also helped legal workers obtain and analyze legal information. For example, as AI technology is widely used in the construction of smart court systems, the function of disclosing legal judgment documents online in some states has been further optimized, lawyers and judges could get access to all these documents expediently. In the future, the more AI technology such as real-time transmission of video signals and big data advanced, the more facility might be gotten by all of us.

3.2 The Application of AI in Substantial Matters of Judicial Adjudication

For the fact-finding and decision-making of the case, AI will also impose certain impact on the final judgment made by judges through the interpretation of legal terms, providing judicial references to judgement and even entering a judgement by itself [4]. For instance, to accord consistent judgments to similar cases, some states are trying to train case robot by promoting the degree of familiarization of robots with the results and legal characters of legal precedents which could assist judges in the process of decision-making [5]. As another example, other states such as Russia are making use of AI as a tool to control judicial discretion. When judges' opinions are confronted with big data, AI will remind judges to consider judgement carefully [6]. Meanwhile, there are some states applying AI in more sensitive areas such as risk assessment. As we all known, the essence of penalty of criminal law is the assessment of whether the defendant is danger to society or not. If the answer is yes, judges need to assault penalty by the assessment of the degree of the danger. When the defendant is highly dangerous, he might get long-term sentence without probation or parole. In Canadian court system, the risk assessment tools were put on trial in some cases have caused controversy because the results of risk assessment might be summit to judges as references to final judgement [7]. Looking forward to the application scenarios of AI technology in the future, various forms of dispute settlement mechanisms including international arbitration will embrace AI technology, but argument about the relationship between AI and judges could not end up in the short run.

3.3 The Controversy of Legal Values Beyond the Application of AI in Legal Practice

There are certain risks or disputes both for the application of AI in procedural and substantial matters of judicial adjudication. On the one hand, if the AI technology is undeveloped, the application of AI in the delivery of legal instruments by big data and court records by automatic speech recognition might record some mistaken information which will delay the process of judicial adjudication. On the other hand, no matter whether AI is a tool to analysis and provide references to judges, or AI is setting up as a judge by itself, the potential risks of "algorithmic black-box" and "algorithmic discrimination" have not been defused thoroughly. In most of the views, acquisition and storage of information by AI is more fairness than the application of AI in the process of decision-making, but the underlying disputes around ethics and fairness is hard to settled. Scholars who support the application of AI widely believe that AI judges are fairer than human beings judges in theory because AI judges get less intervention by the society [8]. Especially in complex cases, AI judges is more friendly to parties who could not afford expensive counsel fee or enjoy high-quality legal services [9]. However, other scholars hold the point that the final judgement must be made and supervised by human beings judges to avoid the wrong interpretation of legal terms by AI [10]. From the perspective of advances in technology, AI might provide a more comprehensive and reasonable basis for judicial decisions, but the debate on scientific and technological ethic will still exist in a long term. Just as other cutting-edge technology, AI is designed by a few experts and social elite, but the results of the application of AI in judicial adjudication are suffered and accepted by vast majority of ordinary people under the circumstances that they do not know well about the working principles of AI. In criminal cases, these results are directly related to the freedom of and property right and interests of citizens. Therefore, to increase human beings' trust for AI and avoid "algorithm discrimination" based on race, sex, or age among others [11], we must make great efforts to promote the disclosure and improvement of algorithm on the basis of democracy.

In addition to debate on ethics of AI, the potential risk of erroneous judgement by AI also deserves chew over. For example, when AI-based self-driving cars cause traffic accidents due to mistake of AI, who are responsible for the accident? Dose the consumer or the producer of the car need to compensate for victims? Under the circumstances of judicial adjudication, when AI judges return a wrong verdict, dose the AI judge owns the copyright of the judgement? Does the government or court need to compensate for suffers affected by the wrong judgement? Is the designer of the AI responsible for wrong judgement just like human being judges? In the near future, all these problems are likely to arise in legal practice with the rapid advances in AI technology.

4 Conclusion

In the ending of Cyber and Justice Holmes with the ending of debate, Professor Neustadt concluded that "neither man nor Cybers should ever replace each other". In recent years, scenarios envisioned in the fiction has integrated into legal practice, but we still need to consider how to take both advantages of high efficiency of AI and the sense of justice of human beings. This paper makes a literature analysis of keywords co-occurrences of papers related to application of artificial intelligence in judicial adjudication, which shows the scenarios of the application of AI in both procedural and substantial matters, while the underlying debate on ethics remains to be solved in a long time.

References

1. Silva, R.A.F.E., Sousa, M.D., et al.: Artificial intelligence and legal careers in Brazil: a review and proposed research agenda. Humanidades Inovacao 8(48), 187–203 (2021)
2. Intahchomphoo, C., Vellino, A., Gundersen, O.E., et al.: References to artificial intelligence in Canada's court cases. Leg. Inf. Manag. 1(20), 39–46 (2020)
3. Nitta, K., Satoh, K.: AI applications to the law domain in Japan. Asian J. Law Soc. 7(3), 471–494 (2020)
4. Sert, M.F., Yıldırım, E., Haşlak, İ: Using artificial intelligence to predict decisions of the Turkish constitutional court. Soc. Sci. Comput. Rev. 6(40), 1416–1435 (2022)

5. Wang, N.: "Black box justice": robot judges and AI-based judgment processes in China's court system. In: 2020 IEEE International Symposium on Technology and Society (ISTAS), pp. 58–65. IEEE, Berkeley (2020)
6. Stepanov, O.A., Basangov, D.A.: On the prospects for the impact of artificial intelligence on judicial proceedings. Tomsk State Univ. J. **2**(475), 229–237 (2022)
7. Chugh, N.: Risk assessment tools on trial: lessons learned for "ethical AI" in the criminal justice system. In: 2021 IEEE International Symposium on Technology and Society (ISTAS), pp. 1–5. IEEE, Berkeley (2021)
8. Fortes, P.R.B.: Paths to digital justice: judicial robots, algorithmic decision-making, and due process. Asian J. Law Soc. **3**(7), 453–469 (2020)
9. Terzidou, K.: The use of artificial intelligence in the judiciary and its compliance with the right to a fair trial. J. Judic. Adm. **3**(31), 154–168 (2022)
10. Shi, J.: Artificial intelligence, algorithms and sentencing in chinese criminal justice: problems and solutions. Crim. Law Forum **2**(33), 121–148 (2022)
11. Buhmann, A., Fieseler, C.: Deep learning meets deep democracy: deliberative governance and responsible innovation in artificial intelligence. Bus. Ethics Q. **1**(33), 146–179 (2023)

Overlapping Fingerprint Segmentation and Separation Based on Deep Learning

Sheng-Gui Su[1], Yu-Hsin Cheng[1], Yih-Lon Lin[1(✉)], and Hsiang-Chen Hsu[2]

[1] National Yunlin University of Science and Technology, Yunlin, Taiwan
{m11017059,m11017050}@gemail.yuntech.edu.tw,
yihlon@yuntech.edu.tw
[2] I-Shou University, Kaohsiung, Taiwan
hchsu@isu.edu.tw

Abstract. This study focuses on the application of semantic segmentation to address the challenge of overlapping fingerprint images with noisy backgrounds. Our approach involves excluding non-fingerprint regions and employing an innovative algorithm for separating and merging the overlapping regions with the non-overlapping areas of two fingerprints. This process allows us to obtain two distinct and independent fingerprint images. To mimic real-world scenarios encountered at crime scenes, we utilize an envelope bag image as the noisy background. We also incorporate artificial fingerprint images generated at various angles, adjust the distance between the two fingerprints, and superimpose them to simulate overlapping. These measures ensure a comprehensive evaluation of our method's effectiveness. Through extensive experimentation, we report impressive results. The fingerprint segmentation model achieves an outstanding accuracy of 0.9894 on the test set. This accuracy confirms the model's ability to robustly segment overlapping fingerprints. Our study contributes to advancing the field of fingerprint analysis by addressing the complex issue of overlapping fingerprints and providing a reliable solution. The achieved accuracy demonstrates the potential of our methodology in practical forensic applications, helping forensic personnel experts extract valuable information from challenging fingerprint datasets.

Keywords: fingerprint image · image segmentation · overlapping fingerprint separation

1 Introduction

Semantic segmentation holds significant importance in the field of computer vision, as it enables the classification of each pixel within an image into distinct categories. Semantic segmentation holds immense application value across numerous fields. In the realm of autonomous driving, it plays a pivotal role in enabling vehicles to perceive their surroundings. By accurately segmenting elements such as roads, pedestrians, and vehicles, semantic segmentation facilitates tasks related to autonomous driving and enhances overall traffic safety. In the domain of medical imaging, semantic segmentation aids doctors in examining abnormal areas of the body by precisely delineating organ boundaries.

F. Zhao and D. Miao (Eds.): AIGC 2023, CCIS 1946, pp. 256–265, 2024.
https://doi.org/10.1007/978-981-99-7587-7_22

This segmentation enables efficient organ segmentation, assists in disease diagnosis, and supports medical professionals in identifying and analyzing regions of interest.

In the study of semantic segmentation, convolutional layers and pooling layers of convolutional neural networks (CNNs) are commonly used to extract features from images. Subsequently, a fully connected layer is employed to predict the category of each pixel.

During a crime scene investigation, forensic personnel meticulously search for potential fingerprints. They employ specialized fingerprint enhancement techniques to enhance the visibility of fingerprints. After capturing images of the enhanced fingerprints at the scene, they proceed to the police station for manual fingerprint identification and comparison. The method used to visualize fingerprints involves spraying Ninhydrin dye onto the paper containing residual fingerprints, followed by placing the paper in an oven to expedite the reaction. The areas where the fingerprints are present typically retain sweat residue. During the fingerprint visualization process, the amino acids in the sweat chemically react with Ninhydrin dye, resulting in the production of a blue-purple product. Ninhydrin dye is prepared by combining Ninhydrin, methanol, and petroleum ether in a specific ratio. While the Ninhydrin method is specifically suitable for dry paper, it exhibits excellent effectiveness in displaying fingerprints. Moreover, the displayed fingerprints can be preserved for an extended period, making it one of the commonly employed and widely used techniques in fingerprint visualization.

When overlapping fingerprints are encountered at a crime scene, forensic personnel often face challenges in direct comparison due to the presence of complex background patterns and the increased difficulty posed by the overlapping nature of the fingerprints. As a result, forensic personnel may opt to forgo immediate comparison in such cases. Indeed, human behavior patterns exhibit a high degree of consistency, and certain locations are frequently touched on a daily basis. Consequently, the objective of this research is to leverage deep learning techniques to effectively separate two overlapping fingerprints from images that contain noisy backgrounds. The primary aim of this separation process is to facilitate the task of forensic personnel in identifying individual fingerprints more accurately. By successfully separating the overlapping fingerprints, the chances of identifying potential criminal suspects can be significantly increased.

2 Related Work

2.1 Fingerprint Image Generation

During the process of collecting fingerprint images, various challenges are often encountered, including the security of personal fingerprint data, inconsistencies in fingerprint images caused by the use of different collection devices, and variations in fingerprint printing methods.

Cappelli et al. [1] proposed a method to generate synthetic fingerprint images. A Gabor-like spatially varying filter is used to iteratively extend an initial empty image containing only one or a few seeds. The input to the orientation image model is the number and location of fingerprint cores and deltas, which are used to adjust the filter according to the underlying ridge orientation. Very realistic fingerprint images are obtained after the final noise and rendering stages.

Xue et al. [2] introduced a rapid fingerprint image generation method that differs from the approach proposed by Cappelli et al. in two key aspects. Firstly, this method incorporates a design element that allows for the adjustment of the fingerprint mask size. By doing so, the generated fingerprint images can more accurately simulate the variations observed in real fingerprints from different fingers. The second aspect the classification of fingerprints into five distinct types based on their overall orientation and the number of singular points. These types include Arch, Tented Arch, Left Loop, Right Loop, and Whorl. By considering these two characteristics, the method enables the generation of fingerprint images in accordance with the specific type and proportion of fingerprints. This approach helps to avoid the generation of an excessive number of fingerprint images of the same type, enhancing the diversity and representativeness of the generated dataset.

2.2 Fingerprint Image Segmentation

In a crime scene, the fingerprint images collected often suffer from significant noise and interference from the background. Consequently, the segmentation of fingerprint images becomes crucial in order to extract clear and intact fingerprint. Tong et al. [3] introduced a method that compares the similarity of each pixel value in an image with its surrounding pixel values. By leveraging the dissimilarity between the ridge and valley regions of a fingerprint, this approach enables the detection of the strength of the fingerprint ridge frequency. After smoothing the image, the resolution of the fingerprint ridge frequency's intensity is improved, leading to enhanced accuracy in subsequent analysis and identification processes. Finally, the fingerprint image is segmented into foreground and background based on the average value of the highest and lowest intensities as a threshold. This method has demonstrated favorable segmentation results when applied to the fingerprint images from the FVC2004 dataset.

Jia et al. [4] proposed an improved fingerprint preprocessing algorithm, including segmentation, enhancement, binarization and thinning. Fingerprint enhancement using Gabor filters based on ridge orientation. The algorithm also addresses segmentation, binarization and thinning techniques, and the efficiency is verified using VC. Experimental results show that the algorithm significantly improves the quality of fingerprint images, and the ridges of fingerprint images are clear, continuous and smooth. The algorithm enhances the reliability of fingerprint identification systems, making it a valuable tool for enhancing fingerprint images.

Stojanović et al. [5] proposed an ANN-based fingerprint ROI segmentation method. The proposed ANN is trained using 10,000 samples from 20 grayscale and binary pattern fingerprint images. Experimental results on a test database containing 200 fingerprint images show excellent performance, supported by three statistical performance metrics. The method accurately and efficiently segments ROIs in fingerprint images, demonstrating strong learning and generalization capabilities. This study provides valuable insights for the development of advanced fingerprint analysis techniques.

Nimkar et al. [6] proposed adaptive (scale) and directional (vector) algorithms inspired by fully variational models and combining scale and vector features. The algorithm decomposes the fingerprint image into noise layer and texture layer, so as to achieve effective segmentation. Experimental analysis of two fingerprint images demonstrates the efficiency of the algorithm using peak signal-to-noise ratio (PSNR) as a metric. A comparative study of adaptive and directional algorithms confirms the superior performance of the proposed method. Adaptive and orientation-based algorithms help improve fingerprint image segmentation and increase the accuracy of recognition systems.

Zhou et al. [7] proposed a fingerprint image segmentation enhancement algorithm based on adaptive dynamic threshold segmentation method. The non-uniform gray level problem is solved by improving the Otsu algorithm, which effectively distinguishes the target and the background in the fingerprint image. The algorithm combines threshold segmentation and region segmentation techniques to improve image quality. Experimental results show that the proposed fingerprint image segmentation algorithm can produce excellent image quality.

El-Hajj-Chehade et al. [8] proposed an enhanced fingerprint segmentation method using a histogram-based thresholding method using a gamma distribution. Unlike conventional methods that assume a Gaussian distribution, our method adapts to the characteristics of the fingerprint image. Experimental evaluation on a large number of real fingerprint images shows promising results. A comparative study with Gaussian-based methods reveals the superiority of our gamma-based technique, resulting in higher quality segmented images. The authors' research demonstrates the efficacy of gamma thresholding in fingerprint image segmentation.

3 Method

3.1 Data Preprocessing

In this study, we employed the Anguli: Synthetic Fingerprint Generator [9] software to generate fingerprint images. This software offers the capability to rapidly generate multiple virtual fingerprint images while providing control over the number and distribution of key fingerprint features, such as ridge endpoints and bifurcation points. By employing this method, it becomes possible to acquire multiple fingerprint images without compromising personal data security or involving complex collection procedures. In this study, we utilized the software to generate a dataset consisting of 10 synthetic fingerprint images.

To simulate fingerprint images collected from real crime scenes, which often exhibit noisy backgrounds, this study employed a synthesis approach using overlapping fingerprint images and noisy background images. The rotation angle of the synthesized fingerprint images in this study ranged from 0° to 360°. The two fingerprints were overlapped with a spacing of 200 pixels to ensure that the overlapping was not excessive, thereby avoiding the loss of reference value in the analysis. For this study, we selected the envelope bag as the noise background image to simulate the size and position of real fingerprints on the envelope bag. By overlapping 10 fingerprint images, a total of 100 overlapping fingerprint images were generated. Subsequently, these images were synthesized onto 6 envelope bag images, resulting in a dataset comprising a total of

600 images. To mitigate any color discrepancies between the simulated fingerprint color and the color displayed by the Ninhydrin dye, the synthesized images were converted to grayscale. The results of the synthetic process, depicting the simulated overlapping fingerprints and background images, are presented in Fig. 1.

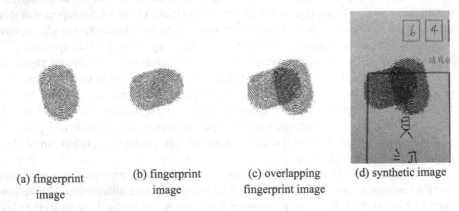

(a) fingerprint
image
(b) fingerprint
image
(c) overlapping
fingerprint image
(d) synthetic image

Fig. 1. Synthetic image of overlapping fingerprints and envelope bag.

3.2 Fingerprint Segmentation Model

In this study, semantic segmentation is employed to conduct binary classification for each pixel within the image, aiming to segment the foreground image (overlapping fingerprint image) and the background image (envelope bag image).

The U-Net model is a deep learning architecture utilized for image segmentation, based on the combination of an Encoder and a Decoder within its architecture. The Encoder employs a series of convolutional layers and pooling layers to extract intricate features from the input image. These features are subsequently reinstated in the Decoder through the integration of upsampling layers and convolutional layers, enabling the generation of segmentation results. The convolutional layers and pooling layers utilized within the U-Net model possess the capability to extract diverse features present in the image, encompassing edges, textures, and colors, all of which are fundamental for effective image segmentation. The Encoder and Decoder of the U-Net model establish a connection via Concatenate, which facilitates the effective combination of features extracted from both components. This mechanism enables the U-Net model to preserve shallow features extracted from the input image while utilizing the deep features derived from additional convolutional layers and pooling layers to enhance the accuracy of segmentation.

The U-Net model was trained using a training set comprising 480 synthetic images, while an additional 120 images from the test set were employed to assess the model's performance after each training iteration. The ratio of the training set to the test set was maintained at 8:2. The model takes a gray-scale image as input, resulting in an input size of $512 \times 512 \times 1$. Since the segmentation task involves binary classification of foreground and background, the model's output size is $512 \times 512 \times 2$.

The process of fingerprint segmentation is illustrated in Fig. 2. In the diagram, the black arrow represents the output of a layer, while the red arrow represents Concatenate. Following the passage of the input composite image through the Encoder and Decoder, two grayscale images are produced as output: the foreground image mask and the background image mask.

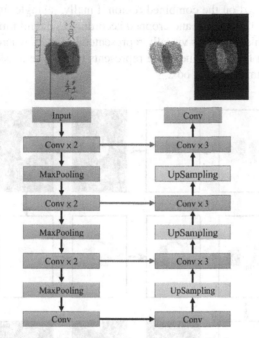

Fig. 2. Fingerprint segmentation process. (Color figure online)

3.3 Overlapping Fingerprint Separation

To facilitate the identification of individual fingerprints by forensic personnel, we have developed an overlapping fingerprint separation algorithm. This algorithm is specifically designed to separate two fingerprints that are overlapped with each other, thereby enabling the distinct identification of each fingerprint. First, Gaussian blurring is applied to the overlapping fingerprint image. Subsequently, binarization is utilized to obtain the image of the overlapping fingerprint area. The next step involves identifying the largest contour area to retain the overlapping fingerprint region. Finally, a Closing morphology operation is performed to enhance the smoothness of the edges the overlapping fingerprint area.

Once the overlapping fingerprint areas are identified, the next step involves separating the non-overlapping fingerprint regions. This process entails several stages. Initially, Gaussian blurring is applied, followed by performing erosion (35 times) and dilation (35 times) morphological operations. Subsequently, binarization is performed to obtain two distinct non-overlapping fingerprint areas. One of the non-overlapping fingerprint areas is selected and combined with the overlapping fingerprint area. An ellipse mask is then generated based on the combined region. Finally, a single fingerprint image is separated using the ellipse mask and cropped according to the minimum bounding box of the fingerprint. This process is visually represented in Fig. 3, where the black arrows depict the algorithm flow, the blue circle represents the ellipse mask, and the red box signifies the minimum bounding box.

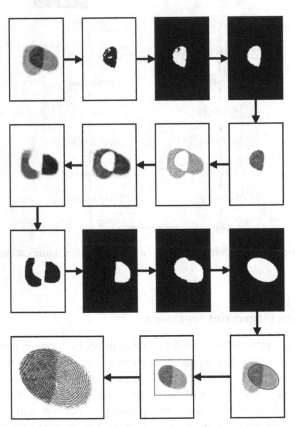

Fig. 3. Overlapping fingerprint separation process. (Color figure online)

4   Experiments and Results

Upon training the U-Net model, we utilize the confusion matrix to assess the disparities between the Ground truth and the Predict image in terms of pixel classification. The confusion matrix comprises True Positive (TP), False Positive (FP), True Negative (TN), and False Negative (FN).

To evaluate the segmentation effect of the U-Net model on both the training set and test set, we employ the following three evaluation indicators. Accuracy (Eq. (1)): This indicator assesses the ratio of correctly classified pixels in the image, providing an overall measure of classification accuracy. Sensitivity (Eq. (2)): evaluates the ratio of correctly predicted foreground regions to the total number of actual foreground regions, indicating the model's ability to accurately identify positive. Specificity (Eq. (3)): measures the ratio of correctly predicted background regions to the total number of actual background regions, indicating the model's ability to accurately identify negative.

$$Accurancy = \frac{TP + TN}{TP + FP + TN + FN} \tag{1}$$

$$Sensitivity = \frac{TP}{TP + FN} \tag{2}$$

$$Specitficity = \frac{TN}{FP + TN} \tag{3}$$

Based on the information presented in Table 1, the training set and test set achieved an accuracy of 0.9894 in the U-Net model. This high accuracy indicates that the model exhibits excellent segmentation performance for both overlapping fingerprints and background images. Additionally, the matching accuracy suggests that there is no overfitting issue present. However, it is worth noting that the Specificity value exceeds the Sensitivity value. This implies that there is a higher probability of misclassifying overlapping fingerprints as background images.

Table 1. Evaluate the performance of U-Net model.

	Training Set	Test Set
Accuracy	0.9894	0.9894
Sensitivity	0.8802	0.8806
Specificity	0.9976	0.9976

Figure 4 demonstrates the effectiveness of our approach in fully isolating fingerprints from images with envelope bag backgrounds. By employing the overlapping fingerprint separation algorithm, we successfully separate the two fingerprints from their respective non-overlapping areas as well as the overlapping region. To prevent potential disruptions to the fingerprint area caused by morphology operations, we utilize an ellipse mask to isolate individual fingerprints. While this ellipse mask may exclude some surrounding portions of the fingerprint, it effectively preserves the primary fingerprint region, facilitating the comparison of fingerprints by forensic personnel.

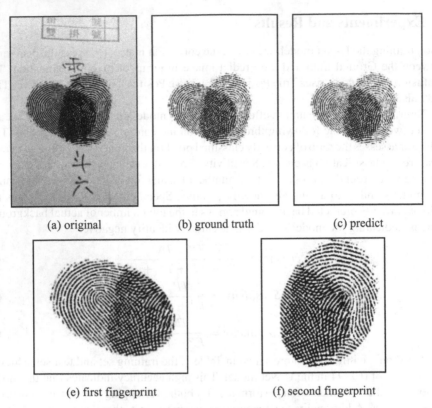

　　　(a) original　　　　　　　(b) ground truth　　　　　　(c) predict

　　　(e) first fingerprint　　　　　　　　　(f) second fingerprint

Fig. 4. Overlapping fingerprint segmentation and separation results

5 Conclusions

This study synthesis of artificially generated fingerprints onto images of envelope bags to simulate scenarios where fingerprints are revealed through dye after fingers come into contact with paper. By utilizing the U-Net model, we were able to successfully separate overlapping fingerprints from images containing envelope bag backgrounds. Furthermore, the overlapping fingerprint separation algorithm enabled the separate extraction of the two individual fingerprints. To further validate the effectiveness of our method, it would be valuable to collect overlapping fingerprint images from real crime scenes in the future.

Acknowledgement. The research was supported by the National Science and Technology Council, Taiwan, under grant no. NSTC 111-2622-E-224-019.

References

1. Cappelli, R., Erol, A., Maio, D., Maltoni, D.: Synthetic fingerprint-image generation. In: 15th International Conference on Pattern Recognition (ICPR), Barcelona, Spain, vol. 3, pp. 471–474 (2000)

2. Xue, J.T., Xing, S.F., Guo, Y., Liu, Z.G.: Fingerprint generation method based on Gabor filter. In: International Conference on Computer Application and System Modeling (ICCASM), Taiyuan, China, pp. V8-115–V8-119 (2010)
3. Tong, X.F., Li, P.F.: Fingerprint image segmentation based on fingerprint ridge intensity. In: International Conference on Machine Learning and Cybernetics, Guilin, China, pp. 1780–1784 (2011)
4. Jia, H., Cao, K.: The research on the preprocessing algorithm for fingerprint image. In: IEEE Symposium on Electrical & Electronics Engineering (EEESYM), Kuala Lumpur, pp. 163–166 (2012)
5. Stojanović, B., Nešković, A., Popović, Z., Lukić, V.: ANN based fingerprint image ROI segmentation. In: 22nd Telecommunications Forum Telfor (TELFOR), Belgrade, Serbia, pp. 505–508 (2014)
6. Nimkar, R., Mishra, A.: Fingerprint segmentation using scale vector algorithm. In: Fifth International Conference on Communication Systems and Network Technologies, Gwalior, India, pp. 530–534 (2015)
7. Zhou, X., Liu, Q., Tan, T.: A study on fingerprint image segmentation algorithm. In: IEEE 2nd Advanced Information Technology, Electronic and Automation Control Conference (IAEAC), Chongqing, China, pp. 2114–2117 (2017)
8. El-Hajj-Chehade, W., Abdel Kader, R., Kassem, R., El-Zaart, A.: Image segmentation for fingerprint recognition. In: IEEE Applied Signal Processing Conference (ASPCON), Kolkata, India, pp. 314–319 (2018)
9. Database Systems Lab Indian Institute of Science. Anguli: synthetic fingerprint generator. https://dsl.cds.iisc.ac.in/projects/Anguli/. Accessed 04 May 2023

Artificial Intelligence for Text Generation: An Intellectual Property Perspective

Huidi Zhang[1], Junming Gong[2], and Wei Wu[3](✉)

[1] Law School, Wuhan University, Wuhan, China
[2] Law School, Leiden University, Leiden, The Netherlands
[3] China Institute of Boundary and Ocean Studies, Wuhan University, Wuhan, China
ww330@whu.edu.cn

Abstract. Purpose: This paper aims at proposing relevant improvement measures for the intellectual property regulation on AI-generated text. Methods: This paper adopts qualitative research methods, collects and analyzes opinions from various sources to illustrate the technical logic of AI-generated text and explore existing intellectual property legal regulations and dilemmas. Results: AI-generated content, especially AI-generated text, follows the process of data collection and processing, models creation and content generation. Regarding intellectual property regulations on the AI-generated text, international law reflects hysteresis while major states' domestic law varies in terms of intellectual property attribution. Conclusion: Based on the research on current situations, this paper proposes corresponding improvement measures, namely self-censorship, self-improvement, international cooperation and strengthening legislation.

Keywords: Artificial Intelligence · AI-generated Text · Original Data · International Cooperation

1 Introduction

In recent years, significant breakthroughs have been made in artificial intelligence technology. Artificial intelligence is a field of science and technology that researches and develops, with the aim of making computer systems capable of simulating human intelligent behavior and decision-making capabilities [1]. It covers several subfields, including machine learning, natural language processing, computer vision, and expert systems, and aims to enable computer systems to learn, understand, reason, and make decisions from data in order to achieve tasks similar to human intelligence [2]. The continuous evolution and innovation of these technologies have not only made artificial intelligence more and more crucial in the field of intellectual property but also triggered a whole new way of thinking about intellectual property law.

When it comes to intellectual property regulation, artificial intelligence technology has brought notable innovations. Artificial intelligence also has the ability to automatically generate abstracts, summaries and analysis reports, greatly simplifying the cumbersome process of literature retrieval and organization and greatly enhancing the efficiency of knowledge management [3]. However, the increasing use of artificial intelligence in creative work also raises new challenges for intellectual property law. In terms of copyright, it is still unclear whether the text generated by artificial intelligence (AI-generated text) should be protected by copyright, which depends on whether the generated text meets the originality standard. At the same time, the law usually attributes copyright to the individual, legal person, or unincorporated organization that created the text. Whether artificial intelligence itself can become the owner of copyright is still controversial [4].

In terms of privacy and data source protection, the application of artificial intelligence in the field of intellectual property may involve a large amount of data collection and analysis, which may lead to the leakage of personal privacy. When using content generated by artificial intelligence (AI-generated content), it is particularly critical to clearly understand the licensed scope and restrictions of the generated content to avoid infringement issues [5].

Nowadays, there are a lot of literary works produced with AI, and the role of this paper is to explore the way of using AI from the intellectual property level, and to contribute to the legal application of AI. First, through a literature review, this paper comprehensively sorts out relevant literature regarding the technical logic of AI generated content and AI-generated text. Secondly, through comparative research, this paper conducts an in-depth analysis of the intellectual property issues of AI-generated content in different fields and mainly discusses the intellectual property regulation and dilemmas of AI-generated texts. Finally, this paper adopts qualitative research methods, collects and analyzes opinions from various parties, and puts forward legislative suggestions in terms of the legitimacy of basic data, strengthening international cooperation, improving the originality and innovation of AI-generated text, and clarifying the ownership of the AI-generated text. Instead of discussing the measures that could be adopted by one category of actors like in other literatures, this paper proposes suggestions for multiple actors, namely individuals, corporations, states and international organizations. These comprehensive suggestions could help balance the relationship between artificial intelligence technology and intellectual property rights at the legal and ethical levels so as to promote the sustainable development of technology and the interests of society.

2 The Technical Logic of the AI-Generated Text

AI-generated text, as a part of all AI-generated content, follows general and specific technical rules. This part briefly summarizes the general working process of AI-generated content and specifically illustrates the operating procedure of AI-generated text.

2.1 General Description

Overall, the process of AI-generated content is divided into four steps. The first step is to check users' needs. Before artificial intelligence software generates content, it usually asks the user what type of content needs to be generated and the requirements for the content [6]. For example, a user can declare that they need to generate text content and specifically request that the topic of the content be higher education.

The second step is to obtain the materials for generation. Designers of artificial intelligence software collect extensive information and turn it into data sets that can be used by artificial intelligence software [7]. This data set is constantly updated as requested by the designers. Artificial intelligence software will capture useful information based on the latest updated data set.

The third step is to run the algorithm. Designers of artificial intelligence software will develop a series of algorithms [7]. These algorithms allow the software to grab the content of the data set and process it according to the needs of users.

The fourth step is to output the content. After the artificial intelligence software understands the user's needs and obtains and processes the relevant information of the data set according to the algorithm, the software outputs the content requested by the user. During this process, technologies like natural language processing, machine learning, deep learning, diffusion models, and transformer neural networks are crucial to the construction of the algorithm [8].

2.2 Specific Illustration

AI-generated text technology uses artificial intelligence and natural language processing to automatically create coherent and contextually relevant written content. The process involves a combination of techniques, including deep learning models, language models, and large datasets. There are three main steps in how AI-generated text works.

First, data collection and processing. The process starts by collecting and preprocessing large amounts of text data. This data includes books, articles, websites, and other sources of text. Preprocessing includes tasks such as tokenization, stemming, and removing stop words in order to prepare the text for analysis [9]. To generate text, the model employs sampling strategies such as greedy sampling [10]. These techniques affect the randomness and creativity of the generated text.

Second, creating models. The core of artificial intelligence to generate text lies in the model architecture. Deep learning models, especially language models, are often used for this purpose. Neural networks, transformers, and generative adversarial networks are all commonly used architectures. Transformers, especially models such as OpenAI's GPT (Generative Pre-Trained Transformers), have attracted attention for their ability to generate coherent and contextually accurate text. The text is tokenized into smaller units, such as words or subwords. Each token is then represented as a high-dimensional vector using embedding techniques like Word2Vec or FastText. These embeddings capture semantic relationships between words. Training the model is also essential. During training, the model learns the statistical patterns and relationships present in the input text data. With the training, the language model may predict the next word or token in a sequence based on previous words. This process is also known as autoregressive language modeling. The model learns to generate text by predicting subsequent words based on the context provided by previous words. Taking GPT-3.5, created by OpenAI, as an example. GPT-3.5 mainly achieves text generation through pre-training and fine-tuning [11]. The pre-training phase mainly includes two activities: data collection and model training. The GPT-3.5 model uses a large amount of text data to learn the statistical regularity, grammatical structure and semantic relationship of language. It analyzes a large number of sentences and predicts what the next word or character will be. This process enables the model to learn general features in terms of vocabulary, syntax, and semantics. After pre-training, GPT-3.5 is fine-tuned for a specific task or application [12]. Fine-tuning is a supervised learning process that uses domain- or task-specific data to adjust the weights of a model to perform better on a specific task. Fine-tuning allows GPT-3.5 to generate content relevant to a specific topic or context. Therefore, the text generated by the GPT-3.5 model is fully in line with the general logic process of AI-generated text.

Third, text generation. Models can generate text based on user needs. The generated text can be post-processed, such as using techniques like grammatical error correction and text summarization to improve readability, coherence and grammatical correctness. Many AI-generated text models are pre-trained on large text corpora and then fine-tuned for a specific task or domain. This transfer learning approach allows models to leverage general language knowledge before specializing in tasks like writing news articles or composing poetry.

AI-generated text can be applied in a wide range of areas, including content creation for articles, blogs, product descriptions, and social media posts; chatbots and virtual assistants, where artificial intelligence generates human-like responses to user queries; language translation; creative writing, poetry, and storytelling; code generation, such as autocompleting code snippets or writing documentation.[13].

3 The Current Situation and Dilemma of Intellectual Property Regulation on AI-Generated Text

Contemporary laws from national and international levels provide certain degrees of regulation while leaving some controversies and dilemmas. This part explores the regulation of states with rapidly developing artificial intelligence technology and the international laws regarding intellectual property, as well as enumerating several drawbacks.

3.1 States' Domestic Law

Contemporary laws from national and international levels provide certain degrees of regulation while leaving some controversies and dilemmas. This part explores the regulation of states with rapidly developing artificial intelligence technology and the international laws regarding intellectual property, as well as enumerating several drawbacks.

The United States

The United States has chosen to expand existing copyright laws to cover AI-generated text. These laws generally grant copyright protection to works that are the product of human creativity, skill and labor. On March 16, 2023, the U.S. Copyright Office published formal guidance for the registration and protection of works of artificial intelligence [14]. The guidance tackles difficult issues in copyright protection for AI-generated content and proposes a new initiative to explore related legal and policy issues. Also, the guidance emphasizes that copyright protects the original works of human beings and clearly states that works generated entirely by automatic and random artificial intelligence programs without human creative input are not eligible for copyright protection [15].

Applicants seeking copyright registration for works containing AI-generated elements must disclose such involvement and briefly describe the human author's contribution. It is worth noting that the applicant must expressly exclude the significant part generated by artificial intelligence through a brief description; otherwise, the validity of the registration in the future may be jeopardized in infringement proceedings. For works involving artificial intelligence, the determination of copyright eligibility will be made on a case-by-case basis [1]. If artificial intelligence independently determines expressive elements such as images or text, the work lacks human authorship and cannot be registered. However, if humans creatively select, arrange, or modify AI-generated material, copyright protection extends to aspects of human creation. Consequently, recent guidance from the U.S. Copyright Office emphasizes that copyright protection relates to human creative works and outlines how it should accommodate AI-generated text while adhering to the nature of human authorship.

The United Kingdom
The UK is considering giving artificial intelligence systems limited "personality," allowing them to be recognized as owners of intellectual property. Computer-generated works of art are eligible for copyright protection under the Copyright, Designs and Patents Act 1988 (CDPA) [16]. Section 178 of the CDPA states that "'computer-generated', in relation to a work, means that the work is generated by computer in circumstances such that there is no human author of the work." [16] Also, "in the case of a literary, dramatic, musical or artistic work which is computer-generated, the author shall be taken to be the person by whom the arrangements necessary for the creation of the work are undertaken." For more than 30 years, the UK has protected computer-generated works, including advanced technologies such as AI-generated text, through the CDPA and has been recognized as a pioneer in promoting and protecting innovative and creative works. In 2021, the United Kingdom Intellectual Property Office (UKIPO) conducted a consultation on artificial intelligence and intellectual property, confirming that existing laws are adequate to protect computer-generated intellectual property, including AI-generated works, and therefore no amendment is necessary [17]. This stance is in line with the UK government's ambition to make the country a global leader in artificial intelligence, reinforced by early applications of artificial intelligence [18].

Since then, however, artificial intelligence has grown in importance, leading Parliament to hold a forensics session in May 2023 on the impact of artificial intelligence on the creative industries [19]. Expert witnesses argue that the CDPA's treatment of "computer-generated" works is outdated as artificial intelligence increasingly plays a major role in creation, not just a secondary one [20]. While the UK currently protects AI-generated works through copyright, the law is still likely to be reviewed, especially given the changing nature of artificial intelligence involvement in the creative process.

Canada
Canada emphasized the concepts of the public domain [21] and fair use [22]. Some argue that AI-generated text is the result of algorithms processing vast amounts of publicly available information and should enjoy limited or no copyright protection. This approach seeks to balance innovation and accessibility, allowing creative endeavors to continue while safeguarding the public interest.

Under Article 13(1) of the Canada Copyright Act, the author is considered the first owner of the copyright [23]. First, the person who writes the code, the developer of the artificial intelligence, can be the author and therefore own the copyright. However, users of artificial intelligence, the people who cause a particular text to be generated, may also be involved. Another possible author could be artificial intelligence itself. If none of these actors should have ownership of the work, then the copyright can remain in the public domain, meaning no intellectual property rights apply [24]. Under Article 5(1)(a) of the Canadian Copyright Act, copyright holders can only be human beings, so a machine that created a work cannot be a copyright holder [23]. However, this does not prevent the Canadian Intellectual Property Office (CIPO) from registering artificial intelligence RAGHAV as a co-author [25]. It can be found that there is still some controversy in Canada on whether artificial intelligence can have intellectual property rights.

China

Article 12 of the Copyright Law of the People's Republic of China stipulates that "the natural person, legal person or unincorporated organization that signs the work is the author, and the corresponding rights exist in the work unless there is evidence to the contrary." [26]. This effectively rules out the possibility of the artificial intelligence itself being the author.

Currently, China has not directly regulated AI-generated texts in the Copyright Law. China adopts a distributed legislative model to regulate artificial intelligence-generated texts in normative documents such as individual laws and regulations. For example, there are regulatory provisions relevant to text generated by artificial intelligence in the Data Security Law [27] and the Personal Information Protection Law [27]. The reason why Chinese legislation adopts this approach instead of simply including AI-generated texts within the protection scope of the Copyright Law or any other legal norms or formulating special laws to regulate them is that China intends to establish a legal system that distinguishes AI-generated texts from the protection of human works. At this stage, the time for independent legislation is not yet ripe, so China only provides distributed regulations.

In judicial practice, with regard to AI-generated texts, China's domestic judgments both recognize AI-generated texts as works and deny AI-generated texts as works. Regarding the ownership of the copyright of AI-generated texts, some courts believe that the rights should belong to the owners of the artificial intelligence software, while others believe that the rights should belong to the users of the artificial intelligence software. The specific definition and standard are not yet clear, but the degree of creative contribution to the final generated text has become a principle guiding the court to make a judgment.

3.2 International Law

International law is of great importance in the regulation of intellectual property regarding AI-generated text. Relevant international conventions and international organizations are the primary source and entities to protect and manage AI-generated text intellectual property.

Fundamental International Conventions

The Berne Convention for the Protection of Literary and Artistic Works and the Agreement on Trade-Related Aspects of Intellectual Property Rights are the pillars of the international intellectual property legal system. Article 3(1) of the Berne Convention for the Protection of Literary and Artistic Works provides that "(1) The protection of this Convention shall apply to: (a) authors who are nationals of one of the countries of the Union, for their works, whether published or not; (b) authors who are not nationals of one of the countries of the Union, for their works first published in one of those countries, or simultaneously in a country outside the Union and in a country of the Union." Article 10(2) of the Agreement on Trade-Related Aspects of Intellectual Property Rights regulates that, "compilations of data or other material, whether in machine readable or other form, which by reason of the selection or arrangement of their contents constitute intellectual creations shall be protected as such. Such protection, which shall not extend to the data or material itself, shall be without prejudice to any copyright subsisting in the data or material itself."

While these two treaties provide the basis for intellectual property regulation worldwide, they were developed before the advent of artificial intelligence and lacked clear provisions for AI-generated works. Although the Agreement on Trade-Related Aspects of Intellectual Property Rights involves the copyright of data collections, it did not take into account the collection and integration of data by non-human subjects when it took effect, which led to difficulties and differences in legal interpretation. Jurisdictional differences in enforcement.

World Intellectual Property Organization

The World Intellectual Property Organization (WIPO) has taken steps to address issues related to artificial intelligence and intellectual property. In September 2019, WIPO hosted the First Session of the WIPO Conversation on IP and artificial intelligence, focusing on the impact of artificial intelligence on intellectual property (IP) policy [28]. The purpose of the dialogue is to jointly raise relevant questions for policymakers. In December 2019, WIPO published an initial issues paper on IP policy and artificial intelligence, launching a public consultation process to gather insights on the issues that IP policymakers will face in light of the growing importance of artificial intelligence. Insights on key issues [28]. The consultation received more than 250 submissions. In May 2020, an updated version of the Issues Paper was released, listing the copyright and ownership of AI-generated texts as one of the issues to be considered [28]. The topics outlined in the revised document were examined in depth at the second and third sessions of the WIPO Dialogue held in July and November 2020 [28].

These two sessions discussed in detail the issues raised in the revised issues paper. WIPO discussions on the impact of artificial intelligence on copyright and patent law highlighted the need for an international dialogue on adapting existing guidelines to

AI-generated texts. However, reaching a consensus on these issues remains a daunting task as member states have diverse interests.

European Union

The current European Union (EU) intellectual property legal framework is silent on intellectual property regulation of AI-generated text. The different interests of member states within the EU have also slowed the EU's legislative process for artificial intelligence.

Although the European Parliament resolution of 20 October 2020 on intellectual property rights for the development of artificial intelligence technologies [29] and the Amendments adopted by the European Parliament on 14 June 2023 on the proposal for a regulation of the European Parliament and of the Council on laying down harmonized rules on artificial intelligence (Artificial Intelligence Act) and amending certain Union legislative acts [29] demonstrates the EU's determination to regulate artificial intelligence, they possess no specific terms of implementation.

3.3 Existing Dilemmas

Under the existing legal framework, there are several dilemmas. Specifically, the dilemmas include the violation of intellectual property rights, deficiency in originality, ambiguity regarding the copyright of the AI-generated text, insufficient international coordination, and the evolution of artificial intelligence.

Violation of Intellectual Property Rights

Training AI models require the use of large amounts of data. However, the use of these data may involve issues of intellectual property rights. The training data for the model may come from publicly available data on the Internet, which may involve the use of original content and may not be properly licensed. Ensuring compliance with intellectual property laws and regulations during data collection is critical. In terms of copyright, if artificial intelligence uses materials such as copyrighted texts to generate new texts, it is necessary to evaluate whether the generated texts infringe the copyrights of the original creators. If artificial intelligence only integrates existing materials and generates new texts, it may lead to intellectual property disputes.

Deficiency in Originality

Generally, only original texts can be granted copyright. The originality of AI-generated text is the basis for its copyright. Intellectual property law generally requires that texts be created with a certain degree of innovation and originality. Whether AI-generated text meets these requirements and whether it is merely imitating existing text is a matter of careful consideration. Although some platforms and tools allow users to generate text, the generated text may be highly similar to other texts, and there may even be meaningless splicing. This raises questions about whether the generated text can obtain independent copyright protection. Since AI-generated texts are often based on pre-existing data, they may lack originality to some extent. This can raise questions related to originality and creativity, especially in areas involving art, literature, etc.

Ambiguity Regarding the Copyright of the AI-generated Text

In many countries, copyright is usually vested in the person or entity that created the text.

However, attribution becomes more complicated when the text is generated by artificial intelligence. There is no generally applicable legal provision that clearly specifies the copyright ownership of AI-generated text. Attribution of authorship becomes complicated when AI-generated text is considered to possess a certain level of creativity. Who should be credited as the author of these texts? Is it the provider of training data, the developer of the model, or the user of the generated text? These issues may require legal and policy clarity to ensure that copyright issues are properly addressed.

In some cases, the law may consider that the text generated by artificial intelligence is the result of human-machine co-creation, and copyright attribution may be shared. This involves determining the relative contributions of artificial intelligence versus human creators in the creative process. In many countries, the issue of authorship of AI-generated texts is not yet clearly regulated by law, as it is a relatively new and complex topic. Copyright is usually assigned to natural or legal persons, while artificial intelligence itself does not fall into these categories. Therefore, there are some controversies and legal gaps in the copyright ownership of AI-generated texts.

Insufficient International Coordination

At present, major international conventions lack regulations on the intellectual property attributes of AI-generated texts. The lack of comprehensive international rules has led to differences in how different jurisdictions deal with AI-generated text. This incoordination complicates cross-border cooperation and increases legal uncertainty for stakeholders. The borderless nature of the internet and AI-generated text further complicates matters. Determining global jurisdiction and applicable law for AI-generated text poses significant challenges.

The Evolution of Artificial Intelligence

The rapid development of artificial intelligence technologies requires an adaptable legal framework. However, the lagging nature of the traditional legal system makes it difficult to quickly incorporate changes and updates to keep pace with the advancement of artificial intelligence. The novelty of AI-generated text has led to a lack of legal precedent to guide decision-making. Courts and legal professionals lack an established framework to draw upon when dealing with disputes related to AI-generated text. That uncertainty could lead to inconsistent rulings and protracted legal proceedings.

In conclusion, IP regulation in the field of AI-generated text is a multifaceted challenge involving a complex interplay between national-level regulation and the international legal framework. As artificial intelligence technologies continue to advance, bridging the gap between evolving technology and legal norms remains an urgent issue. Collaboration between countries, international organizations, and stakeholders is essential to address current deficiencies and create an adaptive legal framework to ensure fair and effective protection of intellectual property rights in the field of AI-generated text.

4 Intellectual Property Regulation of AI-Generated Text: Multi-actor Improvement Measures

Facing the current regulatory systems and the concurrent dilemmas, proposing measures that could be adopted by multiple subjects for improvement is necessary. Specifically, the measures include self-censorship, self-improvement, international cooperation, and new legislation.

4.1 Self-censorship

AI-generated text may use a large amount of existing data as pre-training material. Such data may involve the creative and original work of others, which raises issues of data provenance and usage authorization. Therefore, it is recommended that users and developers should follow applicable laws and regulations to ensure that their actions are legal and compliant [32]. Second, using material from legitimate sources. Developers could avoid using copyrighted material unless properly licensed or within fair use guidelines. Again, use trademarks and branding information discreetly. If the AI-generated text involves trademarks or brand information, developers should follow the corresponding trademark laws to avoid unauthorized use of trademarks. Finally, exploring open source and licenses. When using open-source technologies and models, developers should understand the terms and conditions of the relevant licenses and ensure that the use process complies with the requirements of the licenses. Owners of artificial intelligence text generation software can innovate technical protection measures and develop technologies to identify and track AI-generated text to prevent unauthorized copying and use.

In addition, all parties can specify the ownership of intellectual property rights in the contract for the use of artificial intelligence text generation software. Contracts can stipulate that in specific cases, for example, artificial intelligence generates text for a specific purpose, the intellectual property rights are owned by the user, or that the user gets only limited usage rights rather than full intellectual property rights.

4.2 Self-improvement: Ensuring the Originality of the AI-generated Text

Intellectual property law generally requires creations to be innovative and original. For AI-generated text, the issue of innovation and originality is complicated because it is generated based on large amounts of data and models and may not meet traditional standards of originality. Therefore, it is recommended to expand the training data first, using a rich variety of training data, including various styles, subjects and sources [33]. This may make artificial intelligence models more broadly inspired and innovative. Second, iterative generation. Allowing artificial intelligence software to generate multiple versions of text, and iterates and improves on the basis of generation could helps the model gradually explore more innovative directions. Finally, human intervention and editing can help fine-tune AI-generated text to be more in line with innovative standards.

4.3 Promoting International Cooperation: Balancing Data Processing and Protection

Countries should promote international cooperation to develop intellectual property standards applicable in cross-border situations in order to better address issues on a global scale. First, the formulation of international data protection agreements and standards. Countries can cooperate to formulate common data protection legal frameworks and standards to ensure that data is properly protected globally [34]. For example, an international version like the European General Data Protection Regulation (GDPR) could serve as a reference to promote consistency in data privacy. Second, build data privacy partnerships. Countries can establish data privacy partnerships to share best practices, lessons learned and technical solutions to improve data protection. Third, establish a framework for cross-border data flows, formulate guiding principles for cross-border data flows, and ensure that data flows between different countries can balance the needs of data protection and economic development [35]. This can help avoid unnecessary data barriers. Fourth, ensuring prudent uses of data localization restrictions while avoiding one-size-fits-all data localization restrictions to allow the free flow of data internationally. Data localization measures should be carried out under the premise of ensuring data security and following international guidelines. Fifth, build a cooperative regulatory agency. International cooperative regulatory agencies can coordinate data protection and privacy supervision to ensure that companies comply with national laws and regulations. These efforts may reduce legal conflicts and uncertainties among states.

4.4 Strengthening Legislation

To clarify the intellectual property ownership and protection scope of AI-generated texts, it is necessary to identify when artificial intelligence is considered a "creator" and when AI-generated texts can be considered as possessing intellectual property rights [30]. The following aspects are worth considering. First of all, countries can clarify the authorship rights of AI-generated texts, which could help protect the intellectual property of AI-generated text. Given the collaborative nature of AI-generated text, there is a need to redefine authorship. Countries can modify domestic intellectual property laws and regulations for AI-generated text. For example, it can be stipulated that the intellectual property rights of AI-generated texts belong to artificial intelligence developers, users, training data providers, or multiple subjects [31]. Such provisions will reflect the unique contributions of both parties in the creative process.

In addition, new legislation could address issues related to licensing, duration of protection, and transfer of rights. In view of the rapid development of artificial intelligence technology, the copyright term of AI-generated text can be more flexible. A grading approach may be considered to differentiate creative texts from mechanical texts. The duration of protection may vary according to the degree of human involvement and novelty of the artificial intelligence system. Intellectual property-related laws can force creators and platforms to disclose the contribution of artificial intelligence technology to text generation. This transparency will allow users to assess the reliability and authenticity of AI-generated text. The New laws could also require owners of artificial intelligence software to open-source datasets for inspection, thereby reducing intellectual property violations from the outset. In the legislative process, legislators should consider the impact of AI-generated text on the public interest. The basis of AI-generated text is materials in the public domain. Therefore, in order to promote innovation and social development, the law can limit or adjust the intellectual property protection of AI-generated texts, expand the scope of fair use clauses to accommodate the transformative use of AI-generated texts, and protect the interests of creators.

Finally, the establishment of a dedicated dispute-resolution mechanism is also essential. In order to solve the contradiction between the lag of law and the rapid development of AI-generated texts, countries can set up permanent institutions to track the development of AI-generated content and put forward targeted legislative suggestions or drafts. Given the massive amount of data that artificial intelligence text-generating software collects to create models, related intellectual property infringement disputes are likely to proliferate. Countries can set up their own permanent institutions to deal with intellectual property rights disputes generated by artificial intelligence to speed up litigation procedures and reduce litigation costs.

5 Conclusion

The rapid development of generative artificial intelligence technology has shown great potential in many fields, but there are also complex relationships and various problems with intellectual property protection. There is currently no unified legal framework to address these issues, but countries have begun research on AI-generated content and intellectual property protection. By analyzing the technical logic of AI-generated text and the current national and international legal frameworks, this paper finds that generative artificial intelligence technology faces many challenges and unresolved problems in intellectual property protection. The content generated by artificial intelligence comes from data, and the generated content will be legal only if the method of obtaining data is legal. Also, cross-border cooperation and international coordination are crucial. Although the standards for defining the innovativeness of AI-generated texts and the protection period of AI-generated texts still need to be further determined in practice, existing intellectual property legislation and judicial practice can still provide many references. The intellectual property ownership and protection scope of AI-generated texts need to be allocated and regulated fairly and reasonably through cooperation among legislators, artificial intelligence software owners, data providers, and artificial intelligence software users. In the future, with the innovative multi-actor improvement measures proposed in this paper, individuals, corporations, states and international organizations could delve deeper and develop comprehensive and specific guidelines to ensure that the development of AI-generated content is not only innovative but also ethical and legal and could ultimately have a more positive impact on society.

Acknowledgements. This study is supported by Youth Fund for Research in Humanities and Social Sciences: The Study on International Law Issues of Applying Floating Platforms to Safeguard Rights and Law Enforcement in the South China Sea (20YJC820049); Southern Marine Science and Engineering Guangdong Laboratory (Zhuhai): Research on legal issues related to unmanned shin and marine unmanned equipment (SML2020SP005); Youth Academic Team in Humanities and Social Sciences of Wuhan University (Grant No. 4103-413100001); Major

Projects of National Social Science Fund of China "Researches on China's Positions and Discourse Power regarding International Rules for Cyberspace" (Grant No.: 20&ZD204); 2023 Wuhan University Funded Projects of Fundamental Research Operating Expenses of Central Universities.

References

1. Lee, J., Hilty, R., Liu, K.: Artificial Intelligence and Intellectual Property. Oxford University Press, England (2021)
2. The Chain. https://vocal.media/theChain/artificial-intelligence-development-agency-transf orming-the-future. Accessed 20 Aug 2023
3. Haleem, A., Javaid, M., Qadri, M., Singh, R., Suman, R.: Artificial intelligence (AI) applications for marketing: a literature-based study. Int. J. Intell. Netw. **3**, 119–132 (2022)
4. Herbert Smith Freehills. https://www.herbertsmithfreehills.com/latest-thinking/the-ip-in-ai-does-copyright-protect-ai-generated-works. Accessed 20 Aug 2023
5. Harvard Business Review. https://hbr.org/2023/04/generative-ai-has-an-intellectual-pro perty-problem. Accessed 20 Aug 2023
6. Harvard Business Review. https://hbr.org/2022/11/how-generative-ai-is-changing-creative-work. Accessed 20 Aug 2023
7. McKinsey & Company. https://www.mckinsey.com/featured-insights/mckinsey-explainers/ what-is-generative-ai. Accessed 20 Aug 2023
8. McKinsey Digital. https://www.mckinsey.com/capabilities/mckinsey-digital/our-insights/ the-economic-potential-of-generative-ai-the-next-productivity-frontier. Accessed 20 Aug 2023
9. Deepanshi. https://www.analyticsvidhya.com/blog/2021/06/text-preprocessing-in-nlp-with-python-codes/. Accessed 20 Aug 2023
10. Wu, D., Lin, C., Huang, J.: Active learning for regression using greedy sampling. Inf. Sci. **474**, 90–105 (2019)
11. OpenAI. https://platform.openai.com/docs/models. Accessed 20 Aug 2023
12. Oodles Technologies. https://www.oodlestechnologies.com/blogs/Exploring-the-World-of-Generative-AI:-From-GPT-1-to-GPT-3.5/. Accessed 20 Aug 2023
13. Vocal. https://vocal.media/01/what-is-chat-gpt-5jn3gf0lhz. Accessed 20 Aug 2023
14. McMillan. https://mcmillan.ca/insights/publications/ai-regulatory-roundup-international-developments-in-ai-law-and-policy/. Accessed 20 Aug 2023
15. National Archives. https://www.federalregister.gov/documents/2023/03/16/2023-05321/cop yright-registration-guidance-works-containing-material-generated-by-artificial-intelligence. Accessed 20 Aug 2023
16. GOV.UK. https://assets.publishing.service.gov.uk/government/uploads/system/uploads/att achment_data/file/957583/Copyright-designs-and-patents-act-1988.pdf. Accessed 20 Aug 2023
17. Kretschmer, M., Meletti, B., Porangaba, L.: Artificial intelligence and intellectual property: copyright and patents—A response by the CREATe Centre to the UK Intellectual Property Office's open consultation. J. Intellect. Prop. Law Pract. **17**(3), 321–326 (2022)
18. GOV.UK. https://www.gov.uk/government/publications/uk-international-technology-str ategy/the-uks-international-technology-strategy. Accessed 20 Aug 2023
19. UK Parliament. https://committees.parliament.uk/event/18019/formal-meeting-oral-evi dence-session/. Accessed 20 Aug 2023
20. Lexology. https://www.lexology.com/library/detail.aspx?g=a9b81aa1-7243-4f03-890c-7d2 9f5ccbdd7. Accessed 20 Aug 2023

21. The University of British Columbia. https://copyright.ubc.ca/public-domain/,last accessed 2023/8/20
22. McGill. https://libraryguides.mcgill.ca/c.php?g=710276, last accessed 2023/8/20
23. Government of Canada. https://laws-lois.justice.gc.ca/eng/acts/c-42/. Accessed 20 Aug 2023
24. McGill.   https://www.mcgill.ca/business-law/article/end-creativity-ai-generated-content-under-canadian-copyright-act. Accessed 20 Aug 2023
25. Government of Canada. https://www.ic.gc.ca/app/opic-cipo/cpyrghts/dtls.do?fileNum=118 8619&type=1&lang=eng, last accessed 2023/8/20
26. The State Council The People's Republic of China. https://www.gov.cn/guoqing/2021-10/29/content_5647633.htm. Accessed 20 Aug 2023
27. The National People's Congress of the People's Republic of China. http://www.npc.gov.cn/npc/c30834/202106/7c9af12f51334a73b56d7938f99a788a.shtml. Accessed 20 Aug 2023
28. World Intellectual Property Organization. https://www.wipo.int/about-ip/en/artificial_intelli gence/conversation.html. Accessed 20 Aug 2023
29. European   Parliament.   https://www.europarl.europa.eu/doceo/document/TA-9-2020-0277_EN.html. Accessed 20 Aug 2023
30. Wang, H.: Authorship of artificial intelligence-generated works and possible system improvement in China. Beijing Law Rev. **14**(2), 901–912 (2023)
31. Lee, J.: Artificial Intelligence and International Law. Springer, Singapore (2022). https://doi.org/10.1007/978-981-19-1496-6
32. World Bank. https://id4d.worldbank.org/guide/data-protection-and-privacy-laws. Accessed 29 Aug 2023
33. Data   Map.   https://www.datacamp.com/tutorial/complete-guide-data-augmentation. Accessed 29 Aug 2023
34. The International Association of Privacy Professionals. https://iapp.org/news/a/the-case-for-a-global-data-privacy-adequacy-standard/. Accessed 29 Aug 2023
35. World Economic Forum. https://www3.weforum.org/docs/WEF_A_Roadmap_for_Cross_Border_Data_Flows_2020.pdf. Accessed 29 Aug 2023

Generative Artificial Intelligence Upgrading Design of TCM Medical Diagnosis System Integration Platform

Yan Yang and Peng-fei Bao([✉])

Beijing Jiaotong University, Beijing, China
{21120569,22120622}@bjtu.edu.cn

Abstract. Medical information management systems have great advantages in improving medical convenience and efficiency. With the continuous advancement of traditional Chinese medicine informatization, the prototype framework of traditional Chinese medicine medical diagnostic information systems has gradually taken shape. The emergence of generative artificial intelligence has had a great impact on the original artificial intelligence and informatization development. Therefore, this paper takes the traditional Chinese medicine medical diagnosis system integration platform as the research object, and considers using generative artificial intelligence to generatively analyze the traditional Chinese medicine medical diagnosis system integration platform from three aspects: text, image and audio and video. Artificial intelligence upgrade design, select different types of generative artificial intelligence for corresponding upgrade design based on the characteristics of the subsystems of each TCM medical diagnosis system integrated platform, and compare the advantages with the traditional TCM medical diagnosis system integrated platform, try Provide upgrade design ideas for traditional Chinese medicine information construction and medical decision support.

Keywords: TCM Medical Diagnosis System Integration Platform · AIGC · Upgrading Design

1 Background

1.1 Pain Points of Medical Technical Diagnosis

The traditional modern medical diagnosis method is to use medical examination equipment to check the patient's symptoms, obtain the pathological characteristics through the medical examination equipment, and then analyze the condition by the doctor to obtain the diagnosis result of the patient's condition [1]. This traditional diagnosis method will be affected by the doctor's level, especially in the diagnosis of some difficult and complicated diseases, and the reliability of the results cannot be guaranteed. Moreover, the number of occupational doctors with high professional quality is limited. Compared with traditional modern medical diagnosis methods, TCM medical diagnosis methods rely more on the professional level of doctors and less emphasis on the importance of medical examination equipment, so they are more subjective and are more affected by the level of doctors.

F. Zhao and D. Miao (Eds.): AIGC 2023, CCIS 1946, pp. 280–289, 2024.
https://doi.org/10.1007/978-981-99-7587-7_24

1.2 Artificial Intelligence and Large Number According to Technology

Expert system is an important branch in the field of artificial intelligence. Its main feature is intelligent programming that imitates human expert thinking to solve complex problems in a specific field [2]. On the basis of expert system, TCM diagnosis and treatment expert system pays attention to the design principles and methods of TCM expert system, and can be considered as an auxiliary tool and important reference to help TCM solve complex disease diagnosis [3].

2 The Idea of the Generation of Artificial Intelligence Upgrading of the Integrated Platform of Medical Diagnostic System of Traditional Chinese Medicine

2.1 The Generated Artificial Intelligence of Each System is Combined with the Upgrade Design

As a new use of artificial intelligence in recent years, generative artificial intelligence aims to use artificial intelligence technology to automatically output text, images, videos, audio and other multi-modal data [4]. With the successive launch of applications such as Dall-E2, ChatGPT, and Stable Diffusion, generative artificial intelligence has become the focus of the world's attention. Currently, generative artificial intelligence application scenarios include text generation, image generation, audio and video generation and digital virtual image generation [5]. Therefore, this study, with full consideration, based on the various application scenarios of generative artificial intelligence and is suitable for different types of information systems, obtains corresponding generative artificial intelligence upgrade solutions, such as generative artificial intelligence upgrades for each system. As shown in the overall design diagram, specifically, systems suitable for text-generated artificial intelligence include: electronic medical record systems, electronic medical record systems, medical information systems, health information exchange systems, electronic health record systems, and health information portals. Systems suitable for image generative artificial intelligence include image information systems. Systems suitable for audio and video generative artificial intelligence include telemedicine systems. The system suitable for the joint introduction and upgrade of image generative artificial intelligence and audio and video generative artificial intelligence is the laboratory information system. The system of generative artificial intelligence applicable to the three is the medical decision support system (Fig. 1).

2.2 Detailed Description of the Generated Artificial Intelligence of Each System

Application of Text Generation Artificial Intelligence: Systems suitable for text-generated artificial intelligence include: electronic medical record systems, electronic medical record systems, medical information systems, health information exchange systems, electronic health record systems, and health information portals. The information stored in these systems is mainly in text format, which is suitable for text-generated artificial intelligence upgrades. The application of text generation based on natural language

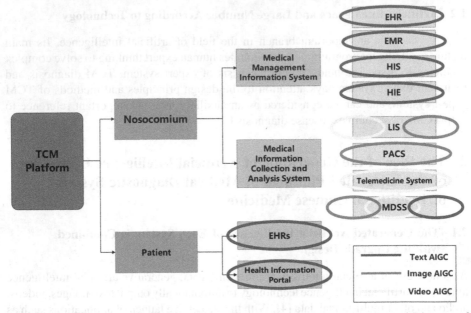

Fig. 1. Generative Artificial Intelligence Combined with Upgrade Design General Diagram of Each System

processing is an earlier application of generative artificial intelligence, which can realize the continuation of text content. A typical application of text generative artificial intelligence is ChatGPT. Due to the characteristics of its large text model, the efficiency of many related natural language processing models cannot be compared with it. Therefore, the large text generative artificial intelligence model can be regarded as interface to upgrade each system. The traditional text question answering system processing framework is as follows (Fig. 2):

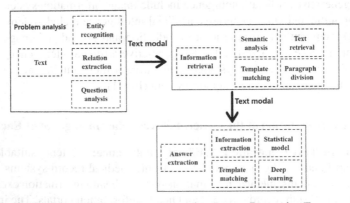

Fig. 2. Traditional Text Question Answering System Processing Framework

The main application of the system architecture is shown in the system architecture (Fig. 3):

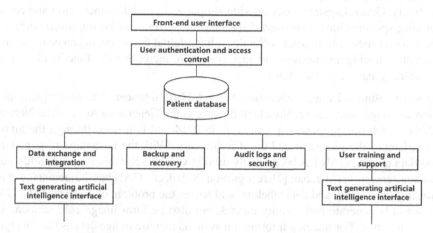

Fig. 3. Mainly Applied at the System Architecture Level

At the client level, users interact with the system through the front-end interface, which mainly includes medical workers entering, viewing and managing patient information, as well as patients accessing and managing personal medical information. In the process of user authentication and access control, the operating procedures of the three parts are fixed, and the generated data is stored in text form. The format is relatively fixed, and there is not much room for upgrading and improving text-generated artificial intelligence. In the application layer, the system records user activities and access in the audit log and security module, tracks access and monitors potential security issues. The backup and recovery module regularly backs up data to prevent data loss. The modules involved in the text mining part are the following three aspects: Data exchange and integration part. This system can perform data exchange and integration with other systems. This generally involves the use of standardized data formats and protocols to ensure data consistency and interoperability. However, electronic health information records represented by electronic medical records mostly exist in the form of text data. The process of data exchange and integration involves text mining and text extraction, including natural language processing such as entity recognition, relationship extraction, and question classification. Technology, text generative artificial intelligence can be introduced as an interface here to replace the original text mining model and NLP technology, achieve text mining effects more efficiently, and extract data suitable for requirements for exchange and integration. Decision support system: the system provides medical assistant decision support about pathological diagnosis and prescription drug interaction when users access patient data. This helps medical staff make informed medical decisions. In related academic research, the academic attempts involved are mainly represented by knowledge recommendation systems and knowledge graph technologies. Although knowledge graph technologies such as Neo4j are widely used, the scale of individual academic research is always small, and common academic research is difficult to compete with large-scale Text-based generative AI techniques for text-trained models. The system may provide training resources and support channels in the user training and support module to help medical staff and patients use the system effectively. General systems only provide simple and fixed documentation and related operating specifications. However, if the original system can be improved using text generation artificial intelligence technology, the original fixed system can be turned into an adaptive intelligent question and answer system, increasing the functionality of the user training and support modules.

The Application of Image Generation Artificial Intelligence: The most typical application of image generative artificial intelligence is the Generative Adversarial Network (GAN), which was proposed by Lan et al. in 2014 and is formed through the mutual comparison of the generator and the discriminator. With the advantage of generating good data quality, GAN has been widely used in various scientific research fields, such as image generation [6], repair [7], recognition [8–10], etc. GAN has the advantage of not requiring supervision and data labeling, and solves the problem of data sets [11]. GAN can not only generate high-quality images, but also perform image enhancement and image migration. For imaging information systems that use image data as the data type, enhanced images can reasonably reduce the technical errors of images and improve doctors' efficiency in processing information data. The following is based on the application fields of GAN as the improvement direction that affects the information system.

Image Generation. As a generation model, GAN can learn unsupervised from a large number of unlabeled data, and can generate generator functions with the ability to process image, voice, language and other forms of data, in order to achieve data augmentation [12]. For the description of multiple image data of the same lesion in a patient, image generation technology can be considered to use GAN as a generative model for a general description, which can minimize the error caused by individual cases and synthesize the most typical pathology as much as possible sexual characteristics.

Image Super Resolution. Image super-resolution is widely used in the academic field. For example, SRGAN (Super-Resolution Generative Adversarial Network) [13], which performs super-resolution on astronomical images and satellite images, can be applied to image super-resolution. It proposes a loss function based on the similarity-aware method [14], which can effectively solve the problem of lost high-frequency details in the restored image. The image super-resolution function can also be migrated to the image repair problem of medical image information data to improve the pathological identification efficiency of medical professionals.

Image Synthesis. The process of creating new images from image descriptions is image composition. For situations where it is inconvenient to collect image data, preliminary image synthesis can be performed through patient descriptions and other image descriptions, which can be used as a preliminary reference, basis and reference for research.

The Application of Video Generation Artificial Intelligence: The application path of audio and video generative artificial intelligence is roughly similar to that of image generative artificial intelligence, the difference lies in the type of audio, video and image data.

3 Functional Structure Description of Generative Artificial Intelligence Upgrade Design of Traditional Chinese Medicine Medical Diagnosis System Integration Platform

Various medical information systems are designed for platform integration, and the generative artificial intelligence upgrade design of the traditional Chinese medicine medical diagnosis system integration platform is obtained (Fig. 4).

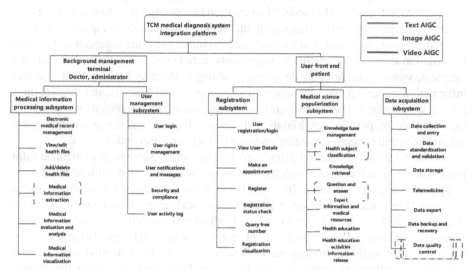

Fig. 4. Generative Artificial Intelligence Upgrading Design of TCM Medical Diagnosis System Integration Platform

As shown in the figure, the TCM medical diagnosis system integration platform for generative artificial intelligence upgrade design consists of a user front end and a background management end. Among them, the user front end is mainly composed of registration subsystem, medical science popularization subsystem and data acquisition subsystem. The registration subsystem is the main system of the entire user frontend interface and the main entrance for patients. It is designed for first-time users to enter the system, including user registration, user login, viewing user details, making an appointment for registration, handling registration, registration status inquiry, and query Functions such as vacant numbers and registration status visualization. The function of the medical science popularization subsystem is mainly to popularize medical and health science knowledge for patient users. Its essence is a knowledge recommendation system. Before the emergence of the large language text training model of generative artificial intelligence, the knowledge recommendation system represented by the knowledge graph was used as a commonly used data model respected by academia. Its functions mainly include: knowledge base management, health topic classification, knowledge retrieval, Q&A community, expert information and medical resources, download of health education materials, and release of health education activity information. The data collection system mainly involves patients' medical and health information and other pathochemical data, including functions such as data collection and entry, data standardization and verification, data storage, data access, data export, data backup and recovery, and data quality control. The technologies involved mainly include database management technology, sensor detection, data backup and recovery technology, and data quality monitoring. The background management system is a data system used by medical information management institutions for patient user medical information management. The users are doctors and medical institution managers. The background management system consists of a medical information processing subsystem and a user management subsystem. The medical information processing subsystem is the core component of medical information management. Medical information management is mainly divided into two functional components: medical information management and medical information analysis. Its functional components include electronic medical record management, viewing or editing health records, adding or deleting health records, medical information extraction, medical information evaluation and analysis, and medical information visualization. The main function of the user management subsystem is to manage users and set user permissions. Mainly includes user login, user rights management, user notifications and messages, security and compliance, and user activity logs.

In the "medical information extraction" function of the subsystem of medical information processing, it can be considered to introduce text generation artificial intelligence to upgrade the extraction of medical text information from electronic medical records. The data type of text information in electronic medical records is suitable for using text generation artificial intelligence to extract medical information A reminder of extraction accuracy and efficiency. Related subsystem functions that are also applicable to text-generating artificial intelligence include health topic classification and question-and-answer community functions, which are similar to text interaction functions in knowledge recommendation systems, and are better than natural language processing in the academic field in terms of text information extraction and entity relationship recognition. Compared with the large text language training model, the model has some shortcomings and is suitable for text generative artificial intelligence. The "data quality control" function module in the data acquisition subsystem is suitable for text generation artificial intelligence, image generation artificial intelligence and audio and video generation artificial intelligence. It involves the data enhancement function module in the data acquisition subsystem. Among them, text generation artificial intelligence is mainly used for the text missing of text medical information such as electronic medical records, health and medical records; image generation artificial intelligence is mainly used for data fuzzy enhancement and missing description of medical filming, such as using confrontation network to realize image missing Partial generation; audio and video generative artificial intelligence is mainly used for the absence of audio and video in the online medical diagnosis process in the telemedicine process or the absence of all audio and video records in the diagnosis record.

4   Functional Upgrade Contrast

Higher Efficiency and Accuracy: The first is text generation artificial intelligence. Compared with traditional NLP models, large-scale pre-training models such as Chat-GPT have many advantages. Large-scale pre-training models are better for longer and more complex context processing, and more accurately reflect user to express wishes. At the same time, large pre-trained models can automatically learn a large number of language features without manual feature engineering, reducing the difficulty and cost of developing complex NLP systems and improving accuracy. The current large-scale pre-training models such as ChatGPT and other models are general-purpose NLP models, which are suitable for multiple tasks and multiple languages. There is no need to train a separate model for each task and prepare multiple corpora in different national languages. Improve development efficiency. And because it is a general model, it can be relatively easily deployed to different applications, shortening the development cycle and improving development efficiency. The knowledge of large pre-trained models can be transferred to specific tasks, and the accuracy on specific tasks can be improved through fine-tuning, reducing the time and resource requirements of model training.

Higher Information Acquisition Rate: Generative artificial intelligence technology helps to improve the information acquisition rate, because it can transform raw data into a form that is easier to parse, retrieve and share, simplify the information processing process, save model running time and resource consumption, and make it easier to obtain all the information. It's easier to find key information. In addition, diversified information presentation methods can be provided to meet the needs and preferences of different users. Among them, text generative artificial intelligence improves information acquisition rate in terms of text summarization and summary, automatic question and answer, document generation and multi-language translation. Image generative artificial intelligence helps to restore missing information in images in terms of image annotation and image generation. By automatically generating image annotations, it makes objects, scenes and content in images easier to understand and improves the searchability of information. At the same time, image generation helps improve image quality and remove noise and blur. The generated data visualization charts and images can help Users understand data and information better. Audio and video generative artificial intelligence improves the accessibility of audio and video content in terms of speech transcription, automatic subtitles, video summarization, music and sound generation, provides a quick preview of key information, saves users time and enhances the expressiveness of information. Strong auxiliary thrust.

5 Feasibility Discussion on Generative Artificial Intelligence Upgrade Design Plan

In recent years, the research on the landing of generative artificial intelligence has been more in-depth, and the research on the function of enabling online learning of generative artificial intelligence has gone deep into three aspects: teacher-student interaction, understanding learners and helping learners to test [15]. From the perspective of man-machine relationship, there are also related studies on the new relationship of generative artificial intelligence governance logic update [16], in the medical system. With the release of WHO "generating evidence for medical devices based on artificial intelligence: training, verification and evaluation framework", this paper discusses the medical devices of artificial intelligence [17]. In terms of the relevant regulations on the landing of generative artificial intelligence and the application of medical devices, China has formulated the "operational Law of generative artificial Intelligence Management", which is helpful to the development of intelligent medicine [18]. There are also calls for restrictions on the medical development of generative artificial intelligence applications, calling for artificial intelligence to limit the generation of prescriptions for AI [19], and the relevant "legal" regulations need to be further optimized in China [20].

6 Prospect

This paper designs the generative artificial intelligence upgrade of the traditional Chinese medical diagnosis system integration platform. The generative artificial intelligence can improve the efficiency and information acquisition rate of the traditional medical diagnosis system integration platform in terms of text, image, audio and video data processing. The improvement has a significant effect. I sincerely hope that the generative artificial intelligence upgrade design of the medical diagnosis system integration platform in this paper can contribute to the development of the medical diagnosis system of traditional Chinese medicine, promote the upgrading of information technology of medical diagnosis of traditional Chinese medicine, and make contributions to the construction of information technology of traditional Chinese medicine.

References

1. Shen, T., Fu, X.: Artificial intelligence in the diagnosis and treatment of malignant tumors. Chin. Tumor J. **40**(12), 881–884 (2018)
2. The application of artificial intelligence in medical expert systems. Technol. Inf.: Acad. (13), 20–22 (2007)
3. Fang, A.: The construction and application of the new Chinese medicine expert system. J. Univ. Tradit. Chin. med. **11**(10), 9–10 (2009). https://doi.org/10.13194/j.jlunivtcm.2009.10.11.zengzhf.126
4. Zheng, H., Guan, M., Tan, Y.: A preliminary study on the application of generative artificial intelligence in central media—Taking the exploration and practice of CCTV News in the field of AI application as an example. Sound Screen World **06**, 98–100 (2023)
5. Zhang, Y., Wang, X., Wang, W.: The impact of generative artificial intelligence on the economy and society. Commun. World **16**, 4344 (2023). https://doi.org/10.13571/j.cnki.cww.2023.16.007
6. Yang, M.: Research on GAN-Based Image Generation algorithm. Anhui University of Science and Technology, Huainan (2021)
7. Gao, J.: Research on Image Repair Algorithm Based on Generative Adversarial Network. Nanjing University of Posts and Telecommunications, Nanjing (2021)
8. Chen, X.: Research on Pose Normalized Face Recognition Method Based on Generative Adversarial Network. Chongqing University of Posts and Telecommunications, Chongqing (2021)
9. Krizhevsky, A., Sutskever, I., Hinton, G.E.: ImageNet classification with deep convolutional neural networks. In: Proceedings of the 25th Annual Conference on Neural Information Processing Systems, pp. 1106–1114. MIT Press, Lake Tahoe (2012)
10. Wu, J., Qian, X.Z.: Application of compact deep convolutional neural network in image recognition. J. Front. Comput. Sci. Technol. **13**(2), 275–284 (2019)

11. Liang, J., Wei, J., Jiang, Z.: Review of generative adversarial networks GAN. Comput. Sci. Explor. (141), 1–17 (2020)
12. Liu, H., Ye, H., Xu, M., et al.: A review of generative adversarial network research. Internet Things Technol. **12**(11), 93–97 (2022). https://doi.org/10.16667/j.issn.2095-1302.2022.11.029
13. Ledig, C., Theis, L., Huszár, F., et al.: Photo-realistic single image super-resolution using a generative adversarial network. In: IEEE Conference on Computer Vision and Pattern Recognition, pp. 4681–4690. IEEE, Piscataway (2017)
14. Dosovitskiy, A., Brox, T.: Generating images with perceptual similarity metrics based on deep networks. In: Advances in Neural Information Processing Systems 29, pp. 658–666. NIPS, [S.l.] (2016)
15. Xiao, J., Bai, Q., Chen, M., et al.: Generative artificial intelligence enables online learning scenarios and implementation paths. Audio-Vis. Educ. Res. (09), 57-6399 (2023). Accessed 08 Sept 2023
16. Han, X.: Logic update and path optimization of generative artificial intelligence governance – from the perspective of man-machine relationship (06), 30–42 (2023). Http://kns.cnki.net/kcms/detail/11.3110.d.20230904.1043.004.html. Accessed 08 Sept 2023
17. Li, Z., Qiu, X., Yang, J., et al.: Who "generating evidence for medical equipment based on artificial intelligence: training, verification and evaluation framework" and its enlightenment to China's medical equipment Industry. Chin. Med. Equip. **19**(07), 157–167 (2022)
18. Qi, X.: Generative AI regulations. Lift Medical treatment. Pharmaceutical Economics, 2023-0720 (001). https://doi.org/10.38275/n.cnki.nyyjj.2023.000600
19. Huibi: It is strictly forbidden to generate prescriptions for AI. Artificial intelligence needs to keep boundaries across borders. Medical Economics, 2023-0824 (001). https://doi.org/10.38275/n.cnki.nyyjj.2023.000645
20. Kong, X.: The governance response of generative artificial intelligence from the perspective of national security – from the perspective of ChatGPT. Gov. Res. 1–10. Https://doi.org/https://doi.org/10.16224/j.cnki.cn33-1343/d.20230905.006. Accessed 08 Sept 2023

Convolutional Neural Network (CNN) to Reduce Construction Loss in JPEG Compression Caused by Discrete Fourier Transform (DFT)

Suman Kunwar[✉]

Selinus University of Sciences and Literature, Ragusa, Italy
sumn2u@gmail.com

Abstract. In recent decades, digital image processing has gained enormous popularity. Consequently, a number of data compression strategies have been put forth, with the goal of minimizing the amount of information required to represent images. Among them, JPEG compression is one of the most popular methods that has been widely applied in multimedia and digital applications. The periodic nature of DFT makes it impossible to meet the periodic condition of an image's opposing edges without producing severe artifacts, which lowers the image's perceptual visual quality. On the other hand, deep learning has recently achieved outstanding results for applications like speech recognition, image reduction, and natural language processing. Convolutional Neural Networks (CNN) have received more attention than most other types of deep neural networks. The use of convolution in feature extraction results in a less redundant feature map and a smaller dataset, both of which are crucial for image compression. In this work, an effective image compression method is purposed using autoencoders. The study's findings revealed a number of important trends that suggested better reconstruction along with good compression can be achieved using autoencoders.

Keywords: JPEG Compression · Discrete Fourier Transform (DFT) · Construction Loss · Convolutional Neural Network (CNN) · Autoencoder

1 Introduction

In today's digital era, human beings are surrounded by digital gadgets. Photographs are now an integral part of a person's daily life, and digital images are widely used in a variety of applications. As digital imaging and multimedia services advance, more and more people can share their data on the Internet. The number of internet users is growing day by day rapidly [1], resulting in increased data transfer, which necessitates efficient image compression. In multimedia, JPEG is one of the most used lossy compression techniques [2]. There

© The Author(s), under exclusive license to Springer Nature Singapore Pte Ltd. 2024
F. Zhao and D. Miao (Eds.): AIGC 2023, CCIS 1946, pp. 290–298, 2024.
https://doi.org/10.1007/978-981-99-7587-7_25

are several variations of JPEG: JPEG 2000, JPEG XS, JPEG Systems, JPEG Pleno, and JPEG XL [3]. Based on the data provided by Web Technology Survey, 74.3% of websites use the JPEG image format [4]. In digital images, pixels have high correlations, and the removal of this correlation will not affect the visual quality of the image [5,6]. To achieve the best quality with the smallest possible size, the low frequency values are preserved as they define the content of the image, and the high frequency values are truncated by a certain amount [7,8].

With the help of DFT, images in the spatial domain can be converted into the frequency domain, and certain frequencies can be ignored or modified to produce a low-information image with adequate quality [9–11]. In practical application, when we compute the DFT of an image, it is impossible to meet the periodic condition that opposite borders of an image are alike, and the image always shows strong discontinuities across the frame border. As a result, this affects the registration accuracy and success rate [12]. To solve this problem, various approaches have been taken. Among them, raised-cosine window [13], blackman window [14], and flap-top window [15] are the most popular ones.

As CNN has advanced rapidly [16], the problem of removing image artifacts from the decoded images has been re-examined. Several state-of-the-art deep learning-based algorithms [17–20] have been developed with great success using this approach. Approaches like using the Hann windows for reducing edge-effects in patch-based image segmentation with CNNs has shown promising results and pointed out the further investigation with different window functions, and with reducing the amount of context needed [21]. While these solutions are promising, still some information is lost in the process.

To minimize the reconstruction loss, an autoencoder network is proposed. The encoder reduces the dimensionality of an image into a number of feature maps. The feature map contains densely packed color values of an image whose original value can be closely reconstructed using the decoder associated with it. To measure the quality of the reconstructed image, Mean Squared Error (MSE), Peak Signal to Noise Ratio (PSNR), and Structural Similarity Index (SSMI) are used. It is also compared with the popular image compression standard JPEG. Here, SSIM measures the similarity in pixels and is used as a second image quality metric.

The remainder of this paper is organized as follows. The proposed method is described in Sect. 2. In Sects. 3 and 4, training and experimental evaluations are described in detail. Section 5 presents the conclusions of this research.

2 Methods

This section gives more information about the implementation, including deep learning network architecture, and the loss function.

2.1 Network Architecture

As autoencoder pipeline claims to provide efficient compression efficiency [22,23], an autoencoder network is constructed. Here, the encoder takes the input to represent the data in the simplest possible way. This is done by extracting the most prominent features and presenting them in a way that the decoder can understand. Whereas the decoder learns to read compressed code representations and generate them based on those representations. The proposed network architecture is depicted in Fig. 1.

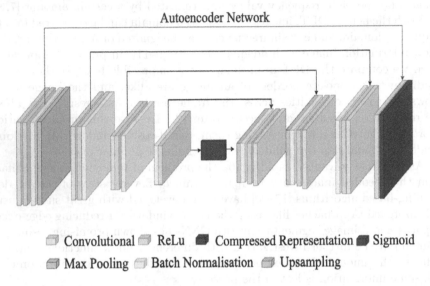

Fig. 1. Proposed Network Architecture that consists of different level of convolutional layer

The encoders are trained along with the decoders without labels. Eight layers of convolution are used in the auto encoder network. Batch Normalization is used to enable faster and more stable training of deep neural networks [24]. The benefit of using a layer is, it allows similar operations to be performed simultaneously. With more convolution kernels, the number of parameters increases linearly. As a result, the number of output channels also increases linearly and helps to reduce the artifacts caused by the gaussian noise [25]. The computation time is also proportional to the size of the input channel and to the number of kernels.

Sigmoid activation function is used in the output layer, whose output is bound between 0 and 1 range, and can be prone to suffering from the vanishing gradient problem [26]. To overcome this, Rectified Linear Unit (ReLU) is used as an activation function to increase the speed up the application and for better results [27]. Upsampling is also done to increase the spatial dimensions of the feature maps [28]. Adam optimizer is used as optimizer which uses momentum and adaptive gradient to compute adaptive learning rates for each parameter [29].

2.2   Loss Function

Loss functions are used to calculate predicted errors created across the training samples. Here, loss is measured in terms of Mean Squared Error (MSE). It calculates the mean of square error of the predicted value and the actual value. It results are derivable and is possible to control the update rate.

$$\sum_{i=1}^{D}(x_i - y_i)^2 \tag{1}$$

3   Training

This section describes the training procedure and parameter settings used in this research.

3.1   Datasets

CIFAR-10 dataset is used here to train and test the model [30]. It consists of 60000 32 × 32 color images and is divided into training and test datasets. Simple random sampling is used here to represent the different sets of images as it gives an equal probability of selecting a particular item [31]. Here, samples are only taken from the training images, the dataset used to train the algorithm. As the method aims to evaluate itself using both real and representative data, no training was done on holdout tests [32].

3.2   Procedure

The training dataset is run through the model in an incremental fashion. While training the model, Adam is used as a optimizer and loss is calculated in terms of Mean Square Error (MSE) as it has better generalization performances than the Cross Entropy (CE) loss [33]. Loss values and accuracy scores while training at different iterations is measured. For error correction, regularization options [34] such as early stopping, and dropout are carried out.

4   Evaluation

This section compares the results that happened while training the model with different settings. First, the model is tested with lower epochs as shown in Fig. 2 and then gradually increased.

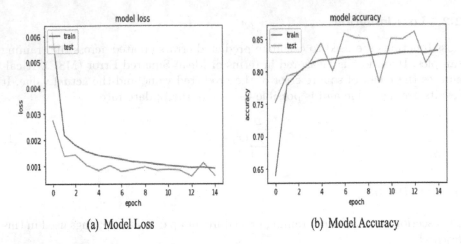

(a) Model Loss   (b) Model Accuracy

Fig. 2. Loss vs Model Loss (a) and Accuracy vs Model Accuracy after 15 epochs of training

Certain spikes can be seen in Fig. 3(a). Here, the loss first decreases, increases, and decreases at last. These spikes are often encountered when training with high learning rates, high order loss functions or small batch sizes [35]. The batch size is reduced, and early stopping is added to the model to avoid overfitting [36]. A checkpoint is also added to save the best value weights obtained while training. The maximum accuracy value has reached up to 0.90104 within 200 epochs of training.

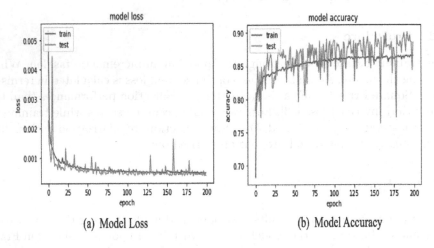

(a) Model Loss   (b) Model Accuracy

Fig. 3. Loss vs Model Loss (a) and Loss vs Model Accuracy (b) after 200 epochs of training

From the comparison shown in Table 1, the model shows high accuracy with regularization and has higher loss with low training. With the increasing number of training variation on loss and accuracy can be seen. The model shows low accuracy with the addition of noise value to the model with normalization.

Table 1. Comparison of proposed model with different settings

Model (variations)	Loss	Accuracy	Epochs
Model	0.162	80.236	15
Model	0.092	85.352	100
Model	0.089	82.307	200
Model with batch size 32	0.13	84.269	200
Model with normalization	0.07	85.94	100
Model with normalization	0.08	87.22	200
Model with normalization and noise	0.111	56.74	30

Image quality can be compromised as a result of distortions during the acquisition and processing of images. Different metrics have been used to measure the quality of compressed results, including Mean Square Error (MSE), Peak Signal to Noise Ratio (PSNR), and Structural Similarity Index (SSMI) [37]. To measure the performance of the reconstructed images of the purposed system, PSNR and SSIM are used. The source code is made available on GitHub.[1]

Table 2. Performance comparison of original and DFT output

Compression Ratio	MSE	PSNR	SSIM	Size
90	0.000194	37.102	0.982	2.613 KB
80	0.000267	35.722	0.97	2.602 KB
70	0.00037	34.309	0.96	2.599 KB
60	0.000521	32.827	0.956	2.577 KB
50	0.000731	31.356	0.938	2.560 KB

Table 3. Size comparison of original and model output image

	Original Image	Model Output
Size	2.594KB	2.543 KB

---

[1] Source code: https://github.com/sumn2u/neuralnetwork-jpeg.

Table 4. Performance comparison of original and model output

	MSE	PSNR	SSIM
Values	0.000199	37.0081	0.986

The results from Tables 2, 3, and 4 show that the purposed system has better compression while preserving the quality. Figure 4 presents a few sample images to compare the subjective quality of these methods.

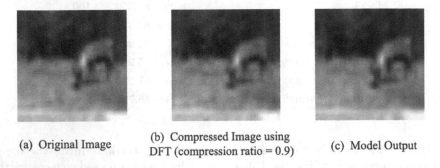

(a) Original Image   (b) Compressed Image using DFT (compression ratio = 0.9)   (c) Model Output

Fig. 4. Original image (a) and the reconstructed image produced by (b) DFT and the (c) purposed model

5   Conclusion

This paper presents a technique for compressing images using an autoencoder to create a lower-dimensional representation of the image. By capturing the most important elements, the technique can reconstruct the image with minimal loss. We trained the model using various settings and quality metrics to measure the outcomes. Our experiments showed that the model performed well, achieving an accuracy of 87.22 on the given dataset. The compressed image produced by the model is smaller than the original image and of superior quality.

Although the model successfully improved the quality of the compressed image with a smaller size, there are still other variations of JPEG that use different compression techniques, which require further research. Additionally, the impact of noise as a regularization method deserves further investigation. In our case, adding noise resulted in moderately good quality images.

References

1. Rahman, M.A., Islam, S.M.S., Shin, J., Islam, M.R.: Histogram alternation based digital image compression using base-2 coding. In: 2018 Digital Image Computing: Techniques and Applications, Canberra, Australia, pp. 1–8. IEEE (2018). https://ieeexplore.ieee.org/document/8615830/

2. Hussain, A.A., AL-Khafaji, G.K., Siddeq, M.M.: Developed JPEG algorithm applied in image compression. IOP Conf. Ser. Mater. Sci. Eng. **928**(3), 032006 (2020). https://doi.org/10.1088/1757-899x/928/3/032006

3. SMPTE: The future of the JPEG standard (2020). https://www.smpte.org/blog/the-future-of-the-jpeg-standard

4. W3Techs: Usage statistics of image file formats for websites (2022). https://w3techs.com/technologies/overview/image_format

5. Gonzalez, R.C., Woods, R.E., Masters, B.R.: Digital image processing (2009)

6. Yuan, S., Hu, J.: Research on image compression technology based on Huffman coding. J. Vis. Commun. Image Represent. **59**, 33–38 (2019)

7. Li, P., Lo, K.T.: Joint image encryption and compression schemes based on 16×16 DCT. J. Vis. Commun. Image Represent. **58**, 12–24 (2019)

8. Rasheed, M.H., Salih, O.M., Siddeq, M.M., Rodrigues, M.A.: Image compression based on 2D Discrete Fourier Transform and matrix minimization algorithm. Array **6**, 100024 (2020). https://linkinghub.elsevier.com/retrieve/pii/S2590005620300096

9. Siddeq, M.M., Al-Khafaji, G.: Applied minimized matrix size algorithm on the transformed images by DCT and DWT used for image compression. Int. J. Comput. Appl. **70**(15), 33–40 (2013)

10. Siddeq, M., Rodrigues, M.: A new 2D image compression technique for 3D surface reconstruction (2014)

11. Siddeq, M., Rodrigues, M.: A novel image compression algorithm for high resolution 3D reconstruction. 3D Res. **5**(2), 7 (2014). https://doi.org/10.1007/s13319-014-0007-6

12. Dong, Y., Jiao, W., Long, T., Liu, L., He, G.: Eliminating the effect of image border with image periodic decomposition for phase correlation based remote sensing image registration. Sensors **19**(10), 2329 (2019). https://www.mdpi.com/1424-8220/19/10/2329

13. Leprince, S., Barbot, S., Ayoub, F., Avouac, J.P.: Automatic and precise orthorectification, coregistration, and subpixel correlation of satellite images, application to ground deformation measurements. IEEE Trans. Geosci. Remote Sens. **45**(6), 1529–1558 (2007)

14. Podder, P., Khan, T.Z., Khan, M.H., Rahman, M.M.: Comparative performance analysis of hamming, hanning and blackman window. Int. J. Comput. Appl. **96**(18), 1–7 (2014). http://research.ijcaonline.org/volume96/number18/pxc3896927.pdf

15. Ge, P., Lan, C., Wang, H.: An improvement of image registration based on phase correlation. Optik **125**(22), 6709–6712 (2014). https://www.sciencedirect.com/science/article/pii/S0030402614010377

16. Pouyanfar, S., et al.: A survey on deep learning: algorithms, techniques, and applications. ACM Comput. Surv. **51**(5), 1–36 (2019). https://doi.org/10.1145/3234150

17. Li, J., Wang, Y., Xie, H., Ma, K.K.: Learning a single model with a wide range of quality factors for JPEG image artifacts removal. IEEE Trans. Image Process. **29**, 8842–8854 (2020). https://doi.org/10.1109/TIP.2020.3020389

18. Svoboda, P., Hradis, M., Barina, D., Zemcik, P.: Compression artifacts removal using convolutional neural networks. arXiv:1605.00366 [cs] (2016). http://arxiv.org/abs/1605.00366

19. Baig, M.H., Koltun, V., Torresani, L.: Learning to inpaint for image compression. arXiv:1709.08855 [cs] (2017). http://arxiv.org/abs/1709.08855

20. Santurkar, S., Budden, D., Shavit, N.: Generative compression. arXiv:1703.01467 [cs] (2017). http://arxiv.org/abs/1703.01467

21. Pielawski, N., Wählby, C.: Introducing Hann windows for reducing edge-effects in patch-based image segmentation. PLoS ONE **15**(3), e0229839 (2020). https://doi.org/10.1371/journal.pone.0229839
22. Cheng, Z., Sun, H., Takeuchi, M., Katto, J.: Deep convolutional autoencoder-based lossy image compression. In: 2018 Picture Coding Symposium (PCS), pp. 253–257 (2018)
23. Alexandre, D., Chang, C.P., Peng, W.H., Hang, H.M.: An autoencoder-based learned image compressor: description of challenge proposal by NCTU (2019). https://arxiv.org/abs/1902.07385
24. Santurkar, S., Tsipras, D., Ilyas, A., Madry, A.: How does batch normalization help optimization? (2018). https://arxiv.org/abs/1805.11604
25. Audhkhasi, K., Osoba, O., Kosko, B.: Noise-enhanced convolutional neural networks. Neural Netw. **78**, 15–23 (2016). https://linkinghub.elsevier.com/retrieve/pii/S0893608015001896
26. Gustineli, M.: A survey on recently proposed activation functions for deep learning (2022). https://arxiv.org/abs/2204.02921
27. Dubey, S.R., Singh, S.K., Chaudhuri, B.B.: Activation functions in deep learning: a comprehensive survey and benchmark (2021). https://arxiv.org/abs/2109.14545
28. Kundu, S., Mostafa, H., Sridhar, S.N., Sundaresan, S.: Attention-based image upsampling (2020). https://arxiv.org/abs/2012.09904
29. Kingma, D.P., Ba, J.: Adam: a method for stochastic optimization (2014). https://arxiv.org/abs/1412.6980
30. Shorten, C., Khoshgoftaar, T.M.: A survey on Image Data Augmentation for Deep Learning. Journal of Big Data 6(1), 60 (Dec 2019), https://journalofbigdata.springeropen.com/articles/10.1186/s40537-019-0197-0
31. Gupta, B.C.: Sampling methods, pp. 89–121 (2021)
32. Yadav, S., Shukla, S.: Analysis of k-fold cross-validation over hold-out validation on colossal datasets for quality classification. In: 2016 IEEE 6th International Conference on Advanced Computing (IACC), pp. 78–83 (2016)
33. Hui, L., Belkin, M.: Evaluation of neural architectures trained with square loss vs cross-entropy in classification tasks (2020). https://arxiv.org/abs/2006.07322
34. Steck, H., Garcia, D.G.: On the regularization of autoencoders (2021). https://arxiv.org/abs/2110.11402
35. Ede, J.M., Beanland, R.: Adaptive learning rate clipping stabilizes learning. Mach. Learn. Sci. Technol. **1**(1), 015011 (2020). https://doi.org/10.1088/2632-2153/ab81e2
36. Rice, L., Wong, E., Kolter, J.Z.: Overfitting in adversarially robust deep learning (2020). https://arxiv.org/abs/2002.11569
37. Deshmukh, K.R.: Image compression using neural networks. Master of science, San Jose State University, San Jose, CA, USA (2019). https://scholarworks.sjsu.edu/etd_projects/666

Research on Rural Feature Classification Based on Improved U-Net for UAV Remote Sensing Images

Rongzhi Lei[✉], Qianguang Tu, and Weihua Kong

School of Geomatics and Municipal Engineering, Zhejiang University of Water Resources and Electric Power, 508 2nd Street, Qiantang District, Hangzhou, China
879429829@qq.com

Abstract. With the rapid development of economy and society, the planning and utilization of agricultural land is also crucial. Aiming at the problems of the current lack of Unmanned Aerial Vehicle Remote Sensing (UAV Remote Sensing) Image Dataset of rural surface features and the low efficiency and low automation of traditional surface feature classification methods, a rural surface feature detection segmentation method based on Improved U-net (U-net+) is proposed based on the semantic segmentation task in Deep Learning. This method integrates the Swin Transformer framework into the U-net infrastructure, which can not only effectively improve the operation efficiency of the model, but also easily extract feature information to improve the accuracy and generalization of the model. At the same time, the UAV Remote Sensing image is used as the data source to produce a general UAV Remote Sensing Image Dataset of rural features: Air Dataset. In order to verify the advantages of U-net+, this study uses two UAV Datasets, the self-made Air Dataset and the famous Semantic Drone Dataset, to operate in U-net+. The experimental results show that the U-net+ proposed in this paper achieves 60.3% of mean Intersection-over-Union (mIoU) in the Air Dataset, which is better than U-net in PA, detection efficiency and model generalization ability; The Semantic Drone Dataset mIoU achieves 39.7%, which is also better than U-net According to the comparative analysis of the two Datasets, the segmentation accuracy of the two models for house buildings and grassland vegetation is relatively high, and the U-net+ is better. Therefore, in actual production and life, the U-net+ is more suitable for the rural terrain scene segmentation task mainly based on "Fine Operation".

Keywords: UAV Remote Sensing Image · Classification of rural features · Deep Learning · U-net+ · Swin Transformer

1 Introduction

"The construction of new socialist countryside" and "The rural revitalization strategy" are inseparable from using and planning of land. The accurate identification and effective classification of land features play a vital role in using and planning of land. At present,

© The Author(s), under exclusive license to Springer Nature Singapore Pte Ltd. 2024
F. Zhao and D. Miao (Eds.): AIGC 2023, CCIS 1946, pp. 299–312, 2024.
https://doi.org/10.1007/978-981-99-7587-7_26

researchers have conducted in-depth research on ground-objects identification, but the research on ground-objects classification is still lacking. Therefore, it's necessary to conduct in-depth research on the accurate classification methods of rural ground-objects.

At present, in the work of ground-objects identification and classification, the original data are mostly from remote sensing images, and the platforms are usually obtained from Satellites, Hot air balloons, small Aircraft and Unmanned Aerial Vehicles. The satellite remote sensing technology has the disadvantage of low spatial and temporal resolution of the image. The image data obtained by small aircraft and hot air balloon have the disadvantages of high cost of human and material resources [1], and they are not suitable for the survey of small areas such as rural towns. The data image collected by UAV Remote Sensing platform has the characteristics of low operation cost, flexible take-off and landing, and high spatial and temporal resolution of the image [2]. Therefore, this study adopts UAV to collect image data.

With the continuous development of Remote Sensing technology, image data has gradually stepped into the era of big data of sea quantification from the previous miniaturization [3–5]. The mass and complexity of image data pose a new challenge to the research of ground-objects classification [6]: how to improve the efficiency and accuracy of ground-objects classification in the face of massive and complex image data [7]. In view of the new problems, the classification methods of ground-objects based on Remote Sensing images are also rapidly updated. From the earliest visual interpretation classification to the computer-based classification of Remote Sensing images, more and more emerging technologies and tools are applied to the research of ground-objects classification. In recent years, as artificial intelligence technology sweeps the scientific and technological field, Deep Learning as a part of artificial intelligence [8–11], obtains more essential features of target categories by constructing Machine Learning models with many hidden layer networks, thus effectively making up for the defect that shallow networks are not deep enough in feature learning. Among them, Convolutional Neural Networks (CNN), as the most effective algorithm to solve image problems in Deep Learning [12–14], is increasingly applied to the identification and classification of Remote Sensing images, and has achieved good research results [15]. Therefore, this study adopts the method of Deep Learning to research the accurate identification and effective classification of rural objects.

Based on the images obtained by the UAV Remote Sensing platform, this study uses Deep Learning technology to establish an improved U-net based feature segmentation model for rural construction land in small areas, and carries out efficient, intelligent and refined automatic identification and classification.

2   Algorithm Principle

2.1   Algorithm Introduction

U-net algorithm was proposed by German scholar Olaf et al. in 2015. Initially, researchers usually applied it to Deep Learning segmentation of medical images [16–18]. Because it has achieved excellent results in the medical field, the U-net algorithm is widely applied to tasks in the fields of Automatic Driving, Remote Sensing image processing and so on, it has a fine performance in Remote Sensing image feature segmentation. The U-net also belongs to the category of CNN, but it's different from CNN in that the model doesn't have Full Connected Layer (FC), and is all composed of Convolutional Layers (Conv), so it is also named Full Convolutional Network (FCN).

2.2 Network Structure

The network structure of the U-net is shown in Fig. 1. The overall structure is U-shaped and symmetry. The structure on the left is Pooling and Subsampling as the compression path, and the structure on the right is Transposed Convolution and Upsampling as the expansion path [19]. The blue box corresponds to the multi-channel feature image, where the channel number is at the top of the box, the gray box represents the copied and clipped feature image, and the arrow represents different operations [20]. First, input the image data, then use four times of Pooling and Subsampling to present the environmental information of the data, and then the structure on the right uses four times of Transposed Convolution and Upsampling to fuse all the information to restore the image information, including the feature information of each layer carried by Pooling Subsampling and the input information of Transposed Convolution and Upsampling. The jump transfer method indicated by the middle arrow fuses the information of up and down sampling with each other, and finally outputs the segmentation map.

In this study, the U-net algorithm is used as the basic network for semantic segmentation experiments. The Swin Transformer framework is introduced to improve the U-net, and established the rural feature segmentation model U-net+. The detection and segmentation results of the two models are compared and analyzed using the same two datasets.

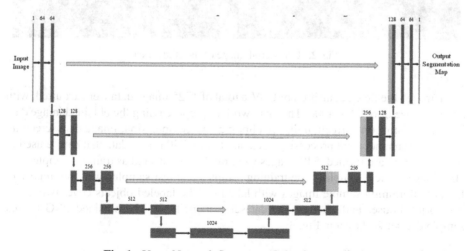

Fig. 1. U-net Network Structure. (Color figure online)

3 Experiment and Analysis

The UAV Remote Sensing image of a village in Fengcheng City, Jiangxi Province is used as the data source, and the low altitude tilt photography of multi rotor UAV is used for real-time aerial photography [21]. The UAV carries a Sony IL CE-6000 digital camera for image acquisition, with a spatial resolution of 0.02 m, a sensor size of 23.5 mm, a lens focal length of 35.3 mm and 25 mm, and the data acquisition time of March 2019. In order to show the overall picture of the research area more comprehensively, the UAV aerial image of the village is shown in Fig. 2. Through the image illustration of the village, it can be found that its main ground-types are basic feature types such as houses, fields, jungles and rivers.

Fig. 2. UAV aerial image of the study area.

Through the field acquisition of UAV, a total of 1725 image data were acquired, with an image size of 6000 × 4000. The main work of preprocessing the obtained image data is: image distortion correction, image clipping and compression, image data screening and image enhancement processing, etc., and finally 740 image data that can be used to make a dataset are formed. 540 images were randomly selected as training samples and 200 images as test samples. The training samples and test samples were not repeated. Label 540 training sample images with labelme. The labeled objects cover two target categories: houses and ground. The dataset is named Air. The obtained PNG format sample dataset is shown in Fig. 3:

Fig. 3. Air Dataset.

In this paper, we use a self-made dataset, in order to verify the generalization and universality of the improved model, the famous UAV Remote Sensing Dataset Semantic Drone Dataset is introduced as the horizontal comparison dataset. This dataset focuses on semantic segmentation of urban scenes. The UAV equipped with a high-resolution camera is used to acquire 600 images with a size of 6000 × 4000. The training contains 400 images and the test set contains 200 images, which cover many target categories, such as paved-area, grass, roof, people, etc. The resulting sample dataset in PNG format is shown in Fig. 4. The semantic drone dataset is very consistent with the self-made Air dataset in terms of the number of image data, application scenarios and target categories.

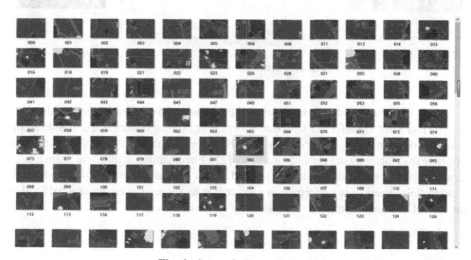

Fig. 4. Semantic Drone Dataset.

Therefore, it is possible to run two different datasets in the same model and compare their segmentation effects.

3.1 Network Improvement

Although the Transformer structure improves the overall performance of the U-net, there are still two shortcomings. First, the category change of different task scenarios will lead to insufficient generalization of the model; Secondly, when the size of the input image is too large, the computing power of the global attention mechanism will be insufficient, which will easily lead to the collapse of the model during operation or the final effect is very poor. Therefore, when improving the U-net, the Swin Transformer framework is adopted. The principle of this framework is to calculate the block attention through sliding window, and reduces the computing power when integrating the block attention, thereby improving the efficiency of the model. At the same time, it can easily extract feature information to improve the accuracy and generalization of the model.

This experiment integrates a global information aggregation module based on Swin Transformer into the Bottleneck structure in the U-net, and its network structure is shown in Fig. 5. Through this improvement, the local high-level features extracted by the CNN are aggregated, and the global information is fused to improve the perception ability of the U-net.

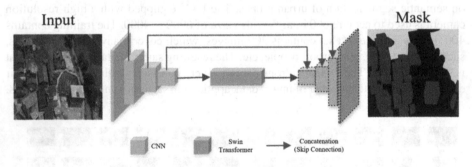

Fig. 5. Network structure of U-net+.

As an important module in the U-net+, the structure of Swin Transformer is shown in Fig. 6. The Swin Transformer structure adopts Hierarchical feature maps to Subsample feature images by 4, 8 and 16 times. The Backbone of this structure can effectively improve the efficiency of segmentation tasks [22]. For example, in the Swin Transformer structure of Fig. 6, the feature image is divided into several disjoint regions, and Multi-headed Self-attention is performed in each region to reduce the computational load of the model and further improve the efficiency of the model.

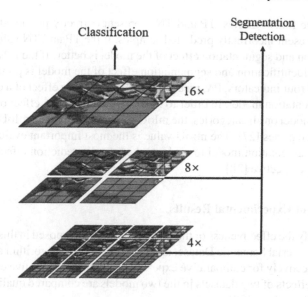

Swin Transformer

Fig. 6. Swin Transformer Structure.

4 Results

4.1 Model Accuracy Evaluation

The evaluation standards used in this experiment include PA, Recall, IoU and mIoU. Its calculation is based on the Confusion Matrix.

The full name of PA is Pixel Accuracy, and Pixel Accuracy is the percentage of pixels with correct marks in the total pixels [23]. The calculation formula is as follows:

$$PA = (TP + TN)/(TP + TN + FP + FN) \tag{1}$$

Recall is the proportion of samples with predicted value of 1 and true value of 1 among all samples with true value of 1 [24]. The calculation formula is as follows:

$$Recall = TP/(TP + FN) \tag{2}$$

The full name of IoU is Intersection over Union. The Intersection ratio is the intersection of the true value and the predicted value of a pixel divided by the union of the true value and the predicted value [25]. The calculation formula is as follows:

$$IoU = TP/(TP + FP + FN) \tag{3}$$

The full name of mIoU is Mean Intersection over Union, and the Mean Intersection Union ratio is the average of all types of IoU [26]. The calculation formula is as follows:

$$mIoU = \frac{1}{k+1} \sum_{i=0}^{k} \frac{TP}{TP + FP + FN} \tag{4}$$

In the above four formulas, TP and TN represent correctly predicted samples, and FP and FN represent incorrectly predicted samples. If the TP and TN values are higher, the identification and segmentation effect of the model is better; If the FP and FN values are higher, the identification and segmentation effect of the model is poor.

Among the four indicators, PA, recall and IoU measure the effect of a category in the semantic segmentation model. In order to measure the prediction effect of the semantic segmentation model on all categories, the mIoU value is derived, and mIoU is the average of IoU of all categories [27]. The mIoU value is the most important evaluation index in the semantic segmentation model to evaluate the overall prediction effect of the model from a macro perspective [28].

4.2 Analysis of Experimental Results

In order to verify the effectiveness of the U-net+ algorithm proposed in this paper, the Air Dataset and the Aerial Semantic Dataset are move in the U-net algorithm and the U-net+ algorithm respectively for comparative experiments. In the comparative experiment, the segmentation effects of two datasets in the two models are compared qualitative analysis and quantitative analysis [29].

First, we will take the Air Dataset as an example. The segmentation effect of the Air dataset in the U-net and the U-net+ is analyzed qualitatively, as shown in Fig. 7, which is a comparison image of the two groups of segmentation effects of the two models. This figure takes two image data as the base image, and compares the segmentation effects of the U-net and the U-net+. In the figure, the blue target area and the green target area are the main ones, in which the blue target area represents house, the green target area represents ground, and other features are represented by the original colors in the image data. Through the comparison of image one, it can be found that the segmentation effects of the two models are not different in the segmentation of houses; In the ground segmentation, the contour line of the ground identified by the U-net is relatively complete, while the contour line of the ground identified by the U-net+ is partially incomplete and some areas are missing. In the comparative image two, it can

Fig. 7. Comparison of segmentation effects between U-net and U-net+ in Air Dataset. (Color figure online)

be found that the segmentation effects of the two are relatively similar in terms of the segmentation of houses; In the ground segmentation, the U-net has some regions missing for the ground segmentation in the right half, while the U-net+ has a comprehensive ground classification. Therefore, in the comparison of the segmentation effect images of the models, the segmentation effect of the U-net+ is slightly better than that of the U-net. In order to more accurately distinguish the advantages and disadvantages of the two models, quantitative methods need to be used to analyze the segmentation effects of the two models, so four numerical indicators, PA, Recall, IoU and mIoU, are introduced to compare the two models.

Table 1 shows the index evaluation results of each category of the two models. First of all, from the macro perspective, the mIoU of the U-net+ is 60.3%, and the mIoU of the U-net is 59.4%, the former is only 0.9% higher than the latter. From the microscopic point of view, first of all, in terms of PA, the U-net+ has lower segmentation accuracy in houses and grounds than the U-net. The difference between the segmentation accuracy of houses is 1%, and the difference between the segmentation accuracy of grounds is 1.2%. Secondly, from the perspective of Recall rate, in the index value of Deep Learning, generally speaking, the higher the accuracy, the lower the Recall rate [30]. Therefore, the Recall rate of the U-net+ is higher than that of the U-net. The Recall rate of the former in the house is 3.8% higher than that of the latter, and the difference between the two in the ground is 4.2%. Finally, on the comparison of IoU indicators, the U-net+ is 1% higher than the U-net in the category of houses, and there is little difference between the two in the category of grounds, with a difference of only 0.8%. Generally speaking, although the U-net+ adds the Swin Transformer module, there is little difference between the U-net+ and the U-net in the segmentation effect of the Air Dataset. The reason for this is that the Swin Transformer module is more inclined to predict more comprehensive positive samples, which leads to an increase in the number of predicted samples and a significant improvement in the Recall rate. On the contrary, the PA shows a downward trend [31]. Despite all this, the U-net+ still has excellent performance in predicting the segmentation effect of each category.

Table 1. Comparison of index values of Air Dataset in U-net and U-net+

Model	Category	PA	Recall	IoU	mIoU
U-net	House	**0.655**	0.789	0.614	0.594
	Ground	**0.612**	0.783	0.574	
U-net+	House	0.645	**0.827**	**0.624**	**0.603**
	Ground	0.600	**0.825**	**0.582**	

Secondly, we will take the Semantic Drone Dataset as an example. The segmentation effect of Semantic Drone Dataset in U-net and U-net+ is also analyzed qualitatively. As shown in Fig. 8, two sets of segmentation effects of the two models are compared. This Figure also uses two image data as the base map to compare the segmentation effect of U-net and U-net+. The Figure is dominated by purple target area and green target area. Purple target area represents Paved-Area, green target area represents Grass, gray target area represents Roof, blue target area represents Lakes, yellow target area represents Wooden Floors, and important feature information is represented by the above colors. Other features are not summarized. By analyzing Image One, we can find that the surface features of Image One are not complex, including Grass, Paved-Area, and the surrounding Flower Beds and Bicycles. Through the segmentation comparison of the two models for Image One, it's found that the U-net+ is superior in Paved-Area segmentation. Although there is little difference in the main road segmentation, the U-net+ is in a regular state in the boundary description of the fence at the lower left corner of the fence; In terms of Grass segmentation, the U-net+ is also more effective, because the U-net+ is very fine for Grass segmentation, and the stones and debris in the Grass are also represented by purple areas; In terms of Bicycle segmentation, neither of them clearly distinguishes three Bicycles, but by comparison, the U-net+ gives a clear description of the rough outline of Bicycles. By analyzing Image One, we can find that the surface features of Image One are diversified, including Lake, Paved-Area, Roof, Grass, Wooden Floors and other surface features. By comparing Image Two, we can find that the U-net+ is obviously better than the U-net in the segmentation of Lake, and the U-net has wrongly segmented Lake; In Paved-Area segmentation, U-net is better than U-net+, and the segmentation result of U-net+ is slightly flawed; In terms of Roof segmentation, both of them do not accurately describe the boundary information. In comparison, the area of Roof described by U-net accounts for a larger proportion; In terms of Grass segmentation, the U-net+ is better. The U-net confuses the Wooden Floor in the lower right corner of the image with the Grass; On the segmentation of Wooden Floor, U-net+ successfully described the boundary information of two Wooden Floors and separated the People lying on the floor. Therefore, the segmentation effect of the U-net+ is slightly better than that of the U-net in the comparison of the segmentation effect images of the model. In order to more accurately distinguish the advantages and disadvantages of the two models, quantitative methods need to be used to analyze the segmentation effect of the two groups of models, and four numerical indicators, PA, Recall, IoU value and mIoU value, are also introduced to compare the two models.

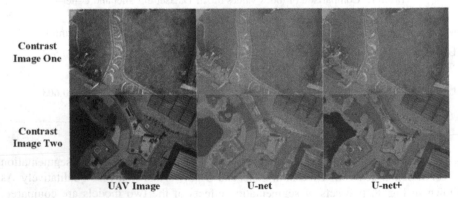

Fig. 8. Comparison of segmentation effects between U-net and U-net+ in Semantic Drone Dataset. (Color figure online)

Table 2 shows the index evaluation results of each category of the two models. First of all, from the macro perspective, the mIoU value of the U-net+ is 39.7%, and that of the U-net is 37.5%. The former is 2.2% higher than the latter. From the micro perspective, first of all, in terms of PA, the segmentation accuracy of U-net+ in Paved-Area and Roof is higher than that of U-net. The difference between Paved-Area and Roof is 2.3%, and the difference between them is the largest, 6%; However, the segmentation accuracy of U-net in Grass is higher than that of U-net+, with a difference of 1.2%. Secondly, from the perspective of Recall rate, the Recall rate of U-net+ in Paved-Area is the same as that of U-net, with a value of 88.5%. There is no difference between them in Grass category, with a difference of only 0.5%. The Recall rate in Roof is 10.2% higher than that of U-net. Finally, on the comparison of IoU indicators, the U-net+ is 2.2% higher than the U-net in Paved-Area category, 2.4% lower than the U-net in Grass category, and 6.8% higher than the U-net in Roof category. In general, because the U-net+ adds the Swin Transformer module, the segmentation effect of the Semantic Drone Dataset is better than the U-net. Although compared with the Air Dataset, the overall segmentation level of the two models for this dataset is low, the U-net+ still performs well.

Table 2. Comparison of index values of Semantic Drone Dataset in U-net and U-net+

Model	Category	PA	Recall	IoU	mIoU
U-net	Paved-Area	0.626	0.885	0.602	0.375
	Grass	0.502	0.519	0.428	
	Roof	0.157	0.143	0.096	
U-net+	Paved-Area	0.649	0.885	0.624	0.397
	Grass	0.496	0.514	0.404	
	Roof	0.217	0.245	0.164	

The running results of the two models for the two Datasets show that: First, from the perspective of the advantages and disadvantages of the model, the overall mIoU value of the U-net+ is higher than that of the U-net. The highest segmentation accuracy of the two models are House and Paved-Area, followed by Ground and Grass. The former group of ground features belong to the same building type, and the latter group of ground features belong to the same grassland and vegetation. Therefore, the model has high segmentation accuracy for buildings and grassland vegetation; Secondly, from the perspective of the applicability of the dataset, the overall accuracy of the Air Dataset is better than that of the Semantic Drone Dataset under the same operating environment of the U-net+; Finally, for the Air Dataset with better segmentation accuracy, its mIoU value reaches 60.3%, House's IoU value is 62.4%, and Ground's IoU value is 58.2%. The overall segmentation accuracy of U-net+ for the Air Dataset is better than that of U-net.

5 Conclusions

At present, with the rapid development of Deep Learning, UAV Remote Sensing as a cutting-edge technology in the field of remote sensing has attracted much attention. Our paper takes the semantic segmentation model U-net as the basic model, uses the UAV Remote Sensing platform to obtain the image data of a village in Fengcheng City, Jiangxi Province, China, to create the Air Dataset, integrates the Swin Transformer structure into the U-net, aggregates the local advanced features extracted by the convolutional neural network, fuses the global information, improves the perception ability of the U-net, and effectively improves the segmentation accuracy and operation efficiency of the model. The results show that:

1. For the Air Dataset, the mIoU value of the U-net is 59.4%, and the mIoU value of the U-net+ is 60.3%, which is only 0.9% higher than the original model. In both models, the segmentation accuracy for House is higher than that for Ground, with IoU values of 62.4% (U-net+) and 61.4% (U-net) respectively. In general, if the overall performance of the model is evaluated by the mIoU index, the improved U-net is slightly better than the U-net.
2. For the Semantic Drone Dataset, the mIoU value of the U-net is 37.5%, and the mIoU value of the U-net+ is 39.7%, which is 2.2% higher than the U-net. Among the two models, Paved-Area has the highest segmentation accuracy, with its IoU values of 60.2% (U-net) and 62.4% (U-net+) respectively, followed by Grass, with its IoU values of 42.8% (U-net) and 40.4% (U-net+) respectively. In general, if the overall performance of the model is evaluated by the mIoU index, the improved U-net is slightly better than the U-net.
3. According to the comparative analysis of the two Datasets, it can be found that the segmentation accuracy of the two models for housing buildings and grassland vegetation is relatively high, and the U-net+ is better. Therefore, in actual production and life, the U-net+ is more suitable for scene segmentation tasks focusing on rural features.

Although the U-net+ has excellent segmentation performance for two Datasets, there are still some shortcomings in this study: there is a lack of comparison with other traditional and cutting-edge algorithms in the selection of comparison algorithms. Traditional and cutting-edge algorithms, as more mature algorithms, have stronger reliability and applicability, making the research more convincing. Therefore, it is necessary to introduce more mature algorithms for comparative analysis in future research work.

Funding. This research was supported by the National Natural Science Foundation of China (41804007, 42104023), and Jiangxi University of Science and Technology High-level Talent Re-search Startup Project (205200100564).

References

1. Liu, W.: Research on ground object classification method based on UAV multispectral remote sensing image. Master's Thesis, Shihezi University, China (2017)

2. Gong, Z.: Discussion on the application prospect of UAVs in marine surveying and mapping. Low Carbon World (01), 72–73 (2016)

3. Qing, J., Yang, Y., Li, D., Du, J.: Research on massive remote sensing image retrieval methods based on distributed elastic search. Mapp. Spat. Geogr. Inf. **42**(06), 64–66+69 (2019)

4. Li, X.: Research on Hyperspectral Database and Data Mining. Graduate School of Chinese Academy of Sciences (Institute of Remote Sensing Applications), China (2006)

5. Ming, D., Luo, J., Shen, Z., Wang, M., Sheng, H.: Research on high resolution remote sensing image information extraction and target recognition technology. Surv. Mapp. Sci. (03), 18–20+3 (2005)

6. Wang, R., Tan, Z., Ren, J.: Image segmentation method for detecting changes in geographical conditions based on vector map constraints. Surv. Mapp. **41**(02), 69–72 (2018)

7. Zhu, C.: Research on surface feature classification of high resolution SAR images based on surveillance and detection. Master's Thesis, Shanghai Jiaotong University, China (2013)

8. Ding, M., Xu, Y., Jiang, C.: Overview of deep belief network research. Ind. Control Comput. **29**(04), 80–81+84 (2016)

9. Yu, K., Jia, L., Chen, Y., Xu, W.: Yesterday, today and tomorrow of deep learning. Comput. Res. Dev. **50**(09), 1799–1804 (2013)

10. Gu, W.: Research on AI development strategy based on big data. Master's Thesis, Tianjin University, China (2019)

11. Peng, T., Sun, L., Liu, C., Wang, Z.: Intelligent computing technology in the age of big data. Sci. Technol. Plaza (03), 4–10 (2016)

12. Ren, H.: Research on convolutional neural network compression algorithm based on depth compression. Master's Thesis, Southeast University, China (2019)

13. Zhang, A.: Research on SAR image classification algorithm based on convolutional neural network and neighborhood correlation. Ph.D. Dissertation, Hefei University of Technology, China (2020)

14. Huang, M.: Research on the application of multilayer convolution neural network in SAR image target detection and recognition. Master's Thesis, Xi'an University of Electronic Science and Technology (2017)

15. Yan, Y.: Overview of research on building extraction from high spatial resolution remote sensing images. Digit. Technol. Appl. (07), 75–76+78 (2012)

16. Li, M.: Research on vessel segmentation method of coronary angiography image based on depth learning. Master's Thesis, Dalian Maritime University, China (2019)

17. Liu, F., Li, H., Zhang, Y., Li, R., Wang, Z., Tang, X.: Application of artificial intelligence in medical imaging diagnosis. Beijing Biomed. Eng. **38**(02), 206–211 (2019)

18. Xia, L., Shen, J., Zhang, R., Wang, S., Chen, K.: Application of deep learning technology in medical imaging. Peking Union Med. J. **9**(01), 10–14 (2018)

19. Wang, T., et al.: A hidden danger detection method for line corridors based on improved U-net semantic segmentation remote sensing images. China South. Power Grid Technol. **13**(08), 67–73 (2019)

20. Guo, Y.: Polarized SAR surface feature type classification based on deep convolution Highway Unit neural network. Master's Thesis, Xi'an University of Science and Technology, China (2018)

21. Gao, B., Zhang, T., Chen, L., Li, J., Qu, M.: Analysis of earthquake emergency applicability of large construction machinery based on UAV photogrammetry technology. J. Inst. Disaster Prev. Sci. Technol. **20**(01), 9–16 (2018)

22. Wang, G.: Research on internet based building intelligent control system. In: Proceedings of the 2018 Annual Academic Conference of the Chinese Society of Civil Engineering, pp. 93–98 (2018)

23. Min, W.: Research and implementation of face correction based on generation countermeasure network. Master's Thesis, University of Electronic Science and Technology of China, China (2019)
24. Liu, Y., Yu, Z., Yang, Y.: Diabetes risk data mining method based on electronic medical record analysis. J. Healthc. Eng. (2021)
25. Zheng, X., Zhong, B.: Overview and evaluation of image linear segment detection algorithm. Comput. Eng. Appl.. Eng. Appl. **55**(17), 9–19 (2019)
26. Zhang, Z., Gao, J., Zhao, D.: MIFNet: segmentation method of gastric cancer pathological image based on multi-scale input and feature fusion. Comput. Appl. **39**(S2), 107–113 (2019)
27. Wei, F.: Research on automatic recognition technology of vehicle pavement cracks based on semantic segmentation. Master's Thesis, Chang'an University, China (2019)
28. Ruan, X., Liu, B.: Review of 3D point cloud data segmentation methods. Int. J. Adv. Netw. Monit. Controls **5**(1) (2020)
29. Wang, J.: Research on key technologies of hand vein recognition. Master's Thesis, China University of Mining and Technology (2015)
30. Pan, B., Pan, W., Liu, Z.: Diamond image semantic segmentation based on cavity convolution neural network. Diam. Abras. Tools Eng. **39**(06), 20–24 (2019)
31. Zhang, T., Geng, H., Cai, Q.: An improved VSM and its application in automatic text classification. Microelectron. Comput. (12), 24–27 (2005)

Exploring the Intersection of Complex Aesthetics and Generative AI for Promoting Cultural Creativity in Rural China After the Post-pandemic Era

Mengyao Guo[1], Xiaolin Zhang[2,6], Yuan Zhuang[3], Jing Chen[2,4], Pengfei Wang[5], and Ze Gao[7(✉)]

[1] University of Macau, Macau 999078, China
yc07330@um.edu.mo
[2] Peking University, Beijing 100091, China
{zhangxiaolin,cjing}@pku.edu.cn
[3] Shandong University, Jinan 250100, China
202020162@mail.sdu.edu.cn
[4] China Agricultural University, Beijing 100091, China
[5] Tsinghua University, Beijing 100091, China
[6] University of Auckland, Auckland 1010, New Zealand
[7] Hong Kong University of Science and Technology, Hong Kong SAR 999077, China
zgaoap@connect.ust.hk

Abstract. This paper explores using generative AI and aesthetics to promote cultural creativity in rural China amidst COVID-19's impact. Through literature reviews, case studies, surveys, and text analysis, it examines art and technology applications in rural contexts and identifies key challenges. The study finds artworks often fail to resonate locally, while reliance on external artists limits sustainability. Hence, nurturing grassroots "artist-villagers" through AI is proposed. Our approach involves training machine learning on subjective aesthetics to generate culturally relevant content. Interactive AI media can also boost tourism while preserving heritage. This pioneering research puts forth original perspectives on the intersection of AI and aesthetics to invigorate rural culture. It advocates holistic integration of technology and emphasizes AI's potential as a creative enabler versus replacement. Ultimately, it lays the groundwork for further exploration of leveraging AI innovations to empower rural communities. This timely study contributes to growing interest in emerging technologies to address critical issues facing rural China.

Keywords: complex aesthetics · generative AI · cultural creativity · post-pandemic era · subjective aesthetic data · machine learning models

---

M. Guo, X. Zhang and Z. Gao—These authors contributed equally to this work.

F. Zhao and D. Miao (Eds.): AIGC 2023, CCIS 1946, pp. 313–331, 2024.
https://doi.org/10.1007/978-981-99-7587-7_27

1   Introduction

With the COVID-19 pandemic devastating rural areas of China, especially tourism, there is an urgent need for innovative approaches to promote economic development and cultural creativity in these regions [1]. This paper explores the potential of leveraging generative artificial intelligence (AI) in conjunction with complex aesthetics to foster cultural creativity in rural China during the post-pandemic period. Through a literature review, case studies, and data analysis, this research aims to bridge the gap in understanding how machine learning algorithms can be used to generate culturally creative outputs when provided with subjective aesthetic data.

The paper begins by establishing the research background, underscoring how the pandemic has impacted rural Chinese communities reliant on tourism. It then identifies a research gap, noting that while generative AI shows promise in this context, machine learning excels at reproducing technical qualities rather than nuanced aesthetic values [2]. Hence, human input is vital for imparting artistic inspiration to machine learning [3]. Next, the paper outlines its contribution - presenting a novel framework utilizing generative AI and complex aesthetics to stimulate cultural creativity.

The study's structure is elaborated, explaining that it employs a mixed methods approach with case studies, a questionnaire, interviews, and text analysis. Contemporary art cases are reviewed, providing examples of art intersecting with rural landscapes. Early examples of using art for rural revitalization in international and Chinese contexts are also summarized. Additionally, existing instances of art and technology collaborating in rural settings are discussed.

Key findings are highlighted, underscoring the importance of site-specific artworks organically woven into local contexts and the potential for generative AI to nurture grassroots artists within villages. In conclusion, this study offers an original perspective on leveraging generative AI and aesthetics to boost cultural creativity in rural China during the post-pandemic recovery. It provides a foundation for further research at the intersection of technology and culture.

1.1   Research Background

The primary objective of this paper is to explore the potential implications of the intersection of complex aesthetics and generative AI for promoting cultural creativity in rural areas of China during the post-pandemic era [4]. The COVID-19 pandemic has devastated rural areas in China, especially in the tourism industry [5]. As a result, there is an urgent need for innovative approaches to promote economic development and cultural creativity [6]. This paper argues that leveraging the intersection of complex aesthetics and generative AI can provide a promising solution to address this need and promote cultural creativity in rural China [7].

1.2   Research Gap

The exploration of using generative AI to foster cultural creativity in rural China presents an intriguing opportunity but faces key challenges. While generative AI shows promise for creating novel cultural content, its machine learning algorithms excel more at reproducing technical qualities than nuanced, subjective aesthetics. Additional human inputs are thus critical for imbuing AI systems with artistic inspiration reflective of local cultural contexts [8]. This research aims to bridge this gap by developing methodologies to incorporate subjective aesthetic data into generative AI models, enabling outputs that authentically enhance rural cultural experiences. However, a review revealed a lack of documented applications of generative AI tailored specifically for rural creativity. Despite the potential benefits for rural communities, this absence represents a missed opportunity. Given generative AI's aptitude for human-guided synthesis, exploring its integration with rural culture merits investigation. This pioneering research can fill the void by proposing techniques to harness generative AI's capabilities for rural-focused cultural innovations that preserve heritage and empower communities [9].

1.3   Contribution

This paper puts forth an original framework utilizing generative AI and complex aesthetics to promote cultural creativity in rural China. Our approach is two-pronged: first, harnessing generative AI's capacity for machine learning and data synthesis to produce novel cultural content reflective of local aesthetics. Second, incorporating nuanced human aesthetic perspectives to imbue the machine-generated content with richness, specificity, and authenticity. This framework represents a pioneering attempt to synergize technological and humanistic approaches to inject new vitality into rural cultural heritage. Beyond proposing a workflow, this research aims to illuminate the current realities of rural communities in China and envision future possibilities for their revitalization. By elucidating the intersection of technology and culture, we seek to foster enlightening dialogues about preserving local traditions while embracing modernity. Our ultimate contribution is presenting a human-centered model of technological integration that elevates rather than diminishes indigenous cultural creativity.

2   Structure of the Research

This study employs a mixed methods approach incorporating qualitative and quantitative techniques to investigate using generative AI and aesthetics for rural cultural creativity. Contemporary art cases situated in rural areas are first reviewed for contextual examples. Early precedents of art applications for rural development internationally and in China are then summarized. Existing collaborations between art and technology in rural settings are also discussed. The core methodology involves three components: case studies of specific Chinese rural art festivals, a questionnaire surveying various stakeholders, and semi-structured interviews with key curators for expert insights. Additionally, textual data is thematically analyzed to identify trends. Together, this multi-faceted approach enables examining the intersection of generative AI, aesthetics, and rural creativity from theoretical and practical lenses. It provides a detailed investigation from reviewing relevant examples to gathering first-hand perspectives. Ultimately, the analysis extracts key findings to inform a proposed framework for applying this intersection to enhance creative outputs. This comprehensive methodology combines literature, cases, surveys, interviews, and textual analysis to holistically explore the potential of generative AI and aesthetics in promoting rural cultural creativity.

3 Related Works

Through artwork events, exhibitions, and exchanges, rural areas can revitalize and gain regional developments by having a higher potential to interact with the crowds and face the market on a larger scale [10]. In this digital new era, content creation (such as NFTs) with AI [11] gives people a great deal of creative space and imagination, especially artists and writers [12,13]. The ways people enjoy rural cultural heritage sites and countryside spots have significantly evolved with the technology developments. Visitors can not only be allowed to look and read but also touch, interact, and even create [14,15]. The following related works chronologically trace the development of the application of complex aesthetics and generative AI in cultural experiences and local heritage. It includes the contemporary art cases that happen on non-urban landscapes, the earlier examples internationally and domestically of the art applications for rural revitalization, and those art and technology intervening in rural areas.

3.1 The Most Earlier Examples Internationally of the Art Applications for Rural Revitalization

The artwork "The Sheepfolds" by Andy Goldsworthy's (1987) intervention in the countryside connects with and enlivens Cumbria's farming traditions and history by constructing an existing old sheepfold. Individuals, communities, churches, and town councils contributed significantly to this sheepfold project[1] [16]. Agnes Denes's "The Tree Mountain" (1996) was completed by tens of thousands of people from all over the world planting eleven thousand trees in the Pincio gravel pit near Ilojavi, Finland, and eventually formed a huge artificial mountain that became part of the artist's land reclamation project. The project's significance is to relieve ecological pressure on the area by planting trees, with the expectation that a primary forest will eventually be created [17]. Another example South Korea's "Gamcheon Culture Village" (2009) was built in the 1920s and

---

[1] Introduction of Sheepfolds: http://www.sheepfoldscumbria.co.uk/html/info/info00.htm.

1930s, and in 2009 the local authorities, under the Empty Homes Residential Preservation Project, transformed the village into a cultural center with art installations as the primary form of presentation and colorful houses. The value of the project is that it improves the poverty and poor living conditions in the area and provides an example of artistic village revitalization [18].

These international examples demonstrate the broader significance of art as a catalyst for rural rejuvenation and community engagement. The integration of artistic interventions in rural areas preserves cultural traditions and addresses environmental, social, and economic challenges, making a remarkable impact on both local and international levels [19].

3.2 The Most Earlier Examples Domestically of the Art Applications for Rural Revitalization

The "Tongling Countryside Public Art Festival", organized by Kegon Leung in 2018, brings together artists, architects, poets, etc., combining theatre, design, installation, and other art forms. The project has not only made Plough Bridge Village the focus of rural tourism but has also driven the development of folklore, culture, and leisure industries, making it more profitable for villagers[2] The "Daojiao New Art Festival", initiated by Li Zhenhua and others, is a fusion of new media art and technology that transforms industrial parks and grain silos into festival venues. The festival has helped local people to gain a better understanding of contemporary art, creating a new industrial system and cityscape, as well as boosting the local economy through the creation of an art town[3] The Lianzhou International Foto Festival annually presented a unique theme since its inception in 2005, showcasing diverse Chinese images and exploring global photography. With over 80 exhibitions and its participation from top international photographers and scholars, it is one of China's most influential international photography festivals [20].

These rural revitalization projects in China have played an essential role in national rural revitalization efforts. However, so far, only one of these projects has maintained its artistic form to this day, and the Daojiao New Art Festival has only been held once. Meanwhile, the once glorious Tongling Countryside Public Art Festival has gradually lost its former splendor. These few cases, on the one hand, have provided a model for revitalizing rural areas through art in China. Still, on the other hand, they have also exposed the shortcomings in its implementation process.

3.3 Latest Application Using AIGC Tools to Boost Rural Tourism and for Rural Revitalization

The "Yunshang Ethnic Villages" project launched in May 2023 uses advanced generative AI to recreate rural cultural heritage, weaving community narratives

---

[2] Tongling Countryside Public Art Festival: https://www.sohu.com/a/514768925_121 106991.

[3] Daojiao New Art Festival: https://www.sohu.com/a/145276407_340420.

into interactive 3D simulations. Digital technology provides new ways of information dissemination and brand-new experience of traveling experiences. Visitors can quickly obtain detailed information on rural tourist attractions, folk culture, and specialty products through virtual reality technology, intelligent portal websites, live broadcast tools, etc. AIGC-powered products are also shown in this festival, aiming to gain more tourists' attention and for inheritance purposes.

30,000 sample images were collected to integrate and extract the cultural characteristics of the She nationality with the help of Boundless AI. Key cultural elements were digitalized and modeled to generate unique She nationality architecture, character images, clothing, and cultural products innovatively. The AIGC-endowed capability of cultural symbol extraction and content generation significantly boosts ethnic minority village cultural derivatives and digital collections, such as symbols and artworks.

With the development of artificial intelligence, new energy, 5G, and other technologies, rural revitalization breaks through the traditional paradigm with the help of technological means to lead rural change [21]. The "Safe Countryside AI Empowerment" project in Baoshan, Yunnan Province, provides AI technology to address social governance issues such as illegal parking and rubbish exposure, helping the local government to modernize its governance transformation. In addition, the most representative AI-enabled rural revitalization is the Digital Countryside Project launched by Shangtang Technology in Songyang, Zhejiang Province. Through the development of AI courses to promote the development of primary and secondary school studies, AR recovery of traditional ancient villages, rural digital museums, and other rural governance.

With the help of generative AI, art has made significant contributions to rural revitalization and the development of rural areas [22]. It has increased cultural awareness in regions inhabited by the She ethnic group and attracted numerous designers to re-imagine local ethnic patterns. For example, the She Pattern 2023 International Pattern Creative Design Competition[4], launched in August 2023, aims to showcase redesigned ethnic patterns specific to the She culture.

4   Content Generation Workflow

The implementation of our system consists of three main stages: initial mockup (3D Rendering), image generation by AI (Midjourney), and AI video generation (Runway). Participants or organizers collect visual information in the initial mockup phase by capturing a 360 local view using a camera and sourcing artwork from artists. These artworks are then digitally placed within the captured local view, rendering the overall images in the 3D environments. This visually rich information serves as input data for a Chat GPT 4.0, which generates text prompts for the subsequent image generation stage.

In the Midjourney stage, which focuses on AI image generation, the prompts generated by the Chat GPT 4.0 are used to select specific images from the 3D

---

[4] Website: https://www.shejijingsai.com/2023/08/994035.html.

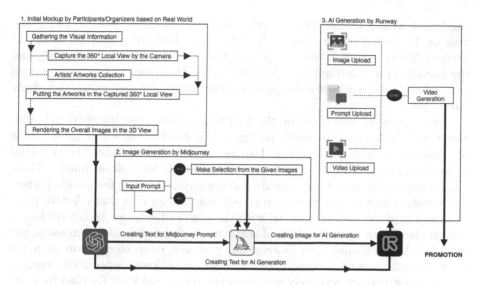

Fig. 1. AI Content Generation Workflow

environments. If the prompts do not yield suitable images, additional descriptions are added to refine the selection process. The chosen images are then transformed and readied for the final stage, in which AI video generation takes place.

Runway's Generative Adversarial Network 2 (GAN 2) model is employed in the AI video generation stage. The images chosen in the Midjourney phase, the prompts generated by the Chat GPT 4.0 model, and any original video content are used as input for the GAN 2 model. The GAN 2 model is capable of generating video content based on these inputs. Users can further influence the output by providing a style reference or uploading a style image, which guides the AI toward a desired aesthetic in the generated videos. This multi-stage implementation results in a novel blend of real-world views, artistically curated content, and AI-enhanced video output, See Fig. 1.

5   Methodology

This study incorporates evidence from a literature review, case studies, semi-structured interviews, and thematic analysis to explore the intersection of complex aesthetics and generative AI and its implications for cultural creativity in rural areas.

5.1   Case Study

We use case studies to explore a specific phenomenon in-depth. This approach yielded detailed insights into the subject matter under investigation. A finite selection of cases was made, and questionnaires and structured interviews were conducted with relevant stakeholders, such as curators, artists, villagers, and visitors. The cases selected for study include the Tongling Country Public Art Festival (2018), Another Wave - the First Daojiao New Art Festival (2016), and the Lianzhou International Foto Festival (2005). These events, each with unique cultural and artistic dimensions, provide a comprehensive and contrasting study landscape.

Our assessment was based on the festival locations, their international influences, and the fame of the participating artists. Festivals based in more remote or rural areas, such as Tongling, Daojiao, and Lianzhou, have the potential to provide unique cultural experiences and stimulate local development. These locales may not be provincial capitals, but the success of these festivals in garnering attention and fostering cultural engagement has proven that effective planning and implementation, coupled with the power of art, can transform lesser-known places into dynamic cultural destinations. Regarding global impacts, the Lianzhou International Foto Festival has garnered more recognition than the other festivals, largely because it has been an annual event since 2005, whereas the other festivals are relatively new. Another Wave - the First Daojiao New Art Festival was a one-off event.

The participating artists in these festivals hail from diverse backgrounds and enjoy varying degrees of fame. For instance, the Lianzhou International Foto Festival's longstanding history has enabled it to attract a broader network of renowned photographers and artists from around the world, thereby enhancing the festival's global standing and influence, which in turn creates a positive feedback loop of growing recognition and talent attraction. In contrast, the newer festivals, Tongling Country Public Art Festival, and Another Wave - the First Daojiao New Art Festival, may feature a combination of budding and established artists. Their artist lineups may predominantly consist of domestic talent, but this does not detract from their cultural relevance or growth potential. Instead, it may offer a unique platform for local artists to gain exposure and for attendees to be introduced to new artistic perspectives. Despite the one-time occurrence of Another Wave - the First Daojiao New Art Festival, it may have left a lasting impression on the local community and the artists involved. Its sole edition may have acted as a catalyst for future projects or left a significant cultural footprint.

5.2   Questionnaire

The setting of the questionnaire was designed to incorporate three question types: apparatus-based questions, five-point scale responses, and open-ended questions. The apparatus-based questions were formulated to categorize interviewees based on their backgrounds. These questions focused on gathering information about the participants' roles, experiences, and qualifications. The survey employed a 5-level scale, where the first level signifies "strongly disagree" and the fifth indicates "strongly agree". This scale is designed to aid researchers and decision-makers in discerning the ways and degree to which art activities bolster rural revitalization. Open-ended questions were also included to capture unconsidered details and fill any gaps within the questionnaire. These questions encouraged participants to provide descriptive and nuanced responses, enabling a more comprehensive understanding of their perspectives.

Our questionnaire, with 329 collected responses, served as an exhaustive instrument for collecting data on various crucial aspects such as the festival's contribution to preserving local culture, how the local culture is incorporated into the festival, the significant challenges in organizing the festival, the festival's impact the local community, the audience, and visitors respond to the festival, the role of generative AI in promoting cultural creativity in future festivals. It investigated the respondents' participation in the festival, their viewpoints on the event's main objectives, and their personal standouts. Furthermore, it assessed the festival's influence on the local community and gauged the audience's reaction.

5.2.1   Data Analyse

A significant portion of this survey revolved around the interviewer's profound understanding of how generative AI could play a potential role in future iterations of the festival. This perspective was pertinent as most respondents felt that the artworks' interactions did not significantly resonate with the local culture, owing to the perceived isolation of these artworks. About 62% of respondents viewed the interaction between artists and locals as only marginally significant, rather than as a means of promoting local culture during festivals.

Another prevalent feedback was the economic constraints not ensuring the sustainable growth of art promotion in rural areas. Respondents who identified these effects as severe or significant constituted more than 50% of all participants. The rationale for this is twofold: the expansion of the pandemic and the demand for artist involvement. We are considering its future potential in applying AI in these rural settings, particularly its utilization by locals. This could reduce the need for artist participation and artwork, thereby streamlining the festival organization.

5.3   Semi-structured Interview

We conducted Semi-structured interviews with 24 stakeholders which include artists, curators, and the general audience (Village workers, villagers, village officials.), and 2 curators in-depth interviews. In this study, we also employed thematic analysis to scrutinize data derived from semi-structured interviews. These interviews encompassed a diverse range of 24 participants' interview content. The insights gleaned from these in-depth interviews, when passed through the lens of thematic analysis, offered a profound understanding of the curators' experiences and perspectives, thereby enriching the overall research findings.

5.3.1 Participants

Table 1 provides information about the participants involved in the study. Here is a summary of the information: We invite 5 curators, 7 artists, and 12 general audiences. The participants are identified by the codes C1, C2, C3, C4, C5, A1, A2, A3, A4, A5, A6, A7, G1, G2, G3, G4, G5, G6, G7, G8, G9, G10, G11, and G12. The participants' ages range from 21 to 44 years old (45 ± 1.25 years). The participants include males (M) and females (F), representing a mix of genders. The "Year of Experience" column indicates the number of years of experience each participant has, ranging from 0 to 9 years. The table provides a snapshot of the participant's demographic information, including their age, gender, and years of experience.

5.3.2 In-Depth Interviews

After the Semi-structured Interview with a total of 24 participants, we selected C1 and C3 for in-depth interviews. Their responses in the questionnaire stood out because of their comprehensive insights and unique perspectives on the discussed topics.

The first respondent C1 was chosen because of his detailed insights into incorporating local culture and rural aesthetics into the festival and thoughtful suggestions on how generative AI could promote cultural creativity. In the interview with C1, we focused on his detailed understanding of integrating local culture into the festival and his thoughts on the role of generative AI in promoting cultural creativity. We asked him to elaborate on his challenges and provide further suggestions for improving the use of generative AI in future festivals.

The second respondent C3 was selected due to her comprehensive understanding of the festival's impact on the local community and her forward-thinking ideas on how generative AI could be used to create digital archives and preserve rural culture. For C3, our interview questions were centered around his perception of the festival's impact on the local community and her innovative ideas on how generative AI could be used to create digital archives and preserve rural culture. We sought to understand her experiences with audience and visitor responses and her views on the potential feasibility of using generative AI in upcoming festivals.

From these interviews, we found that both C1 and C3 shared a belief in the potential of generative AI to enhance the representation of local culture and rural aesthetics in the festival. They also highlighted the challenges involved in integrating advanced technology into a traditional setting, such as the need for resources and gaining acceptance from the local community. Their insights provide valuable direction for future research and planning in integrating generative AI into rural art festivals. We delved deeper into these topics, seeking to understand not just the what but also the how and why of their perspectives.

Table 1. Participants' and their information

Participants	Age	Gender	Year of Experience
C1	38	M	3
C2	34	F	1
C3	37	F	1
C4	28	F	6
C5	31	M	9
A1	29	M	2
A2	30	F	3
A3	44	F	1
A4	37	F	2
A5	22	M	3
A6	34	M	1
A7	37	F	1
G1	21	M	1
G2	24	M	1
G3	27	M	2
G4	31	F	1
G5	44	F	1
G6	37	F	0
G7	31	F	0
G8	24	M	1
G9	27	M	0
G10	21	F	0
G11	24	F	1
G12	27	F	2

This allowed us to better understand the potential for integrating generative AI into future festivals and the possible challenges and opportunities that might arise from this integration.

5.3.3 Thematic Analysis

(See Table 2).

Table 2. Main Themes and relevant codes, and number of code occurrences (OC)

Theme	Relevant Codes	CO
From Urban to Countryside	Enriched creative landscape	28
	Renewed focus on local culture	19
	Sustainable development - Collaboration	30
	Sustainable development - Cultural tourism	22
	Sustainable development - Digital platforms	15
	Total	114
The Form of Art in Countryside	Reflection of traditions and customs	45
	Celebrating cultural identity	12
	Revival of traditional crafts	32
	Pottery, paper-cutting, embroidery	21
	Influence of landscape and nature	18
	Inspiring artistic creations	22
	Total	150
Execution and Development	Enhancing artistic creations	39
	Innovative artistic techniques	26
	Unconventional approaches	14
	Training programs and workshops	10
	Knowledge sharing and interdisciplinary exchange	27
	Promoting local art and sustainable development	7
	Total	123
Interaction and Influences	Community engagement - Input, resources, local knowledge	25
	The social impact of art - Cohesion, cultural pride, well-being	12
	Broadening perspectives, new ideas	36
	Participatory art practices, Co-creation, and community ownership	23
	Influence of traditions Inspiring artistic expression	21
	Empowerment through art - Self-expression, cultural preservation	9
	Total	126
High-tech Involvement	Application of generative AI - Novel artistic outputs	7
	Digital technologies for preservation - Digitization, virtual reality	12
	Immersive rural settings - Virtual reality experiences	33
	Digital platforms - Showcasing, accessibility, cultural exchanges	39
	Challenges and opportunities - Infrastructure, affordability, digital inclusion	17
	Total	108
All		621

This study delves into public attitudes towards combining generative AI and local culture in rural art festivals, uncovering diverse viewpoints. Some respondents believe that generative AI can introduce a variety of art and cultural activities to rural areas, while others express skepticism about the feasibility of this integration. There's also a perception among some participants that rural art festivals do not adequately represent local culture and characteristics. Simultaneously, examining rural artistic practices in China has been conducted, focusing on the high frequency of specific keywords in interviews. This analysis suggests that these terms are significant and relevant in the context of rural art. The respondents appear to perceive urban-rural communication as a catalyst for artistic growth, emphasize the importance of preserving local cultural heritage, and seek strategies for sustainable development, as indicated by the frequent use of these keywords. The high frequency of keywords such as "reflection of traditions and customs" and "revival of traditional crafts" suggests that respondents view rural art as a means of preserving and honoring cultural identity. A renewed interest in traditional artistic practices is indicated, potentially spurred by a desire to reconnect with cultural heritage and generate economic opportunities through craftsmanship. The repeated mention of "enhancing artistic creations" implies a shared aspiration among respondents to improve the quality and impact of rural art. Terms like "innovative artistic techniques", "training programs and workshops", and "knowledge sharing and interdisciplinary exchange" indicate a desire to explore new approaches, develop skills, and foster collaboration and cross-disciplinary interactions. Community engagement is frequently mentioned, highlighting its importance in artistic endeavors by involving local communities and leveraging their resources. Respondents also recognize the potential social impact of art, acknowledging its capacity to foster community cohesion, cultural pride, and overall well-being. The limited mention of the "application of generative AI" suggests that the use of artificial intelligence in artistic creation is still relatively nascent in rural contexts. However, using "digital technologies for preservation" and "digitization, virtual reality" indicates an interest in leveraging digital tools to preserve and present rural art forms.

6   Key Findings

The artworks presented by many artists at art festivals are simply moving their pre-existing creations into another "art museum" space rather than creating organic combinations specific to the local context. This approach only benefits the dissemination of art but fails to promote the rural areas' unique characteristics and cultural heritage. To appropriately promote local cultural experiences, creating works organically integrated with the local context is necessary. Rural areas face a limitation in their revitalization efforts as they rely on artists as external forces who cannot permanently reside in these villages. Hence, rural areas need to cultivate their "artists (villagers)" who better understand the local context and may not necessarily be trained professionals. To this end, generative AI can serve as a tool for these "artists" as machine learning learns from subjective aesthetics (provided by the previously participated artists) and has the potential to contribute to sustainable development while residing in rural areas. This approach can replace the traditional annual rural art festivals, and form a sustainable development cycle [23]. There are two ways to use generative AI to protect local cultural heritage, which has yet to be fully explored in rural revitalization. The first is conducting in-depth academic and inheritance-based preservation while respecting the history and cultural heritage relying on the machine learning of generative AI. The second is to attract audiences by providing interactive generated images/texts/videos that offer multiple visual representations, which can promote tourism and rural development while maintaining entertainment value, like digital storytelling [24].

When looking at the relationship between aesthetics, AI, and cultural creativity from a theoretical perspective, it is first necessary to clarify the intrinsic correlation between these. In terms of the concept of complex aesthetics, which corresponds to generative AI. And complex aesthetics reflects generative art, which uses digital media and information interaction as its hub [25]. Unlike traditional art, which uses physical media and natural language to create, the systematic complexity and typological diversity of complex art reflect different characteristics [26]. The characteristics of complex art can be summarized as three points:

- A greater tendency towards normative language expression, emphasizing the extension of predictable phenomena in both the temporal and spatial dimensions;
- A selective operating mechanism, which sifts out distinguishing and effective data from a large amount of irregular data, which can provide a source of creative consciousness;
- A two-way selectivity of aesthetic value reinforced by the pivot of informational interactions, further narrowing the distance between subjects, which can provide a source of creative awareness; and a two-way selectivity in the interaction of information as a pivot to strengthen aesthetic value, which further reduces the distance between subjects. This further reduces the distance between subjects, which, from the perspective of intersubjectivity, strengthens the construction of the relational network between subjects. To summarize, these three features of complex art highlight the organic, holistic, and processual nature of generative AI. This is compatible with the nature of cultural creativity [27]. After all, for promoting cultural creativity, the correspondence between individual creation and the overall cultural environment, the integration of science and humanities, and the exchange of local and diverse cultures are three essential and necessary paths.

Aiming at it, we can think about the possibility of the use of generative AI in art creation to enhance cultural creativity in Rural China from three levels. First, cultivating art workers in local cultural contexts. This is due to the fact that artists often have difficulty integrating their artworks with local culture when practicing rural art practices. This problem arises mainly due to the existence of cultural differences. Cultural difference is both a challenge and an opportunity [28]. On the one hand, China's vast geography makes rural culture have obvious differences between north and south and east and west, and with the development of society and economy, the differences between urban culture and rural culture are more prominent; on the other hand, the existence of cultural differences provides the necessary conditions to stimulate the imagination of artists, and the fusion of aesthetic experiences in multiple contexts can create more universal works of art for artists. In the process of dealing with cultural differences, "communication" is the most effective method. This exchange is a two-way street. It includes artists' participation in rural art practice, and it also includes local art workers' absorption of external art practice experience, exploration of new ways of artistic expression, and improvement of artistic expression ability. It is worth noting that, for local artists, due to the cultural advantages of the art creation context, they should take the people's artistic expression as the purpose of art creation, focusing on the expression of the aesthetic experience of life. Second, scientific cognitivism, as an aesthetic principle, plays an important theoretical guiding role for generative AI to be embedded in the process of aesthetic appreciation, which in turn generates the judgment of aesthetic value. Scientific cognitivism, as an embodiment of a natural aesthetic stance, is the relationship between aesthetics and the environment that Alan Carlson sought to address when he proposed scientific cognitivism in 1979, in which the issue of fact and value is unavoidable. In China, the natural environment of the countryside provides a good natural aesthetic field, in which the appreciator understands and integrates to find aesthetic value in aesthetic experience, change of aesthetic interest, and even psychological healing. Scientific cognitivism can play a role in generative AI embedded in the process of aesthetic appreciation, because this aesthetic principle contains the basic connotation of aesthetic relevance, that is, through different ways of presenting art, the aesthetic object does not have the knowledge, imagery, and ideology associated with the aesthetic experience. Thirdly, it injects brand-new cultural stimulation points into a rural culture. In terms of the characteristics of rural culture itself, due to its objective conditions and constraints, it will inevitably appear relatively conservative characteristics. This also makes rural culture characterized by relatively single cultural types, slow updating of cultural expressions, and insufficient motivation for artistic creation. Therefore, in the case of insufficient internal development power, external stimuli are essential for the enhancement of cultural creativity. Generative AI, as a new form of cultural expression, can break through the original boundaries of geography, senses, and other dimensions, and promote the renewal and iteration of the original expression and propaganda of rural culture.

7   Research Agenda and Limitation

The research agenda proposed by this study is aimed at pushing the boundaries of generative AI and complex aesthetics to promote cultural creativity within the rural regions of China. This approach seeks to address the economic and cultural challenges faced by these areas due to the aftermath of the COVID-19 pandemic.

7.1   Research Agenda

The primary objective of this research is to explore and implement a framework that leverages the intersection of generative AI and complex aesthetics to stimulate cultural creativity in rural China. By feeding machine learning models with subjective aesthetic data, the goal is to enable them to generate output that is culturally creative and reflective of the local aesthetics. This research also seeks to create a dialogue between technology and culture, hoping to preserve local traditions while embracing the benefits of modernity. In this context, the proposed framework is not meant to replace human creativity but rather to enhance it by infusing it with AI intelligence. As part of this research agenda, the study invites further exploration and experimentation in the following areas:

- Development of methodologies for extracting subjective aesthetic data from pre-existing artistic representations.
- Building machine learning models that can effectively learn from and generate output based on this data.
- Identification of effective ways to weave site-specific artworks into the local context.
- Exploration of how generative AI can nurture grassroots artists within villages.

7.2 Limitations

The lack of a standardized method for content quantification makes it challenging to evaluate artworks, creating obstacles in developing rural cultural creativity. Cultivating "artists (villagers)" is crucial, but preventing subjective bias when training them to use generative AI is challenging. The solution is to let generative AI learn the subjective essence of art from real artists' data and use it to guide villagers to control and implement in real time. Future research could improve generative AI algorithms to better align with cultural experience and local heritage, evaluate its impact on rural communities and cultural heritage, explore new ways to integrate local knowledge, assess social and ethical implications, and compare how generative AI is used in different cultural contexts and rural areas worldwide. Despite the promising potential of this research, there are several limitations to consider:

Cultural Creativity: The subjective nature of aesthetics and cultural creativity may present challenges in defining and extracting subjective aesthetic data that can be processed by machine learning algorithms.

Generative AI: While generative AI has proven effective in various applications, its use in promoting cultural creativity in rural areas is relatively unexplored. As such, there may be unforeseen technical obstacles or limitations to overcome.

Data Availability: The availability and quality of data for both training the machine learning models and evaluating their output may be limited, especially in rural areas.

Socio-cultural Acceptance: There might be resistance to the incorporation of AI and technology in cultural and artistic practices among rural communities. This resistance could stem from fear, misunderstanding, or the desire to maintain traditional practices.

Despite these limitations, the potential benefits of this research in promoting cultural creativity and economic development in rural China should not be underestimated. The proposed framework provides a solid foundation for further research and experimentation in this area.

8 Conclusion Remark

In conclusion, this study provides an original perspective on leveraging the intersection of complex aesthetics and generative AI to promote cultural creativity in rural China during the post-pandemic recovery. Through examining applications in cultural heritage and conducting a multi-method analysis, key findings emerged. First, artworks integrated within local contexts are crucial for effectively promoting cultural experiences versus conventional disconnected rural art festivals. Second, nurturing grassroots "artist-villagers" through generative AI aids sustainable development.

Ultimately, this pioneering research proposes a human-centered framework synergizing AI capabilities and nuanced aesthetics to inject new vitality into rural heritage while fostering enlightening dialogues about tradition and modernity. The core contributions are elucidating generative AI's potential to enable cultural creativity, establishing frameworks to extract subjective aesthetic data, and envisioning AI not as a replacement for human creativity but rather as enhancing it. This conclusion provides a foundation for further exploration of the nexus of technology, culture, and creativity.

Acknowledgments. Acknowledge financial support from the National Social Science Foundation of China (#20CJY045).

References

1. Li, X., Liang, X., Yu, T., Ruan, S., Fan, R.: Research on the integration of cultural tourism industry driven by digital economy in the context of COVID-19-based on the data of 31 Chinese provinces. Front. Public Health **10**, 780476 (2022)
2. Oh, C., et al.: Understanding how people reason about aesthetic evaluations of artificial intelligence. In: Proceedings of the 2020 ACM Designing Interactive Systems Conference, pp. 1169–1181 (2020)
3. Hitsuwari, J., Ueda, Y., Yun, W., Nomura, M.: Does human-AI collaboration lead to more creative art? Aesthetic evaluation of human-made and AI-generated haiku poetry. Comput. Hum. Behav. **139**, 107502 (2023)
4. Banks, M., O'Connor, J.: "a plague upon your howling": art and culture in the viral emergency. Cult. Trends **30**(1), 3–18 (2021)
5. Song, Y., Zhao, P., Chang, H.-L., Razi, U., Sorin Dinca, M.: Does the COVID-19 pandemic affect the tourism industry in China? Evidence from extreme quantiles approach. Econ. Res. Ekonomska Istraživanja **35**(1), 2333–2350 (2022)
6. Gao, Z., Man, S., Wang, A.: AI art and design creation industry: the transformation from individual production to human-machine symbiosis. In: 2022 World Automation Congress (WAC), pp. 52–56. IEEE (2022)

7. Guo, C., Zhang, Y., Liu, Z., Li, N.: A coupling mechanism and the measurement of science and technology innovation and rural revitalization systems. Sustainability 14(16), 10343 (2022)

8. Xie, D., He, Y.: Marketing strategy of rural tourism based on big data and artificial intelligence. Mob. Inf. Syst. 2022 (2022)

9. Che Aziz, R., Nik Hashim, N.A.A., Awang, Z.: Tourism development in rural areas: potentials of appreciative inquiry approach. J. Tourism Hospitality Culinary Arts 10(1), 59–75 (2018)

10. Saxena, G., Ilbery, B.: Integrated rural tourism a border case study. Ann. Tour. Res. 35(1), 233–254 (2008)

11. Wang, A., Gao, Z., Lee, L.H., Braud, T., Hui, P.: Decentralized, not dehumanized in the metaverse: Bringing utility to NFTS through multimodal interaction. In: Proceedings of the 2022 International Conference on Multimodal Interaction, pp. 662–667 (2022)

12. Rezwana, J.: Towards designing engaging and ethical human-centered AI partners for human-AI co-creativity. PhD thesis, The University of North Carolina at Charlotte (2023)

13. Sun, Y., Li, X., Gao, Z.: Inspire creativity with ORIBA: transform artists' original characters into chatbots through large language model. arXiv preprint arXiv:2306.09776 (2023)

14. Pisoni, G., Díaz-Rodríguez, N., Gijlers, H., Tonolli, L.: Human-centered artificial intelligence for designing accessible cultural heritage. Appl. Sci. 11(2), 870 (2021)

15. Gao, Z., Lin, L.: The intelligent integration of interactive installation art based on artificial intelligence and wireless network communication. Wirel. Commun. Mob. Comput. 2021, 1–12 (2021)

16. Manfredi, C.: Things of space: andy goldsworthy's *sheepfolds* and Alec Finlay's *Company of Mountains*, or, materialising as re-siting. In: Nature and Space in Contemporary Scottish Writing and Art. GSLS, pp. 121–148. Springer, Cham (2019). https://doi.org/10.1007/978-3-030-18760-6_6

17. Wildy, J.: A sculpted land: Ecological landscape art of the Harrisons, Patricia Johanson and Agnes denes

18. Choi, Y.J., McNeely, C.L.: A reinvented community: the case of gamcheon culture village. Sociol. Spectr. 38(2), 86–102 (2018)

19. Crespi-Vallbona, M., Richards, G.: The meaning of cultural festivals: stakeholder perspectives in Catalunya. Int. J. Cult. Policy 13(1), 103–122 (2007)

20. Soutter, L.: Notes on photography and cultural translation. Photographies 11(2–3), 329–338 (2018)

21. Sarangi, S.S., Singha, G.: A survey on impact of AI and social media for rural development. JCSE 7, 64–68 (2019)

22. Zhang, L., Zheng, L., Chen, Y., Huang, L., Zhou, S.: CGAN-assisted renovation of the styles and features of street facades-a case study of the Wuyi area in Fujian, China. Sustainability 14(24), 16575 (2022)

23. Qu, M., Cheer, J.M.: Community art festivals and sustainable rural revitalisation. In: Events and Sustainability, vol. 2022, pp. 18–37 (2022). Routledge, ???

24. Bogdanovych, A., Rodríguez-Aguilar, J.A., Simoff, S., Cohen, A.: Authentic interactive reenactment of cultural heritage with 3D virtual worlds and artificial intelligence. Appl. Artif. Intell. 24(6), 617–647 (2010)

25. Arielli, E., Manovich, L.: AI-aesthetics and the anthropocentric myth of creativity (2022)

26. Gao, Z., Zhang, X., Zhu, S., Braud, T.: Metanalysis: a connected-wall video installation based on the movement of traditional Chinese ink painting. In: 2022 Third International Conference on Digital Creation in Arts, Media and Technology (ARTeFACTo), pp. 1–6. IEEE (2022)
27. Donnarumma, M.: Against the norm: othering and otherness in AI aesthetics. Digit. Cult. Soc. **8**(2), 39–66 (2022)
28. Stuhr, P.L.: celebrating pluralism: art, education, and cultural diversity. JSTOR (1999)

Fingerprint Image Enhancement Method Based on U-Net Model

Yu-Hsin Cheng[1], Sheng-Gui Su[1], Yih-Lon Lin[1(✉)], and Hsiang-Chen Hsu[2]

[1] National Yunlin University of Science and Technology, Yunlin, Taiwan
{m11017050,m11017059}@gemail.yuntech.edu.tw,
yihlon@yuntech.edu.tw
[2] I-Shou University, Kaohsiung, Taiwan
hchsu@isu.edu.tw

Abstract. Fingerprint image enhancement is a vital image processing technology that finds applications in fingerprint identification, matching, and biometric authentication. Its objective is to improve the performance and accuracy of fingerprint identification systems by enhancing the quality, clarity, and contrast of fingerprint images while reducing noise. Traditional algorithms often require adjusting parameters based on individual fingerprint images, resulting in inconsistent enhancement results. To address this challenge, this paper proposes a deep learning-based approach for fingerprint image enhancement. Experimental results demonstrate the effectiveness of U-Net architecture with different depths and the impact of incorporating attention gate. Actually, a U-Net model with a depth of six achieves superior performance, surpassing other depths, and the inclusion of attention gate further enhances the results. The proposed method not only retains more fingerprint features but also produces smoother structures and reduces noise surrounding the fingerprint image. These findings contribute to advancing fingerprint image enhancement techniques, bolstering the accuracy and performance of fingerprint identification systems in applications such as forensic analysis and biometric authentication.

Keywords: Fingerprint Enhancement · Attention Gate · Fingerprint Structured

1 Introduction

Fingerprints have long been a significant area of research because they are one of the unique human features. They possess advantages such as uniqueness, immutability, stability, and easy availability. Moreover, they find wide application in fields like crime identification, identity matching, and information security. Fingerprint image enhancement pertains to the fields of fingerprint recognition and fingerprint matching. It employs image processing technology to enhance the quality and clarity of fingerprint images.

During the sampling process, fingerprint images are easily affected by factors such as lighting conditions, hardware equipment, and skin conditions. Consequently, fingerprint images often encounter issues like noise, blur, and unevenness, which subsequently reduce the performance and accuracy of fingerprint recognition systems. To address these issues, fingerprint image enhancement technology emerged. Its primary objective is to eliminate noise, enhance details, and improve contrast. These methods effectively enhance the quality and clarity of fingerprint images, rendering them more suitable for subsequent fingerprint identification and matching tasks.

2 Related Work

Lin et al. [1] proposed an algorithm for enhancing fingerprint images, which effectively improves the clarity of the fingerprint structure. This algorithm involves five steps: normalization, local orientation estimation, local frequency estimation, region masking, and Gabor filtering. Normalization reduces the variation in gray values between ridges and valleys, resulting in a grayscale image for further processing. Local direction estimation utilizes the least average direction estimation algorithm and applies a low-pass filter to correct inaccurate local ridge orientations, generating a fingerprint orientation image. Local frequency estimation leverages the grayscale and orientation maps to identify x-signatures, allowing for the calculation of local frequencies. Area Mask Estimation employs the Directional Map and Frequency to generate a mask image. Finally, contextual convolution is performed using a Gabor filter. The Gabor filter filters the image to obtain accurate pixels, and the discretized direction index is utilized to assemble the enhanced image. The experimental results presented by the authors demonstrate the algorithm's effectiveness in improving the clarity of the fingerprint structure and repairing unclear local fingerprints.

In the AlignNet fingerprint network [2], the authors conducted two preprocessing steps on the fingerprint images before inputting them into the model. The preprocessing involves cartoon-texture decomposition and Ridge Filter. Cartoon-texture decomposition effectively eliminates the background while preserving the ridge and texture information of the fingerprint. Additionally, the Ridge Filter is applied to enhance the sharpness of the ridges. Subsequently, the ConvNetOF model is employed to obtain the regional direction field, and the ridge filter utilizes the classic Gabor filtering method to generate the enhanced fingerprint image. In previous research, numerous scholars and researchers have put forward various methods for enhancing fingerprint images. Among these methods, filtering is one of the most commonly employed techniques as it effectively reduces noise and enhances image clarity. Commonly used filters include the Mean filter, Median filter, Gaussian filter, and Adaptive filter.

In the paper by Han et al. [3], a technology for fingerprint image enhancement utilizing the adaptive median filter is proposed. This method involves dynamically selecting the filter size, allowing the filter to better adapt to different regions within the fingerprint image and achieve superior enhancements. The experimental results presented by the authors demonstrate that this enhancement method outperforms the traditional Median filter method. In the field of fingerprint image processing, there have been studies that employ a series of steps for image enhancement [4]. The first step involves normalization, which reduces the variations in gray scale values between valleys and ridges. Subsequently, an adaptive noise reduction technique called the Edge Direction Total Variation (EDTV) model is applied. Finally, the local frequency is calculated, and the fingerprint is further enhanced using a Gabor filter. The experimental results presented by the author indicate that this fingerprint image enhancement technology enhances the quality of the fingerprint image by approximately 30%.

In the fingerprint recognition algorithm proposed by Szymkowski et al. [5], the author's experiments demonstrate that the optimal results are achieved by utilizing the median filter in conjunction with a 3 x 3 mask on the binarized fingerprint image. The approach involves applying four filters to each input image and subsequently utilizing the KMM algorithm to obtain high-quality fingerprint images. Yusharyahya et al. [6] compared three different algorithms: NBIS, SourceAFIS, and Gabor Filters. NBIS employs Fast Fourier Transform for power transformation in the frequency domain and utilizes the NFIQ algorithm to assess the quality of fingerprint images. On the other hand, SourceAFIS concentrates on image processing techniques in the spatial domain to enhance fingerprint images. The Gabor filter is utilized for contextual filtering in image enhancement, analyzing the local context of fingerprint images. In the final experimental results, the NBIS score is the highest, indicating the effectiveness of the algorithm. The experimental results of Patel et al. [7] demonstrate that the enhanced O'Gorman filter outperforms both the Gabor filter and the original O'Gorman filter in terms of image enhancement. This algorithm effectively eliminates noise, reconstructs broken ridges, and enhances image quality. The evaluation metrics MSE and PSNR validate the superior performance of the enhanced O'Gorman filter compared to the other two algorithms. These findings underscore the potential of the enhanced O'Gorman filter as a reliable and effective solution for fingerprint image enhancement.

Furthermore, various enhancement algorithms, such as histogram equalization, contrast stretching, and edge enhancement, are extensively utilized in fingerprint image enhancement.

3 Dataset

The dataset employed in this experiment is FVC2000 (Fingerprint Verification Competition 2000) [8], which is widely utilized in fingerprint research and algorithm evaluation. This dataset comprises fingerprints sourced from diverse databases, encompassing both real fingerprints and synthetic fingerprint images. It provides a comprehensive platform for testing and evaluating the performance of fingerprint recognition systems.

The FVC2000 dataset comprises various sub-databases, namely DB1, DB2, DB3, and DB4, each representing distinct types of true fingerprints. These real fingerprints are obtained from individuals of different ages, races, and genders. DB1 and DB2 utilize optical and capacitive sensors, respectively, while DB3 provides high-quality optical sensing fingerprint images. DB4, on the other hand, employs synthetic fingerprints. For the purpose of simulating the collection of real fingerprints, this study utilizes DB3 within the FCV2000 dataset. The fingerprint images have a size of 448×478 pixels and a resolution of 500 dpi. The total number of fingerprints is 80, collected from 19 subjects ranging in age from five to seventy-three. Of these fingerprints, 55% belong to males. The collected fingerprints include those from the thumb, index finger, middle finger, and ring finger.

4 Research Methods

This study was divided into two phases, with the first stage depicted in Fig. 1. To generate an enhanced fingerprint image, it is necessary to distinguish the fingerprint region (foreground) from the background. The fingerprint foreground is characterized by streaks and directional patterns, while the background exhibits a uniform color. The second stage involves utilizing semantic segmentation to generate enhanced fingerprint images, which will be discussed in more detail later. The first stage of fingerprint image enhancement encompasses the following five steps:

Fig. 1. Fingerprint Enhanced Image Algorithm.

Step 1: Gradient image as shown in Fig. 2(a), use the grayscale fingerprint image and apply the Sobel filter to obtain the gradient image and gradient size.

Step 2: The mask image is shown in Fig. 2(b). Use a Box filter to filter the gradient image, and multiply the maximum value by 0.2–0.4 as the threshold for image binarization, and then perform image expansion and erosion operations, finally generate a fingerprint mask through the above steps.

Step 3: Ridge direction, use the gradient image obtained in the first step to perform local Box filtering, since the ridge direction is orthogonal to the gradient direction, so take the average in the local area to get the ridge direction.

Step 4: Ridge Frequency, ridge frequency estimation with a small region within the fingerprint. The ridge frequency is estimated as the average number of pixels between consecutive peaks, calculate the distance between all consecutive peaks in the region, and take the average to get the ridge frequency. Finally, it is merged with the ridge direction to obtain the frequency pattern. Refer to Fig. 2(c).

Step 5: Generate a fingerprint enhanced image as shown in Fig. 2(d), use the ridge frequency to create 8 different angle filters, and filter the entire fingerprint image, then obtain the convolution result of the corresponding direction for each pixel, and assembled into an enhanced image. Then use the mask to remove the surrounding noise, and obtain the final image through binarization.

The first architecture for the generation of enhanced fingerprint images, marking the "fingerprint" and "background" areas for 80 original images. Subsequently, the size of the images is expanded to 512 × 512, and the enhanced fingerprint image serves as the label image for semantic segmentation. The second stage employed in this study involves the utilization of the U-Net model and the U-Net with Attention Gate models. These models are both simple and effective. The U-Net model consists of two main components: Encoder and the Decoder. It incorporates skip connections to achieve pixel-level prediction results.

The Encoders typically comprise multiple convolutional and pooling layers, which are responsible for extracting image features and reducing their size. With each convolutional layer, the depth of the feature map increases while the spatial size of the image

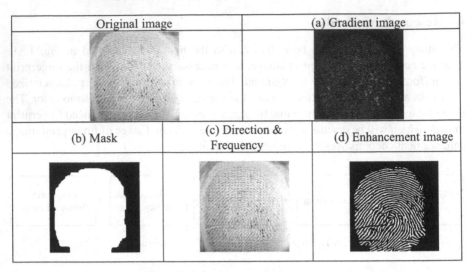

Fig. 2. Fingerprint image enhanced step.

decreases. This enables the model to learn features at different levels, reduce the number of model parameters, and ultimately improve the computational efficiency of the model. The Decoder typically comprises multiple convolution and deconvolution layers, which aim to restore the reduced feature map from the encoder. Each deconvolution layer reduces the depth of the feature map while simultaneously enlarging the spatial size of the image. This allows the model to recover the spatial details of the image and generate prediction results, as shown in Fig. 3.

Furthermore, skip connections are employed to establish connections between corresponding layers in the encoder and decoder. This ensures that feature maps of the same size are connected to their corresponding decoder layers. By utilizing features from different layers, the model can generate more accurate prediction results and enhance overall accuracy.

In the U-Net model, an Attention Gate (as depicted in Fig. 4) is additionally incorporated. This feature enables the model to selectively focus on relevant image features while suppressing irrelevant ones. The Attention Gate mechanism draws inspiration from the concept of attention in human perception, allowing the model to prioritize important image information during the prediction process.

The Attention Gate comprises two main components: Encoder skip connections and Attention Gate. The Encoder skip connections retrieve low-level features from the encoder and directly transmit them to the Attention Gate. The Attention Gate calculates weights for a range of image features, selectively emphasizing specific parts of the image. By highlighting essential regions and suppressing noise or background, the Attention Gate enhances the accuracy of segmentation.

This research primarily employs U-Net models with varying depths and incorporates Attention Gate to compare the performance of deep and shallow layers. The model training methodology is as follows: the fingerprint enhanced dataset is divided into original images and label images, totaling 80 sets of images. These 80 sets of images

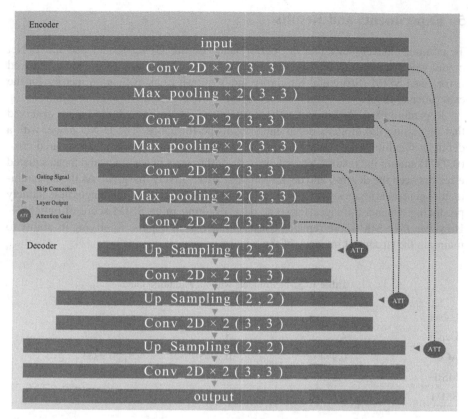

Fig. 3. U-Net+ Attention model.

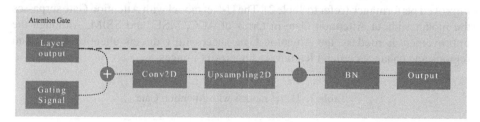

Fig. 4. Attention Gate

are further divided into a training set consisting of 64 groups and a test set consisting of 16 groups.

5 Experiments and Results

To comprehensively evaluate the performance of the U-Net model and the U-Net model with Attention Gate, this study employs ACC (Accuracy), MSE (Mean Squared Error), and SSIM (Structural Similarity Index Measure) as evaluation metrics for the experimental results.

Regarding the U-Net model without Attention Gate (refer to Table 1), it is observed that increasing the model's depth enhances its performance. The U-Net model with a depth of 6 exhibits the highest accuracy (0.9933) and the lowest mean squared error (0.0371) among the results. Nevertheless, the differences in accuracy and mean squared error between the depth-5 and depth-6 models are relatively marginal. As the first architecture generates images without manual markings, it encounters challenges in accurately producing enhanced images for some data, resulting in an SSIM score of only 55%. However, the SSIM score remains consistent across all depths, signifying its ability to maintain the structural integrity of the fingerprint image.

Table 1. U-Net models without Attention Gate.

U-Net			
Depth	4	5	6
ACC	0.9746	0.9863	0.9933
MSE	0.4156	0.0414	0.0371
SSIM	0.5518	0.5521	0.5524

The U-Net model is enhanced by incorporating the Attention Gate, resulting in improved performance (refer to Table 2). The U-Net model with Attention Gate surpasses the model without Attention Gate in terms of ACC, MSE, and SSIM. Although the improvement is modest, the Attention Gate successfully directs attention to relevant regions and enhances crucial features in fingerprint images.

Table 2. U-Net models with Attention Gate.

U-Net + Attention Gate			
Depth	4	5	6
ACC	0.9912	0.9929	0.9939
MSE	0.3850	0.0379	0.0363
SSIM	0.5494	0.5454	0.5496

In the experimental results (as depicted in Fig. 5), the U-Net model with Attention Gate and a depth of 6 demonstrates superior performance in enhancing fingerprint images. Notably, the image generated by the U-Net model with Attention Gate and a depth of 6 exhibits clearer details compared to other depths (the red area is the magnified area). The experimental findings provide evidence that incorporating the Attention Gate into the U-Net model effectively enhances the quality and clarity of fingerprint images.

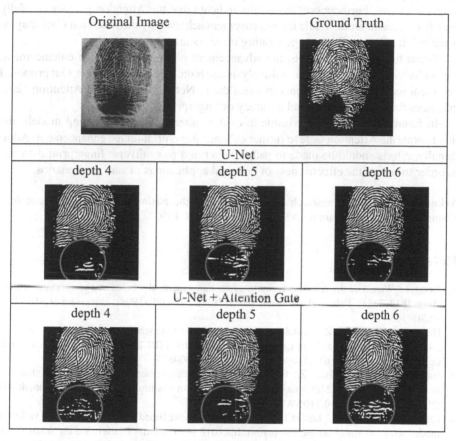

Fig. 5. Different model and depth of fingerprint result.

6 Conclusions

The experimental results prove that the U-Net model with different depths has a good effect on fingerprint image enhancement. Increasing the depth of the U-Net architecture improves performance, as evidenced by higher accuracy and lower mean square error. Increasing the depth of U-Net can improve performance, this is evidenced by higher accuracy and lower mean square error. Deeper architectures are better suited to capturing intricate details and improving the quality and clarity of fingerprint images. In addition, the consistency of the SSIM at different depths, it shows that the model preserves the structural integrity of the fingerprint image during the enhancement process.

The addition of the Attention Gate to the U-Net model offers further enhancements. Models incorporating Attention gate outperform those without Attention gate, as demonstrated by the Accuracy, MSE, and SSIM. The results highlight the effectiveness of the Attention gate in focusing on relevant regions and enhancing important features in fingerprint images. Furthermore, the results indicate that the Attention gate successfully prioritizes crucial areas. While the improvement achieved by the Attention Gate may be marginal, it still contributes to generating more accurate images.

These findings contribute to the advancement of fingerprint image enhancement, particularly in the realms of criminal analysis and biometric authentication. Our proposed deep learning method, which incorporates the U-Net architecture and Attention Gate, enhances the quality, clarity, and accuracy of fingerprint images.

In future research, it is advisable to explore alternative deep learning models and incorporate the Attention Gate to further enhance fingerprint image enhancement. Additionally, efforts should be made to gather larger and more diverse fingerprint datasets, in order to ensure the effectiveness of practical applications in various scenarios.

Acknowledgement. The research was supported by the National Science and Technology Council, Taiwan, under grant no. NSTC 111–2622-E-224–019.

References

1. Hong, L., Wan, Y., Jain, A.: Fingerprint image enhancement: algorithm and performance evaluation. IEEE Trans. Pattern Anal. Mach. Intell. **20**(8), 777–789 (1998). https://doi.org/10.1109/34.709565
2. He, Z., Liu, E., Xiang, Z.: Partial fingerprint verification via spatial transformer networks. In: 2020 IEEE International Joint Conference on Biometrics (IJCB), Houston, TX, USA, 2020, pp. 1–10. https://doi.org/10.1109/IJCB48548.2020.9304877
3. Han, K., Wang, Z., Chen, Z.: Fingerprint image enhancement method based on adaptive median filter. In: 2018 24th Asia-Pacific Conference on Communications (APCC), pp. 40–44 (2018).https://doi.org/10.1109/APCC.2018.8633498
4. Liban, A., Hilles, S.M.: Latent fingerprint enhancement based on directional total variation model with lost minutiae reconstruction. In: 2018 International Conference on Smart Computing and Electronic Enterprise (ICSCEE), pp. 1–5 (2018). https://doi.org/10.1109/ICSCEE.2018.8538417
5. Szymkowski, M., Saeed, K.: A novel approach to fingerprint identification using method of sectorization. In: 2017 International Conference on Biometrics and Kansei Engineering (ICBAKE), Kyoto, Japan, pp. 55–59 (2017). https://doi.org/10.1109/ICBAKE.2017.8090613
6. Yusharyahya, K.A., Nugroho, A.S., Purnama, J., Galsinium, M.: A comparison of fingerprint enhancement algorithms for poor quality fingerprint images. In: 2014 International Conference of Advanced Informatics: Concept, Theory and Application (ICAICTA), Bandung, Indonesia, pp. 342–347 (2014).https://doi.org/10.1109/ICAICTA.2014.7005966
7. Patel, M.B., Patel, R.B., Parikh, S.M., Patel, A.R.: An improved O'Gorman filter for fingerprint image enhancement. In: 2017 International Conference on Energy, Communication, Data Analytics and Soft Computing (ICECDS), Chennai, India, pp. 200–209 (2017). https://doi.org/10.1109/ICECDS.2017.8389784
8. Maio, D., Maltoni, D., Cappelli, R., Wayman, J.L., Jain, A.K.: FVC2000: fingerprint verification competition. IEEE Trans. Pattern Anal. Mach. Intell. **24**(3), 402–412 (2002). https://doi.org/10.1109/34.990140

Applications and Implication of Generative AI in Non-STEM Disciplines in Higher Education

Tao Wu[1] 🆔 and Shu hua Zhang[2]([✉])

[1] Zhujiang College of South China Agricultural University, Guangzhou 510900, China
wutao@scauzj.edu.cn
[2] Tianjin University of Finance and Economics, Tianjin 300222, China
szhang@tjufe.edu.cn

Abstract. There has been considerable research on the use of generative artificial intelligence techniques to support teaching and learning in science, technology, engineering, and mathematics (STEM) subjects in higher education. However, few studies have explored the role of such technologies in non-STEM subjects in higher education. This paper reviews the relevant literature on the application of generative AI in higher education and proposes the application and implications of using generative AI tools to support student and instructors work in non-STEM higher education disciplines. An assessment of the role of AI in complex student tasks in non-STEM subjects is provided. Several considerations for the effective use of generative AI in non STEM higher education are suggested. Faculty and students should focus on: 1) ensuring that ethical and moral implications are addressed; 2) using AI to augment rather than replace human intelligence; 3) using AI as an instructional tool rather than a fully automated system; 4) using AI to improve academic assessment and self-assessment methods; 5) critically reviewing the results of generative AI systems.

Keywords: Large language models · Natural language processing · academic assessment · critical thinking · Source bias · non-STEM disciplines

1 Introduction

With the development of Artificial Intelligence (AI), the potential application of AI technology has brought promising innovations in various fields, such as the medical and agricultural industries, which require highly specialized technology [1, 2]. As a result, the demand for AI-trained talent will continue to grow across multiple industries. Teaching the use of AI tools as a means of solving problems is on the higher education agenda. Although the application of AI in the education sector is not a novel idea, relevant exploration research has already begun decades ago [3]. In the early days, the application of artificial intelligence in assisting education was expected to be achieved by two modules just as follows; 1. An expert system that analyzes and processes the input data and outputs an ideal result model [4]. Also, the system should be able to receive and interpret the comments and explanations of the students and be able to make its own

© The Author(s), under exclusive license to Springer Nature Singapore Pte Ltd. 2024
F. Zhao and D. Miao (Eds.): AIGC 2023, CCIS 1946, pp. 341–349, 2024.
https://doi.org/10.1007/978-981-99-7587-7_29

comments and explanations as appropriate [5], 2. In early 1982, Sleeman and Brown [6] reviewed the state of the art in instructional technology and first defined the Intelligent Tutor System (ITS). ITS is described as a computer program that uses artificial intelligence techniques to represent knowledge and interact with students. Auto Tutor was proposed as a digital learning environment that tutors students through conversations in natural language, which is also considered to be a learning mode of the scene dialog[7]. However, the development and capabilities of AI technology are still a bottleneck to the depth and breadth of the application of artificial intelligence in education (AIEd) [8]. In the UNSECO 2020 report[1] on the Digital Transformation of Education project, it was proposed that artificial intelligence will play a pivotal role in personalized learning. Subsequently, a promising revolution in artificial intelligence that will impact the next few decades has quietly emerged since Large Language Model succeeded. On November 30, 2022, launched OpenAI ChatGPT, a generative AI technology that has taken universities, colleges, and K-12 schools by storm. It was believed to have the potential to revolutionize much of our higher education teaching and learning. ChatGPT is regarded as an impressive, easy-to-use, publicly accessible system demonstrating the power of large language model. More recently, text processing, image, audio, video, and other output have become available with similar generative models [9–11]. The educational revolution wrought by generative AI will have a profound impact on both teachers and students in the classroom and on campus [12–14]. It is worth noting that the skeptics claimed that the digital computation and logical reasoning of ChatGPT based on the LLM model are obviously not ideal, and these models will pose challenges in solving practical problems[2]. In an empirical study of Currie and Barry [15], they found: "While ChatGPT is considered a risk to academic integrity, its ability to solve complex problems is far inferior to that of humans. Thus, higher-order taxonomies may limit its usefulness as a cheating tool." With the introduction of GPT4.0, the shortcomings of Generative AI's ability in numerical computation and logical reasoning have been greatly improved [16]. Therefore, according to the technical characteristics and blemish of the LLM model itself, it is necessary to propose an appropriate framework for the scope of application and warning of generative AI in the education industry especially in non-STEM disciplines. Based on the technical characteristics of generative AI and some of its inherent flaws, this paper proposes the scope and key points of application for non-STEM disciplines. In summary, Sect. 5 proposes the following six recommendations.

2 Large Language Model and Generative AI

A large language model (LLM) is an artificial intelligence model trained on large amounts of text data. This model is capable of understanding language, generating text, and performing tasks that require the understanding of natural language. The function of this

---

[1] AI in Education: Change at the Speed of Learning, UNESCO Institute for Information Technologies in Education, https://iite.unesco.org/publications/ai-in-education-change-at-the-speed-of-learning/

[2] Why the 'intelligence' of ChatGPT does not know how to solve this problem? Vincenti Botti, https://valgrai.eu/2023/04/12/why-the-intelligence-of-chatgpt-does-not-know-how-to-solve-this-problem/

model is to predict the next word in a sentence or phrase based on the context provided by previous words or character. LLM is characterized by their large scale, containing billions of parameters, which help them learn complex patterns in linguistic data. These models are often based on deep learning architectures such as Pre-trained Transformers, which helps them achieve impressive performance on various NLP tasks [17]. To put it simply, the system uses large-scale data training to find the best combination of a character (letter) and a word. Initially, these are sorted based on probability and frequency, and then processed into an understandable language [18].

Generative AI is a branch of artificial intelligence that focuses on the creation of new data instances based on patterns learned from existing data. Generative AI can be considered as a description of the final product of various algorithms, models, and datasets are its critical components. Currently, the most well-known AI chat and creative tools represented by generative AI include DALL-E (OpenAI), Stable Diffusion (Stability AI), cohere (GitHub), Claude (Anthropic), GPT series (OpenAI). Generative AI output is a carefully calibrated combination of data used to train the algorithm, and sometimes models can appear "creative" in generating output due to the sheer volume of data used to train these algorithms (for example, GPT-3 was trained on 45TB of text data). What's more, these models often have a stochastic element, meaning they can produce multiple outputs based on a single input request, which is why generative AI has been criticized for erratic output. The technical characteristics of the above algorithm design have resulted in the three most distinctive features of generative AI. 1). In a black-box mode based on outcome-oriented rules. 2). Traces of cognitive tendencies under manual intervention. 3). The source and amount of underlying data affect the results.

3   Generative Artificial Intelligence in Education

The application prospects of educational technology, especially artificial intelligence, in higher education are drawing more and more attention. Before the advent of generative AI, Crompton and Burke found that in a systematic review of AI in higher education from 2016 to 2022, 72% of studies focused primarily on students, 17% on instructors, and 11% on school administration. After coding 138 qualified full papers, it was found that the 5 most common keywords were ": 1) assessment/evaluation, 2) prediction, 3) AI assistant, 4) intelligent tutoring system (ITS), and (5) managing student learning [19]. While ChatGPT emerged as a tool for creative invention, generative AI is disrupting traditional educational technology pathways, strengthening, and deepening the functional applications described above, and opening up new prospects for new functional applications, such as assisting people with disabilities to learn. In general, however, generative AI in higher education focuses on two aspects: scientific writing and the application of learning practices in various disciplines [20–22].

3.1   Generative AI Aid in Scientific Writing

Current research suggests that Generative AI can excel in these two aspects of scientific writing: first, to help researchers in universities with scientific writing, such as grant applications; second, to help students and researchers write reports or posters, optimizing and simplifying their expression to make it concise and easy to understand [23–25]. However, critical voices are still loud. Arguments and related references against over-reliance on GPT as an ideal writing partner, as shown in Table 1.

Table 1. Various critical opinions on generative AI in scientific writing

Opposing Viewpoint	Description	Reference
Lack of accuracy and reliability	Generative AI may not always produce accurate or reliable information, potentially leading to bias, misuse, and misinformation	[26–28]
Limited knowledge	Generative AI may lack the level of knowledge required to write a paper on a particular topic. When you type in a prompt, it may use very short, filler words to write your response, rather than actually using specific terms	[28, 29]
Inability to provide accurate references or evidence	The references and sources provided are questionable, because generative AI such as GPT does not have the retrieval function	[28]
Ethical concerns	Concerns about unfair competition such as these encourage some forms of opportunism, such as disguised plagiarism	[28, 30, 31]
Loss of critical and creative thinking	Generative AI does not seem to help develop critical and creative writing skills due to lacking transparency, and creativity	[32–34]

3.2 The Potential Benefits of Generative Artificial Intelligence in Various of Disciplines in Higher Education

This section discusses three aspects of the benefits that the application of generative artificial intelligence in higher education brings to students and teachers. The first is academic support that improves students' self-regulation and concentration through generative artificial intelligence connected to other devices. The second is to help students use tools to create knowledge maps and conceptual summaries. The third is to help students practice the virtual reality mode so they can master the ability to apply what they have learned. While embracing these innovative applications, the potential risks of generative artificial intelligence cannot be ignored. For this reason, the Sect. 4. Offers suggestions for vigilance.

3.3 Large Language Model's Application on Finance and Economics Disciplines

Demir and Zaremba proposed that GPT based on the LLM model be applied to three major aspects of finance [35]. a) The LLM model can help users in financial document classification, sentiment analysis, and named entity recognition in financial documents; b) Another area of LLM application is natural language generation of financial reports and summaries; c) Text-based financial analysis, such as financial sentiment analysis, news analysis, and social media analysis. GPT have been used to extract insights from unstructured financial data such as financial news articles, social media posts, and investor communications to detect market sentiment and predict trends. A valuable study found that those who can really benefit the most from the tools of generative artificial intelligence dominated by LLM models are those with weak skill abilities [36]. In higher education, instructors can inspire students to use LLM-based functional tools to summarize, classify, and abbreviate financial and economic documents, and generative AI tools can play a more reliable role in these areas. Generative AI also has an extended function as a tutoring tool for courses and assignments, providing users with programming explanations, examples, and guidance to help them understand complex concepts and techniques, find related resources, and diagnose and solve technical problems [37]. However, its stability and reliability are still in question. However, with future development and improvement, more comprehensive programming skills guidance will be provided. In this way, people who do not understand programming in non-STEM fields can also use this tool to program data analysis and data mining.

4 Recommendation

Although the new technology of generative artificial intelligence can bring an exciting and inspiring innovation to teaching and learning in higher education, what teachers and students need to realize is that its underlying application is still a large language model, just a probability-based expression generation. Therefore, just because a generative AI tool can answer questions doesn't mean it really understands. Therefore, in response to the skeptics' views in the third part, the following five application concerns are proposed for the use of generative AI in non-STEM disciplines to benefit all sides.

4.1 Ethics and Morality Challenge

Academic ethics risks include falsified references in ChatGPT research and potential intellectual property infringements in lesson plans or homework answers. The reliance on papers for evaluation, despite the shortcomings of anti-fraud software, exacerbates these risks. To mitigate this, assessment modes should be diversified beyond solely graded work. We suggest integrating peer review in collaborative projects, incorporating student learning trajectories, and considering tool usage into evaluation. This would include non-textual content, the study plan design, reading volume, generative AI use, and understanding the limits of AI tool use in academic assessment.

4.2 The Scaffolding Function of Generative AI in Non-STEM Disciplines

The meta-analysis showed computer-based scaffolding in STEM disciplines boosted achievement [38, 39]. Similarly, technology that enhances memory and documents speeds up reading. In non-STEM disciplines, however, generative AI should only support systemic thinking, not replace it. The shortcomings of generative AI tools: reliance on outdated data, lack of creative exploration, and generalized, non-specialized problem solving. Students and teachers must understand its limitations as a tool, not a panacea, given its technical limitations such as pre-trained databases based on prior 2021, sentence probabilities, and lack of trained human reasoning skills.

4.3 Generative AI as Assistive Tool not Autonomous Force

First, generative AI contains inherent biases from its LLM foundation [40]. Second, it lacks true understanding, producing superficial content without logical connections that challenge academic rigor. Third, over-reliance on these tools inhibits creativity and critical thinking in users. Given these risks, schools must establish policies that disclose limitations and define any work generated entirely by AI as academic misconduct. However, universities should continue to advance work in this transformative technology within an ethical framework. In summary, while generative AI holds promise, its biases, lack of depth, and inhibition of original thought pose risks. Schools must guide appropriate use while supporting continued innovation.

4.4 Generating Better Academic Assessment with AI

Competency-based learning in STEM allows flexible academic assessment [41]. With generative AI emerging, non-STEM fields can also reform evaluation have based on competency models. However, studies testing ChatGPT in accounting, management, and sociology reveal unsatisfactory answer accuracy [42]. Instead, generative AI can provide formative assessments through conversational agents connected to databases via speech, text, or graphics. This assists teachers in programming personalized, project-based learning. For students, interacting via voice, text, and images motivates self-directed learning. Instructors can thus measure effort for formative grades [43]. Although requiring some AI mastery, this process-focused assessment relieves teaching burdens while encouraging student diligence and enabling comprehensive, equitable evaluation. Overall, generative AI promises to expand competency-based, personalized assessment but accuracy limitations mean it should support, not drive, non-STEM assessment.

4.5 Exercise Prudence with Generative AI's Imaginative Outputs

The concern of potential moral and ethical hazards of LLM is extend to sociology, such as insults to different types of people or inaccurate language that could be harmful to minorities and due to differences in education and ability to properly understand and use, while amplifying socioeconomic class differences through the LLM [44]. This concern about discrimination is not unfounded, as the current training dataset is still in English. Therefore, the views and opinions of a small number of non-English languages may not necessarily be taken into account. In addition, LLM cannot respond to language that lacks detailed context and scenes [45], such as asking about the weather yesterday and the day before yesterday, or speaking in a teasing manner. In academic research, generative AI cannot find the latest research directions and propose the most innovative ideas to researchers, because it takes time to train the database, and what can be seen and found by generative AI is not the latest, and because the algorithm is judged by probability. In addition, there is no guarantee that the content produced by LLM will have a strong bias [40], and it is impossible to take a total neutral position on one side's opinions on debates. Therefore, for academic research in non-STEM subjects, especially sociology, we suggest that Industry Association and government agencies provide some kind of specialized database platform as a training purpose to ensure neutrality and reduce prejudice.

5 Conclusion

This paper is a contribution to the literature expansion of LLM-based general artificial intelligence in non-STEM disciplines in higher education. Due to differences in interests and undergraduate education, instructors and students of non-STEM disciplines may have weak skills in the use of programming or mathematical, physical, and chemical tools. To this end, non-STEM disciplines can take advantage of several implications of the accessibility features of generative AI. For students: 1) Summarize and classify data into visual images or simple text; 2) For complex expressions, abbreviate, improve the efficiency of interaction and the clarity and convenience of team interaction in collaborative learning;3) Bring programming assistance to novices who do not know programming, help them understand and perform data mining and data analysis tasks to clearly understand the results; 4) Formulate basic study plans and study arrangements for students to improve their learning ability and self-regulation; 5)Fragmented knowledge is rearranged and summarized by generative AI tools, and a conceptual roadmap is set to improve learning efficiency. For instructors: 1) Assist teachers in assessing the learning process and mastering problem-solving tools in designing instruction. 2) Use generative AI as an example to help students generate critical thinking. For example, instructors inform students about the inherent flaws and usage frameworks of generative AI, encourage students to find the flaws of generative AI tools in explaining and defining technical terms and finding solutions to problems, and find inappropriate generative AI literature and fake information and explain its flaws coming.

In terms of public interest, the establishment of a global international research organization for generative artificial intelligence, equipped with the enormous computing power required to run the models, and to provide specific and clear guidelines on the scope and operation of generative artificial intelligence, and to provide a joint effort to further develop the technology, will be a good choice to invest in preventing the erosion of excessive commercialization.

Acknowledgement. This project was supported in part by Guangdong Higher Education Association (23GYB118).

References

1. Khan, A.A., et al.: Internet of Things (IoT) assisted context aware fertilizer recommendation. IEEE Access **10**, 129505–129519 (2022)
2. Kujur, A., et al.: Data complexity based evaluation of the model dependence of Brain MRI images for classification of Brain Tumor and Alzheimer's disease. IEEE Access **10**, 112117–112133 (2022)
3. Welham, D.: AI in training (1980–2000): foundation for the future or misplaced optimism? Br. J. Educ. Technol. **39**, 287–296 (2008)
4. Davis, R.: Interactive transfer of expertise: acquisition of new inference rules. Artif. Intell. **12**(2), 121–157 (1979)
5. Ligęza, A.: Expert systems approach to decision support. Eur. J. Oper. Res. **37**(1), 100–110 (1988)
6. Sleeman, D., Brown, J.S.: Intelligent Tutoring Systems, 345 pp. Academic Press, London (1982)
7. Graesser, A., et al.: AutoTutor: a tutor with dialogue in natural language. Behav. Res. Methods **36**, 180–192 (2004)
8. Luckin, R., Holmes, W.: Intelligence Unleashed: An argument for AI in Education (2016)
9. Liu, N., et al.: Unsupervised Compositional Concepts Discovery with Text-to-Image Generative Models (2023)
10. Hinz, T., Heinrich, S., Wermter, S.: Semantic object accuracy for generative text-to-image synthesis. IEEE Trans. Pattern Anal. Mach. Intell. **44**, 1552–1565 (2022)
11. Tang, T., et al.: Learning to Imagine: Visually-Augmented Natural Language Generation (2023)
12. Gimpel, H., et al.: Unlocking the Power of Generative AI Models and Systems such as GPT-4 and ChatGPT for Higher Education A Guide for Students and Lecturers Unlocking the Power of Generative AI Models and Systems such as GPT-4 and ChatGPT for Higher Education A Guide for Students and Lecturers (2023)
13. Kasneci, E., et al.: ChatGPT for Good? On Opportunities and Challenges of Large Language Models for Education. Learn. Individ. Diff. **103**, 102274 (2023)
14. Olga, A., et al.: Generative AI: Implications and Applications for Education (2023)
15. Currie, G., Barry, K.: ChatGPT in nuclear medicine education. J. Nucl. Med. Technol. (2023). https://doi.org/10.2967/jnmt.123.265844
16. Orrù, G., et al.: Human-like problem-solving abilities in large language models using ChatGPT **6**, 1 (2023)
17. Cambria, E., White, B.: Jumping NLP curves: a review of natural language processing research [review article]. Comput. Intell. Mag. **9**, 48–57 (2014)
18. Li, J., et al.: Visualizing and understanding neural models in NLP. arXiv preprint arXiv:1506.01066 (2015)
19. Crompton, H., Burke, D.: Artificial intelligence in higher education: the state of the field. Int. J. Educ. Technol. High. Educ. **20**(1), 22 (2023)
20. Baidoo-Anu, D., Owusu Ansah, L.: Education in the era of generative artificial intelligence (AI): Understanding the potential benefits of ChatGPT in promoting teaching and learning (2023). SSRN 4337484
21. Bär, K., Hansen, Z., Khalid, W.: Considering Industry 4.0 aspects in the supply chain for an SME. Prod. Eng. **12** (2018)
22. Ifenthaler, D., Schumacher, C.: Reciprocal Issues of Artificial and Human Intelligence in Education. Taylor & Francis. p. 1–6 (2023)
23. Salvagno, M., et al.: Can artificial intelligence help for scientific writing? Crit. Care (London, England) **27**, 75 (2023)

24. Aljanabi, M., et al.: ChatGPT: open possibilities. Iraqi J. Comput. Sci. Math. **4**(1), 62–64 (2023)
25. Chen, T.-J.: ChatGPT and other artificial intelligence applications speed up scientific writing. J. Chin. Med. Assoc.: JCMA (2023, Publish Ahead of Print)
26. Dwivedi, Y., et al.: "So what if ChatGPT wrote it?" Multidisciplinary perspectives on opportunities, challenges and implications of generative conversational AI for research, practice and policy. Int. J. Inf. Manage. **71** (2023)
27. Dwivedi, Y., et al.: So what if ChatGPT wrote it? Multidisciplinary perspectives on opportunities, challenges and implications of generative conversational AI for research, practice and policy. **71**, 1–18 (2023)
28. Katar, O., et al.: Evaluation of GPT-3 AI language model in research paper writing (2022)
29. Negrini, D., Lippi, G.: Generative Artificial Intelligence in (laboratory) medicine: friend or foe? Biochimica Clinica (2023)
30. Sok, S., Heng, K.: ChatGPT for education and research: a review of benefits and risks. SSRN Electron. J. (2023)
31. Dehouche, N.: Plagiarism in the age of massive Generative Pre-trained Transformers (GPT-3): 'The best time to act was yesterday. The next best time is now. Ethics Sci. Environ. Polit. **21**, 1721 (2021)
32. Ahmad, N., Murugesan, S., Kshetri, N.: Generative artificial intelligence and the education sector. Computer **56**(6), 72–76 (2023)
33. Marr, B.: The Top 10 Limitations Of ChatGPT, March 2023. https://www.forbes.com/sites/bernardmarr/2023/03/03/the-top-10-limitations-of-chatgpt/?sh=454cab428f35
34. Chan, C., Tsi, L.: The AI Revolution in Education: Will AI Replace or Assist Teachers in Higher Education? (2023)
35. Zaremba, A., Demir, E.: ChatGPT: Unlocking the future of NLP in finance (2023). SSRN 4323643
36. Noy, S., Zhang, W.: Experimental evidence on the productivity effects of generative artificial intelligence (2023). SSRN 4375283
37. Biswas, S.: Role of ChatGPT in Computer Programming.: ChatGPT in Computer Programming. Mesopotamian J. Comput. Sci. **2023**, 8–16
38. Belland, B.R., et al.: A pilot meta-analysis of computer-based scaffolding in STEM education. J. Educ. Technol. Soc. **18**(1), 183–197 (2015)
39. Kim, N.J., Belland, B., Walker, A.: Effectiveness of computer-based scaffolding in the context of problem-based learning for stem education: bayesian meta-analysis. Educ. Psychol. Rev. **30** (2018)
40. Ntoutsi, E., et al.: Bias in data-driven artificial intelligence systems—an introductory survey. Wiley Interdisip. Rev.: Data Min. Knowl. Discov. **10**(3), e1356 (2020)
41. Henri, M., Johnson, M., Nepal, B.: A review of competency-based learning: tools, assessments, and recommendations: a review of competency-based learning. J. Eng. Educ. **106**, 607–638 (2017)
42. Tenakwah, E., et al.: Generative AI and Higher Education Assessments: A Competency-Based Analysis (2023)
43. Sarsa, S., et al.: Automatic Generation of Programming Exercises and Code Explanations Using Large Language Model, pp. 27–43 s (2022)
44. Weidinger, L., et al.: Ethical and social risks of harm from language models. arXiv preprint arXiv:2112.04359 (2021)
45. Mitchell, M., Krakauer, D.: The debate over understanding in AI's large language models. Proc. Natl. Acad. Sci. U.S.A. **120**, e2215907120 (2023)

Industry 4.0 Technology-Supported Framework for Sustainable Supply Chain Management in the Textile Industry

Ding Chen[1] ⓘ, Umar Muhammad Gummi[1,2] ⓘ, Jia Lei[3] ⓘ, and Heng Gao[4(✉)] ⓘ

[1] School of Economics and Management, Xi'an Shiyou University, Xi'an, China
[2] Department of Economics, Sokoto State University, Sokoto, Nigeria
[3] Xinjiang Institute of Technology, Aksu, China
[4] Dalian Maritime University, Dalian, China
13804076717@163.com

Abstract. The ever-growing degradation of the environment has led to the building and implementation of various strategies to achieve sustainable development. The textile industry is not left out in this struggle to reach sustainability through sustainable supply chain management. This paper proposes a sustainable closed-loop supply chain for the textile industry, integrating some industry 4.0 Technologies to facilitate data-driven decision-making within the supply chain and ensure sustainability. The study reveals that block-chain application in the closed-loop supply chain will support tracking and enable supply chain traceability and provide secure means of recording information about the actors and their actions in the supply chain. The paper further recommends the use of Big Data technologies throughout the life cycle stages for data retrieval, processing, storage, mining and application in the textile industry. Also, Industry 4.0 technologies will facilitate sustainable supply chain in the textile industry as it helps in data collection, analysis and information transmission.

Keywords: Industry 4.0 technology · Sustainable Supply Chain · Textile Industry

1 Introduction

In the wake of the global call for environmental sustainability due to the mounting danger of climate change due to carbon emissions, the governments of different countries have been implementing various carbon-reduction strategies. These strategies have ranged from macroeconomic policies to financial attempts. More recently, the transformation of the supply chain has come to mind, as it is believed that most activities that involve the manufacturing and distribution of goods to their final destinations, as well as their usage and disposal to adversely impact the environment, mostly through the use of fossil fuels that are the major sources of carbon emissions. Thus, Preuss [1] argues that the global sustainability agenda cannot be fully actualized without considering the supply chain. This is further enhanced by the assertion of Fahimnia et al. [2] and Tseng et al. [3] that

F. Zhao and D. Miao (Eds.): AIGC 2023, CCIS 1946, pp. 350–361, 2024.
https://doi.org/10.1007/978-981-99-7587-7_30

admitting sustainable practices in the supply chain is a healthy way to cause a revolution in the effectiveness of firms with respect to environmental and social performance.

Sustainable Supply Chain Management (SSCM) indicates a management practice incorporates environmental consciousness, economic awareness, and social performance into the supply chain system [4]. The implication of this is that the idea of sustainability must not be excluded in the central functions of the supply chain. Therefore, to achieve sustainability in the supply chain, some guidelines have been identified. For instance, studies such as Brandenburg et al. [5] and Tsai et al. [6] suggest that it is essential for firms in the supply chain to declare their environmental and social performances, alongside their profits, in their financial reports. This comes with the belief that by additionally declaring their environmental and social performances, the degree of their sustainability practices can be gauged. Also, Panigrahi et al. [4] points that the attainment of sustainability requires that all the firms cooperate to uphold the required standards, which can be ensured by enforcing specific laws.

In light of the highlighted challenges, especially the associated complexity, advanced technologies are required, particularly in data collection, analysis, and information transmission. Thus, analysts have begun to look in the direction of the technologies that followed the fourth industrial revolution. They include artificial intelligence, the Internet of Things (IoT), and advanced robotics. These technologies have proven their relevance and worth by collecting data on a large scale, thereby circumventing the complexity of the modern supply chain. While the high volume of data is called big data, the analytic mechanism is called bid data analytics (BDA). Thus, BDA has been known for its several advantages, including the generation of solutions and new insights from complex problems in the supply chain, and the enhancement of effectiveness and efficiency in the supply chain management.

Despite the rising attention on SSCM and the overwhelming influence of the fourth industrial revolution technologies, the literature is still limited in terms of how these technologies can aid sustainability practices in supply chain management. Therefore, this study focuses on the textile industry, relying on this gap. The textile industry accounts for a unique space in the entire manufacturing industry. Amutha [7] disclosed that it is one of the industries with the marks on the environment, as it engages the use of dangerous chemical substances, impairs water quality, and arbitrarily disposes of used clothing without much recycling. All these sum up to significant carbon and water footprints. Unfortunately, being long and complicated, implementing sustainable practices along its entire chain is taxing, and few studies have delved into this dimension. This study thus covers this empirical gap by examining the application of Industry 4.0 technologies to the SSCM in the textile industry.

2 Textile Industry and Sustainability Issues

The supply chain covers both the downstream and upstream. The upstream segment in the textile industry consists of all activities related to the manufacturing and processing of textile products, while the downstream segment relates to all distribution activities through which the textile products reach the final consumer.

The typical activities in the textile supply chain include fibres production, ginning processes, spinning and extrusions activities, processing weaving and knitting, and garment production. This is further expatiated in the simplified supply chain flow. The life cycle of a textile product from the raw material production to the waste level is emphasized. According to the study, the life cycle of a textile product begins from the production of natural fibres by farmers and growers or synthetic fibres by the chemicals industry, through to the yarn formation, which occurs through fibre preparation, texturizing, and spinning then to fabric formation through knitting, weaving, slashing, to the wet processing through dyeing, finishing and printing, to the fabrication stage through cutting and sewing, to consumption and finally to waste.

At different stages in the life cycle, other than fibres, several other resources are utilized, which are of significant environmental concern, such as dyes, additives, chemicals, water, and energy. Gardetti and Torress [8] highlighted several ways through which activities in the supply chain create environmental and social impacts. The study emphasized that through the use of pesticides during the obtaining of fibre as raw materials, health problems are created for workers, soils are degraded, and there is a loss of biodiversity. The study further revealed that the use of chemicals throughout the entire upstream activities could create carcinogenic and neurological effects, allergies, and sometimes cause fertility-related issues.

Excessive water consumption is one of many issues identified with activities in the textile industry's supply chain. Water serves as an important resource used at nearly every stage in the textile product's life cycle. It is used at every step in the wet-processing sequence to convey chemicals in and out of the fabric materials, cotton tagged 'a thirsty crop' requires a lot of water for processing. After Agriculture, the textile industry is considered the highest consumer and polluter of clean water. The excessive use of water in the environment raises concern for two reasons, clean water globally is of finite supply, and it's becoming increasingly scarce as increasing population and economic growth continue to increase demand for clean water relative to finite water supply. According to the study, the discharge of wastewater that has been charged with chemical additives in the processing stage in the supply chain activities also poses a danger to the water bodies if released untreated.

3 Proposed Framework

This study proposes an integration of the Industry 4.0 technologies to facilitate data-driven decision-making within the supply chain and ensure sustainability within the supply chain. The applications of IoT devices, blockchain, big data, etc., are used to address these issues relating to sustainability. According to Tortorella and Fettermann [10], digitization of processes and implementing practices that use smarter equipment is a crucial success factor that is beneficial to productivity, the efficiency of resources, and the reduction of waste. The development of the industry 4.0 has been employed to restructure the supply chains towards creating digital networks in improving sustainability [11].

The IoT's ability to capture real-time information from the different stages of the supply chain through the use of Radio Frequency Identification (RFID) tags and sensors-based data communication networks was highlighted by Pal and Yasar [12]. The need for big data-driven analysis to evaluate trust and behaviors in the supply chain was empha-sized by Tseng et al. [13]. Also, the capability of the block chain to secure information collected at each stage of the supply chain gives the need for the application [15]. We start by elucidating different types of data that can be generated at different stages of the supply change and then talk about the data collection and management process and the technologies involved.

3.1 Nature of Data Collected at Different Stages of the Supply Chain

In tracking these activities in the supply chain, certain information about what is being done is needed at each stage of the textile production. The nature of the information to be collected varies with each actor and the activities performed in the supply chain. The first actor is the raw materials or fibre manufacturer. Information relating to the origin of the raw materials, type of fibre, the location of the materials, the treatment, and the performance characteristics of fibre flexibility, durability, and elasticity should be col-lected at this first stage. At the process of fabric manufacture, the detailed information of the supplier of the fibres, the date it was provided, a description of the materials and the types of fibres, blend ratio, chemical composition, and the distribution records are pro-vided. At the garment manufacturers' stage, information ranging from grading, pattern making, nesting, marking, cutting, sewing, quality inspection, pressing, and packaging is collected. The clothing distribution and sales take place after the manufacturer has packaged the garments, and at this stage, the wholesalers and retailers are the actors involved. It is important to note that logistics activities are required at every stage of the supply chain in transporting the raw materials fibres producers to the point of trans-porting the garments to the final consumers. Information regarding the freight operators, drivers, importers, and transportation means are then collected at the logistics stage.

3.2 Data Collection, Storage and the Technologies Involved

As highlighted by Pal and Yasar [12] the IoT has the ability to capture real-time infor-mation in the supply chain. The internet of things is also defined by Gorcun [16] as the communication between machines, devices, and equipment in which their virtual personalities and capabilities are obtained due to advancements in technologies. Data relating to movement and activities in the supply chain are detected by sensors and then sent to the system for processing automatically. In collecting the information discussed in the previous section, several tools can be used to achieve this aim, for example, RFID. This process can continue at all the supply chain stages in collecting data. Through smart phones, information about consumption patterns and preferences are also obtained.

As depicted in Fig. 1, the cloud computing technologies and blockchain provide storage capability for the data collected across all stages. The blockchain also helps to manage and distribute the information and at the same time provides security to the data by ensuring that no one tampers with the data collected, which is possible through the immutability property of the blockchain [14]. At each actor's level in the supply chain, there are humans who manage the entire process (data collection, storage, processing and distribution). The humans collect information which may not have been collected through the automated process, manage the database and also function as agent within the supply chain. Mastos et al. [9] propose that through an agent ecosystem, an agent will be able to negotiate on behalf of their company, register their company assets in the blockchain and also track assets.

The aim of the data collection, storage, and security is for the end-users, which could be the governments, private individuals, companies, among others. The availability of these data makes it possible for the end-user to know the supply chain management of the textile industry and identify firms that are not contributing to environmental, social, and economic sustainability. The data could also help governments formulate policies to achieve sustainability goals.

The complexity and fragmentation involved in the supply chain of the textile industry and the aim to reduce the cost of operations for manufacturers and ensure environmental, economic, and social sustainability gave rise to transparency and accuracy in the supply chain process. Consumers are becoming more aware of the environmental and social impact of textile production, and the government is seeking active ways to solve the issues in the industry. As a result, the need for tracing and tracking the supply chain process of the industry increases, and advances in technologies have provided traceability solutions. The application of blockchain in this closed-loop supply chain depicted is designed to support tracking and enable supply chain traceability.

Blockchain is referred to as a Distributed Ledger Technology (DLT) through which transactions are recorded securely. It is a protocol that allows users to group data into blocks and link into a chain. Each block depends on the previous one and contains the transactions' identification, the transaction's actual content, and a pointer to the preceding block in the chain, hence the uniqueness of each block. The immutability property of the blockchain is made possible because no modification by any single element can be done to the block's content that was previously agreed by the nodes. Blockchain technology also helps to discover the status of the whole chain anytime, so it is easy to detect if any of the blocks in the chain is being manipulated. The nodes in the chain are likened to the actors involved in the upstream and downstream stages of the textile industry. Each node of the DLT comprises the basic information and supply-demand information, including historical information, demand information, and supply ability. The activities can be categorized into primary and secondary in the textile industry. The primary activities relate to sales and operation of products, collecting, receiving and storing and disseminating of products. On the other hand, the secondary activities relate to technological development, accounting, planning, and human resource management in the industry [17].

Also, in the closed-loop supply chain, information regarding the origin of the material, composition, length, and color of the fibre can be gotten from the fibre producer, while information relating to the thickness, strength, and humidity level is provided by the yarn spinning factory. Similarly, information about the production network and garment assembling is also provided, while information regarding the packaging and merchandising, transit, and distribution are given at the wholesaling and retailing stage. To ensure sustainability, all the collection of information must begin from the fibre producer who has information about the source of the raw materials to the last actor in the downstream stage. The blockchain can provide a secure means of recording information about the actors and their actions in the supply chain in an immutable distributed ledger which includes the time and locations of supply chain processes, operations, and transactions [14].

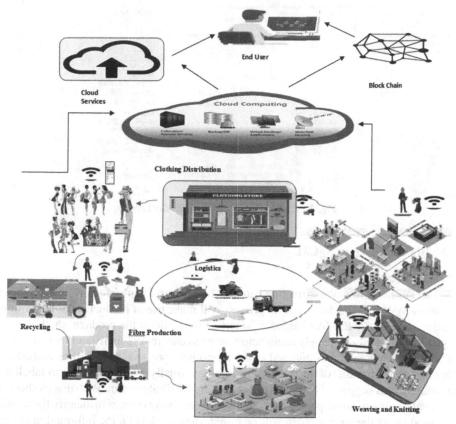

Fig. 1. Closed-Loop Supply Chain in Textile Industry with Integration of Big Data Technologies. Source: Author's compilation

As depicted in Fig. 2, product tracking is possible by blockchain technology. All in the closed-loop supply chain are provided with the tracking application that enables easy tracking and tracing of the product from the first to the last stage of the supply chain. The drivers and trucks involved in transportation, the type, and details about the exchanged products are logged as immutable transactions, which provide full visibility. Mastoset al. [9] identified three components of the tracking application, the first is the decentralized applications (dApp) [18] used for interacting within the blockchain ledger using a mobile interface or web-based interface. The second component is called the consortium permissioned blockchain based back end network and ledger system. Smart contracts are the third component that helps achieve required automation and permission handling on the blockchain system [19]. Registered users can only have access to the system and are only allowed to perform pre-defined actions [9].

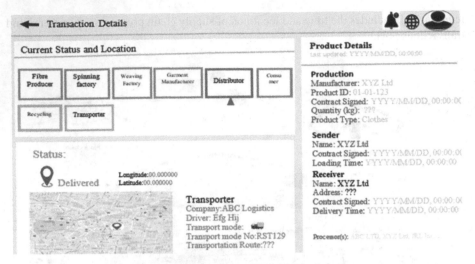

Fig. 2. Product Tracking UI with Blockchain

3.3 Big-Data-Driven SSCM

(1) Data Retrieval

Data retrieval involves extracting data that has already been stored in already chosen database systems. The data retrieval process will make use of product embedded information devices (PEID) like barcodes, smart sensors, albeit the products shall also be registered manually by supply chain actors so as to ensure originality and inclusiveness of information gathered as outlined in the blockchain process PEID will be embedded for example in the label of the textile, a technology usually referred to as smart labelling or smart tags as suggested by Feldhofer [20] using RFID algorithm. With this method of retrieval, the use of smart phones and other smart censoring devices to identify the textile at any stage of the supply chain will be easier, making data on the information of the textile available so as to be able to track it, reducing inefficiency in lifecycle operations [21] and ensuring environmental sustainability practices are upheld.

(2) Computational Framework.

In big data, a computational framework is a reference model that big data analysts can utilize in big data analytics [22]. This framework is studied, understood and used as a model for current analytics across different stages and levels in the supply chain and proper utilization of the computational framework will help to optimize results obtained from the analytics process [22]. A computational framework is required to ensure that data processing, despite how endless and relentless it has become, is context-dependent [23]. This statement means that the data is required to be processed according to the needs and requirements of the data users and not abstractly or generally, but according to the information required to make decisions or recommendations based on the evidence available which can be gotten through the information garnered concerning the textiles. There are many frameworks that can be utilized in the supply chain, some of which include Hadoop, Storm, MapReduce, Flink amongst others but due to the need for sustainability in the supply cycle process, there is a need for real-time data-driven decision making, so as to be able to reduce and negate if possible, the effect of wrong environmental practices in the textile supply chain cycle. Hadoop tends to process data in batches, requiring clustering, split into smaller processing jobs all in an indexed and arranged manner amongst many other processes before results can be gotten. This framework is greatly employed due to its simplicity as it is less complex and requires next to no technical knowledge in its usage. Despite the advantages of this framework, this paper recommends Apache Storm for the textile supply chain.

Storm is a real-time computational framework that is used to process high velocity of data in real time, boosting its ability to process over a million records per second per node using a cluster of modest size. The supply chain in the textile industry consists mainly of resourcing materials, processing or manufacturing the garments, transporting these products to warehouses, selling points or consumers and finally storing the textiles. The storm framework has both preventive and optimized objectives in supply chain activities for example: concerning manufacturing, it considers quality assurance, embeds preventative measures in manufacturing, reduces plant downtime and at the same time optimizes the supply chain. Storing and selling activities that Storm contributes to preventing shrinkage of the textile and eliminates stock-outs with its advanced algorithm at the same time optimizing pricing, offers and advertisements. Concerning transportation, Storm monitors the drivers and riders, supports vehicle health by recommending preventive maintenance while offering optimized routes and more economic pricing through optimization of the pricing system utilized.

The proper utilization of this framework will ensure that there is transparency and proper control systems at each stage of the supply chain. In the event of environmental unhealthy practices, such activities are then visible to supply chain actors and data users and decisions on corrections are evidence-based and not hypothetical, making the whole supply chain process and management both effective and efficient in delivering environmental sustainable practices and optimizing supply chain management processes concurrently.

(3) Data Processing.

The data processing covers the cleaning, integration, reduction, and transformation of textile information according to the preference of the data user. The data integration process requires the data processors to unify and amalgamate information that has come from diverse sources to give data users a unified view of the activities in the supply chain. After the integration of information obtained from data collectors into a uniform view, there is a need for this information to be processed through the data reduction process. Data reduction in the supply chain is the conversion of information of the textiles from raw form (numbers, letters, etc.) to a corrected, ordered and simplified form, usually a graphical representation. This requires the data processor to develop an easier to understand depiction of the result of the analysis carried out at various levels. At this stage, the data processor examines all information gathered, identifies incomplete, incorrect, inaccurate or irrelevant information and replaces, modifies or deletes the bad data found. After processing, a data cube is recommended, as seen in Wang [2] for proper and easier interpretation of the trends, patterns and results of the data processed before being transferred to the data warehouse for easy accessibility for data users and supply chain actors.

(4) Data Storage.

Data storage uses technologically advanced tools to record information gathered at the data mining stage [8]. Data storage methods to be utilized in this framework include structured, unstructured and semi-structured methods. For the structured storage method, the Distributed Database System (DDBS) is recommended as the location of the different actors in the supply chain is diverse, cutting across not just countries but also continents. This database system improves performance as information may be entered and processed from the different textile factories, stores, manufacturers, and users irrespective of their locations, all in real-time, making decision making faster and more effective. In storing unstructured data, NoSql is to be utilized as it is more flexible due to the type of information that is extracted. Information such as user and session data on textiles, messages between supply chain actors and log data; time-series data such textile information; and large objects such as video and images of the textile, which can't be structured into columns and roles are better stored using NoSql as proposed by Chauhan and Bansal [25] and Nayak, Poriya and Poojary [26].

Semi-structured data will utilize XML. XML is an abbreviation for extensible markup language, and it can be used to store semi-structured datasets that can't be structured into columns and rows but still contain vital information that when based on important factors (e.g., time, dates, etc.), are of very vital use to decision-makers and information users. Examples of semi-structured data that XML can store include mails sent concerning production quantity and HTML data like tracking the delivery location using maps, amongst other examples. The use of this data storage system helps information users to understand the timing of the movement of the textiles among actors in the life cycle due to the labelling and hierarchical structure of this model of storage as seen in (see Haw and Lee [27]). This will ensure that information stored is always available to supply chain actors and users, guiding decisions to be taken concerning environmental sustainability. (5) Data Mining.

From the framework developed above, the data mining process embeds various procedures of analysis of the available data, all utilizable in extracting different patterns, similarities and differences most suitable in analyzing the volume of data available. Data mining is concerned with the extraction and segregation of large datasets using machine learning, statistics, and various database systems or methods that are particularly applicable to the data extracted.

Data clustering is concerned with the grouping of information about textiles in such a way that they are either similar or different amongst themselves so that similar textiles are gathered at a point while the different textiles are gathered at another point. Another data mining procedure utilized is the anomaly detection procedure which examines textile types that can be considered as 'noise' [21] due to the deviation of the textile type from the established pattern of textile that is considered in the chain. Sequential patterning is then used to find possible statistical trends that are relevant to the textiles under review then the association rule is applied to consider unique, new, and unknown facts concerning the textiles. An example is seen in Wu et al. [28].

In the decision-making tree procedure of the data mining process, tests on attributes of the textiles are attached to the internal nodes of the decision tree structure, and the outcome of the test is denoted by each branch, as seen in Mengash [29]. The decision tree will be used to group textiles based on differences recognized after a test is carried out on the qualities of that textile. In prediction, historical data is considered and utilized in forecasting possible rules applicable to future textiles that may be in the supply chain, as seen in Tuffery [30]. The use of different data mining procedures ensures that supply chain actors can choose whichever is most economical, efficient, and effective in line with environmental sustainability goals in the supply chain management process and make decisions using evidence-based findings hence a more efficient process (Fig. 3).

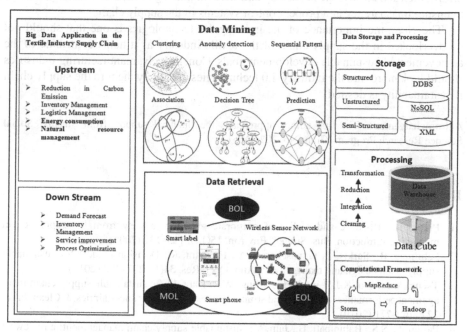

Fig. 3. Big-Data Driven Product Life Cycle Management

4 Conclusion

The SSCM has become a major concern for end consumers, industries, governments and policy makers. End consumers are becoming more aware of the impact of the textile industry supply chain on their environment and society and they also seek to know the origin and sustainability of the products they purchase. In the same vein, the government is actively seeking ways to solve the sustainability issues of the textile industry.

The application of the industry 4.0 technologies in the upstream and downstream segment of the textile industry show ways in which sustainability can be achieved through data collection, analysis, and information transmission. These technologies also show the ability to solve the complexity involved in the textile industry which makes it difficult to track the activities at each stage of the supply chain. The internet of things devices makes it possible to capture real-time information in the supply chain. Information like the origin of the raw materials, location of the materials, type of fibre, chemical composition, pattern making, cutting and logistics activities are being captured through the internet of things. The cloud services and blockchain prove the data storage processes of this information. The blockchain shows the ability to improve the traceability, security and transparency of the data stored at each stage of the textile supply chain. The information collected at each stage is converted to decision-support knowledge if there is proper utilization of big data. The big data application for sustainable supply chain management provides the real-time information and knowledge needed by end-users in making decisions. Similarly, it helps to ensure transparency and proper control systems in the supply chain.

Therefore, the importance of the industry 4.0 technologies is one that should not be overlooked in the supply chain of the textile industry as it helps contribute to the achievement of sustainable development goals. Conclusively, manufacturing industries should look more into the industry 4.0 technologies and apply them to the supply chain if sustainability is desired.

Acknowledgement. The first author appreciates the financial support from Shaanxi Social Science Research Program (Grant No. 2021D051).

References

1. Preuss, L.: Rhetoric and reality of corporate greening: a view from the supply chain management function. Bus. Strateg. Environ. **15**(14), 123–139 (2005)
2. Fahimnia, B., Sarkis, J., Gunasekaran, A., Farahani, R.: Decision models for sustainable supply chain design and management. Ann. Oper. Res. **34**(1), 277–278 (2017)
3. Tseng, M.L., Wu, K.J., Lim, M.K., Wong, W.P.: Data-driven sustainable supply chain management performance: a hierarchical structure assessment under uncertainties. J. Clean. Prod. **227**, 760–771 (2019)
4. Panigrahi, S.S., Bahinipati, B., Jain, V.: Sustainable supply chain management: a review of literature and implications for future research. Manage. Environ. Qual. Int. J. **30**, 1001–1049 (2019)
5. Brandenburg, M., Gruchmann, T., Oelze, N.: Sustainable supply chain management - a conceptual framework and future research perspectives. Sustainability **11**, 7239 (2019)
6. Tsai, E.M., Bui, T.D., Tseng, M.L., Ali, M.H., Lim, M.K., Chiu, A.S.: Sustainable supply chain management trends in world regions: a data-driven analysis. Resour. Conserv. Recycl. **167**, 105421 (2021)
7. Amutha, K.: Sustainable Fibres and Textiles. The Textile Institute Book Series, pp. 347–366 (2017)
8. Gardetti, M.A., Torress, A.L.: Sustainability in Fashion and Textiles: Values, Design, Production and Consumption. Greenleaf Publishing Limited, UK (2013)
9. Mastos, T.D., et al.: Introducing an application of an Industry 4.0 solution for circular supply chain management. J. Clean. Prod. **300**(15), 126886 (2021)
10. Tortorella, G.L., Fettermann, D.: Implementation of Industry 4.0 and lean production in Brazilian manufacturing companies. Int. J. Prod. Res. **56**(8), 2975–2987 (2018)
11. Atzori, L., Lera, A., Morabito, G.: The Internet of Things: a survey. Comput. Netw. **54**(15), 2787–2805 (2021)
12. Pal, K., Yasar, A.: Internet of Things and blockchain technology in apparel manufacturing supply chain data management. Procedia Comput. Sci. **170**, 450–457 (2020)

13. Tseng, M., Tan, R. R., Chiu, A. S. F., Chien, C., Kuo, C. T.: Circular economy meets industry 4.0: Can big data drive industrial symbiosis? Resour. Conserv. Recycl. **131**, 146–147 (2018)
14. Kshetri, N., Voas, J.: Blockchain in developing countries. IEEE IT Prof. **20**, 11–14 (2018)
15. Peng, G.C., Gala, C.J.: Cloud ERP: a new dilemma to modern organisations? J. Comput. Inf. Syst. **54**, 22–30 (2014)
16. Gorcun, O.F.: The rise of smart factories in the fourth industrial revolution and its impacts on the textile industry. Int. J. Mater. Mech. Manufact. **6**, 136–141 (2018)
17. Cai, W., Wang, Z., Ernst, J.B., Hong, Z., Feng, C., Leung, V.C.M.: Decentralized applications: the blockchain-empowered software system. IEEE **6**, 53019–53033 (2018)
18. Wang, Y., Han, J.H., Beynon-Davies, P.: Understanding blockchain technology for future supply chains: a systematic literature review and research agenda. Supply Chain Manage. Int. J. **24**, 62–84 (2019)
19. Koo, D., Hur, J., Yoon, H.: Secure and efficient data retrieval over encrypted data using attribute-based encryption in cloud storage. Comput. Electr. Eng. **39**, 34–46 (2013)
20. Feldhofer, M.: An authentication protocol in a security layer for RFID smart tags. IEEE **2**, 759–762 (2004)
21. Zhang, Y., Ren, S., Liu, Y., Si, S.: A big data analytics architecture for cleaner manufacturing and maintenance processes of complex products. J. Clean. Prod. **142**, 626 (2017)
22. Ning, C., You, F.: Data-driven stochastic robust optimization: general computational framework and algorithm leveraging machine learning for optimization under uncertainty in the big data era. Comput. Chem. Eng. **111**, 115–133 (2018)
23. Drigas, A.S., Leliopoulos, P.: The use of big data in education. Int. J. Comput. Sci. Issues **11**, 58 (2014)
24. Wang, Z., Chu, Y., Tan, K. L. D., Abbadi, A.E., Xu, X. Scalable data cube analysis over big data. arXiv, 1311, 5663 (2013)
25. Chauhan, D., Bansal, K.L.: Using the advantages of NoSQL: a case study on MongoDB. Int. J. Rec. Innov. Trends Comput. Commun. **5**, 90–93 (2017)
26. Nayak, A., Poriya, A., Poojary, D.: Type of NOSQL databases and its comparison with relational databases. Int. J. Appl. Inf. Syst. **5**, 16–19 (2013)
27. Haw, S.C., Lee, C.S.: Data storage practices and query processing in XML databases: a survey. Knowl.-Based Syst. **24**, 1317–1340 (2011)
28. Wu, T., Chen, Y., Han, J.: Association mining in large databases: A re-examination of its measures. In: Kok, J.N., Koronacki, J., Lopez de Mantaras, R., Matwin, S., Mladenič, D., Skowron, A. (eds.) Knowledge Discovery in Databases: PKDD 2007. PKDD 2007. Lecture Notes in Computer Science, vol. 4702, pp. 621–628. Springer, Heidelberg (2007). https://doi.org/10.1007/978-3-540-74976-9_66
29. Mengash, H.A.: Using data mining techniques to predict student performance to support decision making in university admission systems. IEEE Access **8**, 55462–55470 (2020)
30. Tufféry, S.: Data Mining and Statistics for Decision Making. John Wiley & Sons, UK (2011)

Author Index

Printed in the United States
by Baker & Taylor Publisher Services

Printed in the United States
by Baker & Taylor Publisher Services